Advanced PIC Microcontroller Projects in C

Advanced PIC Microcontroller Projects in C

From USB to RTOS with the PIC18F Series

Dogan Ibrahim

ELSEVIER

AMSTERDAM · BOSTON · HEIDELBERG · LONDON
NEW YORK · OXFORD · PARIS · SAN DIEGO
SAN FRANCISCO · SINGAPORE · SYDNEY · TOKYO

Newnes is an imprint of Elsevier

Newnes

Newnes is an imprint of Elsevier
The Boulevard, Langford Lane, Kidlington, Oxford, OX5 1GB, UK
30 Corporate Drive, Suite 400, Burlington, MA 01803, USA

First edition 2008
Reprinted 2008

Notice
No responsibility is assumed by the publisher for any injury and/or damage to persons
or property as a matter of products liability, negligence or otherwise, or from any use
or operation of any methods, products, instructions or ideas contained in the material
herein. Because of rapid advances in the medical sciences, in particular, independent
verification of diagnoses and drug dosages should be made

British Library Cataloguing in Publication Data
A catalogue record for this book is available from the British Library

Library of Congress Cataloging-in-Publication Data
A catalog record for this book is available from the Library of Congress

ISBN: 978-0-7506-8611-2

For information on all Newnes publications
visit our website at www.elsevierdirect.com

Printed and bound in *China*

08 09 10   10 9 8 7 6 5 4 3 2

Contents

Preface

A microcontroller is a microprocessor system which contains data and program memory, serial and parallel I/O, timers, and external and internal interrupts—all integrated into a single chip that can be purchased for as little as two dollars. About 40 percent of all microcontroller applications are found in office equipment, such as PCs, laser printers, fax machines, and intelligent telephones. About one third of all microcontrollers are found in consumer electronic goods. Products like CD players, hi-fi equipment, video games, washing machines, and cookers fall into this category. The communications market, the automotive market, and the military share the rest of the applications.

This book is written for advanced students, for practicing engineers, and for hobbyists who want to learn more about the programming and applications of PIC18F-series microcontrollers. The book assumes the reader has taken a course on digital logic design and been exposed to writing programs using at least one high-level programming language. Knowledge of the C programming language will be useful, and familiarity with at least one member of the PIC16F series of microcontrollers will be an advantage. Knowledge of assembly language programming is not required since all the projects in the book are based on the C language.

Chapter 1 presents the basic features of microcontrollers, discusses the important topic of numbering systems, and describes how to convert between number bases.

Chapter 2 reviews the PIC18F series of microcontrollers and describes various features of these microcontrollers in detail.

Chapter 3 provides a short tutorial on the C language and then examines the features of the mikroC compiler.

Chapter 4 covers advanced features of the mikroC language. Topics such as built-in functions and libraries are discussed in this chapter with examples.

Chapter 5 explores the various software and hardware development tools for the PIC18F series of microcontrollers. Various commercially available development kits as well as development tools such as simulators, emulators, and in-circuit debuggers are described with examples.

Chapter 6 provides some simple projects using the PIC18F series of microcontrollers and the mikroC compiler. All the projects are based on the PIC18F452 micro-controller, and all of them have been tested. This chapter should be useful for those who are new to PIC microcontrollers as well as for those who want to extend their knowledge of programming PIC18F microcontrollers using the mikroC language.

Chapter 7 covers the use of SD memory cards in PIC18F microcontroller projects. The theory of these cards is given with real working examples.

Chapter 8 reviews the popular USB bus, discussing the basic theory of this bus system with real working projects that illustrate how to design PIC18F-based projects communicating with a PC over the USB bus.

The CAN bus is currently used in many automotive applications. *Chapter 9* presents a brief theory of this bus and also discusses the design of PIC18F microcontroller-based projects with CAN bus interface.

Chapter 10 is about real-time operating systems (RTOS) and multi-tasking. The basic theory of RTOS systems is described and simple multi-tasking applications are given.

The CD-ROM that accompanies this book contains all the program source files and HEX files for the projects described in the book. In addition, a 2K size limited version of the mikroC compiler is included on the CD-ROM.

Dogan Ibrahim
London, 2007

Acknowledgments

The following material is reproduced in this book with the kind permission of the respective copyright holders and may not be reprinted, or reproduced in any other way, without their prior consent.

Figures 2.1–2.10, 2.22–2.36, 2.37, 2.38, 2.41–2.55, 5.2–5.4, 5.17, 5.20, 8.8, and 9.13, and Table 2.2 are taken from Microchip Technology Inc. data sheets PIC18FXX2 (DS39564C) and PIC18F2455/2550/4455/4550 (DS39632D).

Figure 5.5 is taken from the web site of BAJI Labs.

Figures 5.6–5.8 are taken from the web site of Shuan Shizu Ent. Co., Ltd.

Figures 5.9, 5.13, 5.18 are taken from the web site of Custom Computer Services Inc.

Figures 5.10, 5.19, and 6.43 are taken from the web site of mikroElektronika Ltd.

Figure 5.11 is taken from the web site of Futurlec.

Figure 5.21 is taken from the web site of Smart Communications Ltd.

Figure 5.22 is taken from the web site of RF Solutions.

Figure 5.23 is taken from the web site of Phyton.

Figures 5.1 and 5.14 are taken from the web site of microEngineering Labs Inc.

Figure 5.16 is taken from the web site of Kanda Systems.
Thanks is due to mikroElektronika Ltd. for their technical support and for permission to include a limited size mikroC compiler on the CD-ROM that accompanies this book.

PIC®, PICSTART®, and MPLAB® are all registered trademarks of Microchip Technology Inc.

CHAPTER 1

Microcomputer Systems

1.1   Introduction

The term *microcomputer* is used to describe a system that includes at minimum a microprocessor, program memory, data memory, and an input-output (I/O) device. Some microcomputer systems include additional components such as timers, counters, and analog-to-digital converters. Thus, a microcomputer system can be anything from a large computer having hard disks, floppy disks, and printers to a single-chip embedded controller.

In this book we are going to consider only the type of microcomputers that consist of a single silicon chip. Such microcomputer systems are also called *microcontrollers*, and they are used in many household goods such as microwave ovens, TV remote control units, cookers, hi-fi equipment, CD players, personal computers, and refrigerators. Many different microcontrollers are available on the market. In this book we shall be looking at programming and system design for the PIC (programmable interface controller) series of microcontrollers manufactured by Microchip Technology Inc.

1.2   Microcontroller Systems

A microcontroller is a single-chip computer. *Micro* suggests that the device is small, and *controller* suggests that it is used in control applications. Another term for microcontroller is *embedded controller*, since most of the microcontrollers are built into (or embedded in) the devices they control.

A microprocessor differs from a microcontroller in a number of ways. The main distinction is that a microprocessor requires several other components for its operation, such as program memory and data memory, input-output devices, and an external clock circuit. A microcontroller, on the other hand, has all the support chips incorporated inside its single chip. All microcontrollers operate on a set of instructions (or the user program) stored in their memory. A microcontroller fetches the instructions from its program memory one by one, decodes these instructions, and then carries out the required operations.

Microcontrollers have traditionally been programmed using the assembly language of the target device. Although the assembly language is fast, it has several disadvantages. An assembly program consists of mnemonics, which makes learning and maintaining a program written using the assembly language difficult. Also, microcontrollers manufactured by different firms have different assembly languages, so the user must learn a new language with every new microcontroller he or she uses.

Microcontrollers can also be programmed using a high-level language, such as BASIC, PASCAL, or C. High-level languages are much easier to learn than assembly languages. They also facilitate the development of large and complex programs. In this book we shall be learning the programming of PIC microcontrollers using the popular C language known as mikroC, developed by mikroElektronika.

In theory, a single chip is sufficient to have a running microcontroller system. In practical applications, however, additional components may be required so the microcomputer can interface with its environment. With the advent of the PIC family of microcontrollers the development time of an electronic project has been reduced to several hours.

Basically, a microcomputer executes a user program which is loaded in its program memory. Under the control of this program, data is received from external devices (inputs), manipulated, and then sent to external devices (outputs). For example, in a microcontroller-based oven temperature control system the microcomputer reads the temperature using a temperature sensor and then operates a heater or a fan to keep the temperature at the required value. Figure 1.1 shows a block diagram of a simple oven temperature control system.

The system shown in Figure 1.1 is very simple. A more sophisticated system may include a keypad to set the temperature and an LCD to display it. Figure 1.2 shows a block diagram of this more sophisticated temperature control system.

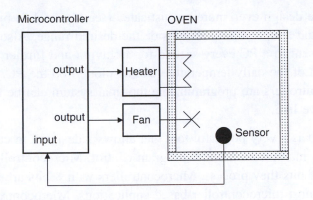

Figure 1.1: Microcontroller-based oven temperature control system

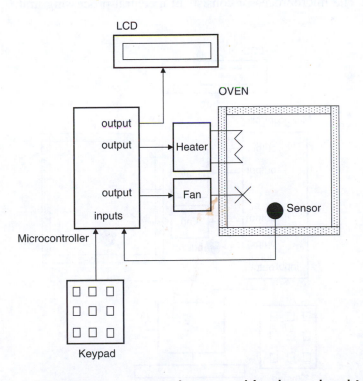

Figure 1.2: Temperature control system with a keypad and LCD

We can make the design even more sophisticated (see Figure 1.3) by adding an alarm that activates if the temperature goes outside the desired range. Also, the temperature readings can be sent to a PC every second for archiving and further processing. For example, a graph of the daily temperature can be plotted on the PC. As you can see, because microcontrollers are programmable the final system can be as simple or as complicated as we like.

A microcontroller is a very powerful tool that allows a designer to create sophisticated input-output data manipulation under program control. Microcontrollers are classified by the number of bits they process. Microcontrollers with 8 bits are the most popular and are used in most microcontroller-based applications. Microcontrollers with 16 and 32 bits are much more powerful, but are usually more expensive and not required in most small- or medium-size general purpose applications that call for microcontrollers.

The simplest microcontroller architecture consists of a microprocessor, memory, and input-output. The microprocessor consists of a central processing unit (CPU) and a

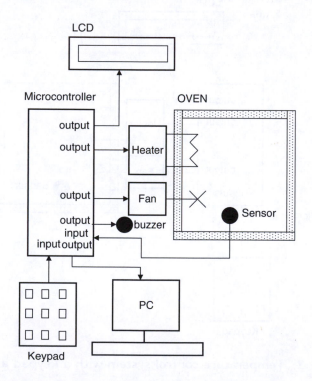

Figure 1.3: A more sophisticated temperature controller

control unit (CU). The CPU is the brain of the microcontroller; this is where all the arithmetic and logic operations are performed. The CU controls the internal operations of the microprocessor and sends signals to other parts of the microcontroller to carry out the required instructions.

Memory, an important part of a microcontroller system, can be classified into two types: program memory and data memory. Program memory stores the program written by the programmer and is usually nonvolatile (i.e., data is not lost after the power is turned off). Data memory stores the temporary data used in a program and is usually volatile (i.e., data is lost after the power is turned off).

There are basically six types of memories, summarized as follows:

1.2.1   RAM

RAM, random access memory, is a general purpose memory that usually stores the user data in a program. RAM memory is volatile in the sense that it cannot retain data in the absence of power (i.e., data is lost after the power is turned off). Most microcontrollers have some amount of internal RAM, 256 bytes being a common amount, although some microcontrollers have more, some less. The PIC18F452 microcontroller, for example, has 1536 bytes of RAM. Memory can usually be extended by adding external memory chips.

1.2.2   ROM

ROM, read only memory, usually holds program or fixed user data. ROM is nonvolatile. If power is removed from ROM and then reapplied, the original data will still be there. ROM memory is programmed during the manufacturing process, and the user cannot change its contents. ROM memory is only useful if you have developed a program and wish to create several thousand copies of it.

1.2.3   PROM

PROM, programmable read only memory, is a type of ROM that can be programmed in the field, often by the end user, using a device called a PROM programmer. Once a PROM has been programmed, its contents cannot be changed. PROMs are usually used in low production applications where only a few such memories are required.

1.2.4   EPROM

EPROM, erasable programmable read only memory, is similar to ROM, but EPROM can be programmed using a suitable programming device. An EPROM memory has a small clear-glass window on top of the chip where the data can be erased under strong ultraviolet light. Once the memory is programmed, the window can be covered with dark tape to prevent accidental erasure of the data. An EPROM memory must be erased before it can be reprogrammed. Many developmental versions of microcontrollers are manufactured with EPROM memories where the user program can be stored. These memories are erased and reprogrammed until the user is satisfied with the program. Some versions of EPROMs, known as OTP (one time programmable), can be programmed using a suitable programmer device but cannot be erased. OTP memories cost much less than EPROMs. OTP is useful after a project has been developed completely and many copies of the program memory must be made.

1.2.5   EEPROM

EEPROM, electrically erasable programmable read only memory, is a nonvolatile memory that can be erased and reprogrammed using a suitable programming device. EEPROMs are used to save configuration information, maximum and minimum values, identification data, etc. Some microcontrollers have built-in EEPROM memories. For instance, the PIC18F452 contains a 256-byte EEPROM memory where each byte can be programmed and erased directly by applications software. EEPROM memories are usually very slow. An EEPROM chip is much costlier than an EPROM chip.

1.2.6   Flash EEPROM

Flash EEPROM, a version of EEPROM memory, has become popular in microcontroller applications and is used to store the user program. Flash EEPROM is nonvolatile and usually very fast. The data can be erased and then reprogrammed using a suitable programming device. Some microcontrollers have only 1K flash EEPROM while others have 32K or more. The PIC18F452 microcontroller has 32K bytes of flash memory.

1.3   Microcontroller Features

Microcontrollers from different manufacturers have different architectures and different capabilities. Some may suit a particular application while others may be totally unsuitable for the same application. The hardware features common to most microcontrollers are described in this section.

1.3.1   Supply Voltage

Most microcontrollers operate with the standard logic voltage of +5V. Some microcontrollers can operate at as low as +2.7V, and some will tolerate +6V without any problem. The manufacturer's data sheet will have information about the allowed limits of the power supply voltage. PIC18F452 microcontrollers can operate with a power supply of +2V to +5.5V.

Usually, a voltage regulator circuit is used to obtain the required power supply voltage when the device is operated from a mains adapter or batteries. For example, a 5V regulator is required if the microcontroller is operated from a 5V supply using a 9V battery.

1.3.2   The Clock

All microcontrollers require a clock (or an oscillator) to operate, usually provided by external timing devices connected to the microcontroller. In most cases, these external timing devices are a crystal plus two small capacitors. In some cases they are resonators or an external resistor-capacitor pair. Some microcontrollers have built-in timing circuits and do not require external timing components. If an application is not time-sensitive, external or internal (if available) resistor-capacitor timing components are the best option for their simplicity and low cost.

An instruction is executed by fetching it from the memory and then decoding it. This usually takes several clock cycles and is known as the *instruction cycle*. In PIC microcontrollers, an instruction cycle takes four clock periods. Thus the microcontroller operates at a clock rate that is one-quarter of the actual oscillator frequency. The PIC18F series of microcontrollers can operate with clock frequencies up to 40MHz.

1.3.3   Timers

Timers are important parts of any microcontroller. A timer is basically a counter which is driven from either an external clock pulse or the microcontroller's internal oscillator. A timer can be 8 bits or 16 bits wide. Data can be loaded into a timer under program control, and the timer can be stopped or started by program control. Most timers can be configured to generate an interrupt when they reach a certain count (usually when they overflow). The user program can use an interrupt to carry out accurate timing-related operations inside the microcontroller. Microcontrollers in the PIC18F series have at least three timers. For example, the PIC18F452 microcontroller has three built-in timers.

Some microcontrollers offer capture and compare facilities, where a timer value can be read when an external event occurs, or the timer value can be compared to a preset value, and an interrupt is generated when this value is reached. Most PIC18F microcontrollers have at least two capture and compare modules.

1.3.4   Watchdog

Most microcontrollers have at least one watchdog facility. The watchdog is basically a timer that is refreshed by the user program. Whenever the program fails to refresh the watchdog, a reset occurs. The watchdog timer is used to detect a system problem, such as the program being in an endless loop. This safety feature prevents runaway software and stops the microcontroller from executing meaningless and unwanted code. Watchdog facilities are commonly used in real-time systems where the successful termination of one or more activities must be checked regularly.

1.3.5   Reset Input

A reset input is used to reset a microcontroller externally. Resetting puts the microcontroller into a known state such that the program execution starts from address 0 of the program memory. An external reset action is usually achieved by connecting a push-button switch to the reset input. When the switch is pressed, the microcontroller is reset.

1.3.6   Interrupts

Interrupts are an important concept in microcontrollers. An interrupt causes the microcontroller to respond to external and internal (e.g., a timer) events very quickly. When an interrupt occurs, the microcontroller leaves its normal flow of program execution and jumps to a special part of the program known as the interrupt service routine (ISR). The program code inside the ISR is executed, and upon return from the ISR the program resumes its normal flow of execution.

The ISR starts from a fixed address of the program memory sometimes known as the interrupt vector address. Some microcontrollers with multi-interrupt features have just one interrupt vector address, while others have unique interrupt vector addresses, one for each interrupt source. Interrupts can be nested such that a new interrupt can suspend the execution of another interrupt. Another important feature of multi-interrupt capability is that different interrupt sources can be assigned different levels of priority. For example, the PIC18F series of microcontrollers has both low-priority and high-priority interrupt levels.

1.3.7 Brown-out Detector

Brown-out detectors, which are common in many microcontrollers, reset the microcontroller if the supply voltage falls below a nominal value. These safety features can be employed to prevent unpredictable operation at low voltages, especially to protect the contents of EEPROM-type memories.

1.3.8 Analog-to-Digital Converter

An analog-to-digital converter (A/D) is used to convert an analog signal, such as voltage, to digital form so a microcontroller can read and process it. Some microcontrollers have built-in A/D converters. External A/D converter can also be connected to any type of microcontroller. A/D converters are usually 8 to 10 bits, having 256 to 1024 quantization levels. Most PIC microcontrollers with A/D features have multiplexed A/D converters which provide more than one analog input channel. For example, the PIC18F452 microcontroller has 10-bit 8-channel A/D converters.

The A/D conversion process must be started by the user program and may take several hundred microseconds to complete. A/D converters usually generate interrupts when a conversion is complete so the user program can read the converted data quickly.

A/D converters are especially useful in control and monitoring applications, since most sensors (e.g., temperature sensors, pressure sensors, force sensors, etc.) produce analog output voltages.

1.3.9 Serial Input-Output

Serial communication (also called RS232 communication) enables a microcontroller to be connected to another microcontroller or to a PC using a serial cable. Some microcontrollers have built-in hardware called USART (universal synchronous-asynchronous receiver-transmitter) to implement a serial communication interface. The user program can usually select the baud rate and data format. If no serial input-output hardware is provided, it is easy to develop software to implement serial data communication using any I/O pin of a microcontroller. The PIC18F series of microcontrollers has built-in USART modules. We shall see in Chapter 6 how to write mikroC programs to implement serial communication with and without a USART module.

Some microcontrollers (e.g., the PIC18F series) incorporate SPI (serial peripheral interface) or I^2C (integrated interconnect) hardware bus interfaces. These enable a microcontroller to interface with other compatible devices easily.

1.3.10   EEPROM Data Memory

EEPROM-type data memory is also very common in many microcontrollers. The advantage of an EEPROM memory is that the programmer can store nonvolatile data there and change this data whenever required. For example, in a temperature monitoring application, the maximum and minimum temperature readings can be stored in an EEPROM memory. If the power supply is removed for any reason, the values of the latest readings are available in the EEPROM memory. The PIC18F452 microcontroller has 256 bytes of EEPROM memory. Other members of the PIC18F family have more EEPROM memory (e.g., the PIC18F6680 has 1024 bytes). The mikroC language provides special instructions for reading and writing to the EEPROM memory of a PIC microcontroller.

1.3.11   LCD Drivers

LCD drivers enable a microcontroller to be connected to an external LCD display directly. These drivers are not common since most of the functions they provide can be implemented in software. For example, the PIC18F6490 microcontroller has a built-in LCD driver module.

1.3.12   Analog Comparator

Analog comparators are used where two analog voltages need to be compared. Although these circuits are implemented in most high-end PIC microcontrollers, they are not common in other microcontrollers. The PIC18F series of microcontrollers has built-in analog comparator modules.

1.3.13 Real-time Clock

A real-time clock enables a microcontroller to receive absolute date and time information continuously. Built-in real-time clocks are not common in most microcontrollers, since the same function can easily be implemented by either a dedicated real-time clock chip or a program written for this purpose.

1.3.14 Sleep Mode

Some microcontrollers (e.g., PICs) offer built-in sleep modes, where executing this instruction stops the internal oscillator and reduces power consumption to an extremely low level. The sleep mode's main purpose is to conserve battery power when the microcontroller is not doing anything useful. The microcontroller is usually woken up from sleep mode by an external reset or a watchdog time-out.

1.3.15 Power-on Reset

Some microcontrollers (e.g., PICs) have built-in power-on reset circuits which keep the microcontroller in the reset state until all the internal circuitry has been initialized. This feature is very useful, as it starts the microcontroller from a known state on power-up. An external reset can also be provided, where the microcontroller is reset when an external button is pressed.

1.3.16 Low-Power Operation

Low-power operation is especially important in portable applications where microcontroller-based equipment is operated from batteries. Some microcontrollers (e.g., PICs) can operate with less than 2mA with a 5V supply, and around 15µA at a 3V supply. Other microcontrollers, especially microprocessor-based systems with several chips, may consume several hundred milliamperes or even more.

1.3.17 Current Sink/Source Capability

Current sink/source capability is important if the microcontroller is to be connected to an external device that might draw a large amount of current to operate. PIC microcontrollers can source and sink 25mA of current from each output port pin. This current is usually sufficient to drive LEDs, small lamps, buzzers, small relays, etc. The current capability can be increased by connecting external transistor switching circuits or relays to the output port pins.

1.3.18   USB Interface

USB is currently a very popular computer interface specification used to connect various peripheral devices to computers and microcontrollers. Some PIC microcontrollers provide built-in USB modules. The PIC18F2x50, for example, has built-in USB interface capabilities.

1.3.19   Motor Control Interface

Some PIC microcontrollers, for example the PIC18F2x31, provide motor control interface capability.

1.3.20   CAN Interface

CAN bus is a very popular bus system used mainly in automation applications. Some PIC18F-series microcontrollers (e.g., the PIC18F4680) provide CAN interface capability.

1.3.21   Ethernet Interface

Some PIC microcontrollers (e.g., the PIC18F97J60) provide Ethernet interface capabilities and thus are easily used in network-based applications.

1.3.22   ZigBee Interface

ZigBee, an interface similar to Bluetooth, is used in low-cost wireless home automation applications. Some PIC18F-series microcontrollers provide ZigBee interface capabilities, making the design of such wireless systems very easy.

1.4   Microcontroller Architectures

Two types of architectures are conventional in microcontrollers (see Figure 1.4). *Von Neumann* architecture, used by a large percentage of microcontrollers, places all memory space on the same bus; instruction and data also use the same bus.

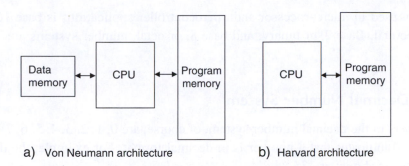

a) Von Neumann architecture   b) Harvard architecture

Figure 1.4: Von Neumann and Harvard architectures

In *Harvard* architecture (used by PIC microcontrollers), code and data are on separate buses, which allows them to be fetched simultaneously, resulting in an improved performance.

1.4.1 RISC and CISC

RISC (reduced instruction set computer) and CISC (complex instruction computer) refer to the instruction set of a microcontroller. In an 8-bit RISC microcontroller, data is 8 bits wide but the instruction words are more than 8 bits wide (usually 12, 14, or 16 bits) and the instructions occupy one word in the program memory. Thus the instructions are fetched and executed in one cycle, which improves performance.

In a CISC microcontroller, both data and instructions are 8 bits wide. CISC microcontrollers usually have over two hundred instructions. Data and code are on the same bus and cannot be fetched simultaneously.

1.5 Number Systems

To use a microprocessor or microcontroller efficiently requires a working knowledge of binary, decimal, and hexadecimal numbering systems. This section provides background information about these numbering systems for readers who are unfamiliar with them or do not know how to convert from one number system to another.

Number systems are classified according to their bases. The numbering system used in everyday life is base 10, or the decimal number system. The numbering system most commonly used in microprocessor and microcontroller applications is base 16, or hexadecimal. Base 2, or binary, and base 8, or octal, number systems are also used.

1.5.1   Decimal Number System

The numbers in the decimal number system, of course, are 0, 1, 2, 3, 4, 5, 6, 7, 8, 9. The subscript 10 indicates that a number is in decimal format. For example, the decimal number 235 is shown as 235_{10}.

In general, a decimal number is represented as follows:

$$a_n \times 10^n + a_{n-1} \times 10^{n-1} + a_{n-2} \times 10^{n-2} + \ldots\ldots + a_0 \times 10^0$$

For example, decimal number 825_{10} can be shown as:

$$825_{10} = 8 \times 10^2 + 2 \times 10^1 + 5 \times 10^0$$

Similarly, decimal number 26_{10} can be shown as:

$$26_{10} = 2 \times 10^1 + 6 \times 10^0$$

or

$$3359_{10} = 3 \times 10^3 + 3 \times 10^2 + 5 \times 10^1 + 9 \times 10^0$$

1.5.2   Binary Number System

The binary number system consists of two numbers: 0 and 1. A subscript 2 indicates that a number is in binary format. For example, the binary number 1011 would be 1011_2.

In general, a binary number is represented as follows:

$$a_n \times 2^n + a_{n-1} \times 2^{n-1} + a_{n-2} \times 2^{n-2} + \ldots\ldots + a_0 \times 2^0$$

For example, binary number 1110_2 can be shown as:

$$1110_2 = 1 \times 2^3 + 1 \times 2^2 + 1 \times 2^1 + 0 \times 2^0$$

Similarly, binary number 10001110_2 can be shown as:

$$10001110_2 = 1 \times 2^7 + 0 \times 2^6 + 0 \times 2^5 + 0 \times 2^4 + 1 \times 2^3$$
$$+ 1 \times 2^2 + 1 \times 2^1 + 0 \times 2^0$$

1.5.3 Octal Number System

In the octal number system, the valid numbers are 0, 1, 2, 3, 4, 5, 6, 7. A subscript 8 indicates that a number is in octal format. For example, the octal number 23 appears as 23_8.

In general, an octal number is represented as:

$$a_n \times 8^n + a_{n-1} \times 8^{n-1} + a_{n-2} \times 8^{n-2} + \dots + a_0 \times 8^0$$

For example, octal number 237_8 can be shown as:

$$237_8 = 2 \times 8^2 + 3 \times 8^1 + 7 \times 8^0$$

Similarly, octal number 1777_8 can be shown as:

$$1777_8 = 1 \times 8^3 + 7 \times 8^2 + 7 \times 8^1 + 7 \times 8^0$$

1.5.4 Hexadecimal Number System

In the hexadecimal number system, the valid numbers are: 0, 1, 2, 3, 4, 5, 6, 7, 8, 9, A, B, C, D, E, F. A subscript 16 or subscript H indicates that a number is in hexadecimal format. For example, hexadecimal number 1F can be written as $1F_{16}$ or as $1F_H$.

In general, a hexadecimal number is represented as:

$$a_n \times 16^n + a_{n-1} \times 16^{n-1} + a_{n-2} \times 16^{n-2} + \dots + a_0 \times 16^0$$

For example, hexadecimal number $2AC_{16}$ can be shown as:

$$2AC_{16} = 2 \times 16^2 + 10 \times 16^1 + 12 \times 16^0$$

Similarly, hexadecimal number $3FFE_{16}$ can be shown as:

$$3FFE_{16} = 3 \times 16^3 + 15 \times 16^2 + 15 \times 16^1 + 14 \times 16^0$$

1.6   Converting Binary Numbers into Decimal

To convert a binary number into decimal, write the number as the sum of the powers of 2.

Example 1.1

Convert binary number 1011_2 into decimal.

Solution 1.1

Write the number as the sum of the powers of 2:

$$1011_2 = 1 \times 2^3 + 0 \times 2^2 + 1 \times 2^1 + 1 \times 2^0$$
$$= 8 + 0 + 2 + 1$$
$$= 11$$

or, $1011_2 = 11_{10}$

Example 1.2

Convert binary number 11001110_2 into decimal.

Solution 1.2

Write the number as the sum of the powers of 2:

$$11001110_2 = 1 \times 2^7 + 1 \times 2^6 + 0 \times 2^5 + 0 \times 2^4$$
$$+ 1 \times 2^3 + 1 \times 2^2 + 1 \times 2^1 + 0 \times 2^0$$
$$= 128 + 64 + 0 + 0 + 8 + 4 + 2 + 0$$
$$= 206$$

or, $11001110_2 = 206_{10}$

Table 1.1 shows the decimal equivalent of numbers from 0 to 31.

1.7   Converting Decimal Numbers into Binary

To convert a decimal number into binary, divide the number repeatedly by 2 and take the remainders. The first remainder is the least significant digit (LSD), and the last remainder is the most significant digit (MSD).

Example 1.3

Convert decimal number 28_{10} into binary.

Table 1.1: Decimal equivalent of binary numbers

Binary	Decimal	Binary	Decimal
00000000	0	00010000	16
00000001	1	00010001	17
00000010	2	00010010	18
00000011	3	00010011	19
00000100	4	00010100	20
00000101	5	00010101	21
00000110	6	00010110	22
00000111	7	00010111	23
00001000	8	00011000	24
00001001	9	00011001	25
00001010	10	00011010	26
00001011	11	00011011	27
00001100	12	00011100	28
00001101	13	00011101	29
00001110	14	00011110	30
00001111	15	00011111	31

Solution 1.3

Divide the number into 2 repeatedly and take the remainders:

```
28/2  →  14   Remainder 0   (LSD)
14/2  →  7    Remainder 0
7/2   →  3    Remainder 1
3/2   →  1    Remainder 1
1/2   →  0    Remeinder 1   (MSD)
```

The binary number is 11100_2.

Example 1.4

Convert decimal number 65_{10} into binary.

Solution 1.4

Divide the number into 2 repeatedly and take the remainders:

```
65/2   →   32   Remainder 1   (LSD)
32/2   →   16   Remainder 0
16/2   →   8    Remainder 0
8/2    →   4    Remainder 0
4/2    →   2    Remainder 0
2/2    →   1    Remainder 0
1/2    →   0    Remainder 1   (MSD)
```

The binary number is 1000001_2.

Example 1.5

Convert decimal number 122_{10} into binary.

Solution 1.5

Divide the number into 2 repeatedly and take the remainders:

```
122/2   →   61   Remainder 0   (LSD)
61/2    →   30   Remainder 1
30/2    →   15   Remainder 0
15/2    →   7    Remainder 1
7/2     →   3    Remainder 1
3/2     →   1    Remainder 1
1/2     →   0    Remainder 1   (MSD)
```

The binary number is 1111010_2.

1.8   Converting Binary Numbers into Hexadecimal

To convert a binary number into hexadecimal, arrange the number in groups of four and find the hexadecimal equivalent of each group. If the number cannot be divided exactly into groups of four, insert zeros to the left of the number as needed so the number of digits are divisible by four.

Example 1.6

Convert binary number 10011111_2 into hexadecimal.

Solution 1.6

First, divide the number into groups of four, then find the hexadecimal equivalent of each group:

```
10011111 = 1001 1111
              9    F
```

The hexadecimal number is $9F_{16}$.

Example 1.7

Convert binary number 1110111100001110_2 into hexadecimal.

Solution 1.7

First, divide the number into groups of four, then find the hexadecimal equivalent of each group:

```
1110111100001110 = 1110 1111 0000 1110
                      E    F    0    E
```

The hexadecimal number is $EF0E_{16}$.

Example 1.8

Convert binary number 111110_2 into hexadecimal.

Solution 1.8

Since the number cannot be divided exactly into groups of four, we have to insert, in this case, two zeros to the left of the number so the number of digits is divisible by four:

```
111110 = 0011 1110
            3    E
```

The hexadecimal number is $3E_{16}$.

Table 1.2 shows the hexadecimal equivalent of numbers 0 to 31.

Table 1.2: Hexadecimal equivalent of decimal numbers

Decimal	Hexadecimal	Decimal	Hexadecimal
0	0	16	10
1	1	17	11
2	2	18	12
3	3	19	13
4	4	20	14
5	5	21	15
6	6	22	16
7	7	23	17
8	8	24	18
9	9	25	19
10	A	26	1A
11	B	27	1B
12	C	28	1C
13	D	29	1D
14	E	30	1E
15	F	31	1F

1.9   Converting Hexadecimal Numbers into Binary

To convert a hexadecimal number into binary, write the 4-bit binary equivalent of each hexadecimal digit.

Example 1.9

Convert hexadecimal number $A9_{16}$ into binary.

Solution 1.9

Writing the binary equivalent of each hexadecimal digit:

$A = 1010_2 \quad 9 = 1001_2$

The binary number is 10101001_2.

Example 1.10

Convert hexadecimal number $FE3C_{16}$ into binary.

Solution 1.10

Writing the binary equivalent of each hexadecimal digit:

$F = 1111_2 \quad E = 1110_2 \quad 3 = 0011_2 \quad C = 1100_2$

The binary number is 1111111000111100_2.

1.10   Converting Hexadecimal Numbers into Decimal

To convert a hexadecimal number into decimal, calculate the sum of the powers of 16 of the number.

Example 1.11

Convert hexadecimal number $2AC_{16}$ into decimal.

Solution 1.11

Calculating the sum of the powers of 16 of the number:

$$2AC_{16} = 2 \times 16^2 + 10 \times 16^1 + 12 \times 16^0$$
$$= 512 + 160 + 12$$
$$= 684$$

The required decimal number is 684_{10}.

Example 1.12

Convert hexadecimal number EE_{16} into decimal.

Solution 1.12

Calculating the sum of the powers of 16 of the number:

$$
\begin{aligned}
EE_{16} &= 14 \times 16^1 + 14 \times 16^0 \\
&= 224 + 14 \\
&= 238
\end{aligned}
$$

The decimal number is 238_{10}.

1.11   Converting Decimal Numbers into Hexadecimal

To convert a decimal number into hexadecimal, divide the number repeatedly by 16 and take the remainders. The first remainder is the LSD, and the last remainder is the MSD.

Example 1.13

Convert decimal number 238_{10} into hexadecimal.

Solution 1.13

Dividing the number repeatedly by 16:

```
238/16  →  14  Remainder 14 (E)   (LSD)
14/16   →   0  Remainder 14 (E)   (MSD)
```

The hexadecimal number is EE_{16}.

Example 1.14

Convert decimal number 684_{10} into hexadecimal.

Solution 1.14

Dividing the number repeatedly by 16:

```
684/16  →  42  Remainder 12 (C)   (LSD)
42/16   →   2  Remainder 10 (A)
2/16    →   0  Remainder 2        (MSD)
```

The hexadecimal number is $2AC_{16}$.

1.12   Converting Octal Numbers into Decimal

To convert an octal number into decimal, calculate the sum of the powers of 8 of the number.

Example 1.15

Convert octal number 15_8 into decimal.

Solution 1.15

Calculating the sum of the powers of 8 of the number:

$$
\begin{aligned}
15_8 &= 1 \times 8^1 + 5 \times 8^0 \\
&= 8 + 5 \\
&= 13
\end{aligned}
$$

The decimal number is 13_{10}.

Example 1.16

Convert octal number 237_8 into decimal.

Solution 1.16

Calculating the sum of the powers of 8 of the number:

$$
\begin{aligned}
237_8 &= 2 \times 8^2 + 3 \times 8^1 + 7 \times 8^0 \\
&= 128 + 24 + 7 \\
&= 159
\end{aligned}
$$

The decimal number is 159_{10}.

1.13   Converting Decimal Numbers into Octal

To convert a decimal number into octal, divide the number repeatedly by 8 and take the remainders. The first remainder is the LSD, and the last remainder is the MSD.

Example 1.17

Convert decimal number 159_{10} into octal.

Solution 1.17

Dividing the number repeatedly by 8:

```
159/8   →   19   Remainder 7   (LSD)
19/8    →   2    Remainder 3
2/8     →   0    Remainder 2   (MSD)
```

The octal number is 237_8.

Example 1.18

Convert decimal number 460_{10} into octal.

Solution 1.18

Dividing the number repeatedly by 8:

```
460/8   →   57   Remainder 4   (LSD)
57/8    →   7    Remainder 1
7/8     →   0    Remainder 7   (MSD)
```

The octal number is 714_8.

Table 1.3 shows the octal equivalent of decimal numbers 0 to 31.

1.14   Converting Octal Numbers into Binary

To convert an octal number into binary, write the 3-bit binary equivalent of each octal digit.

Example 1.19

Convert octal number 177_8 into binary.

Solution 1.19

Write the binary equivalent of each octal digit:

$1 = 001_2$ $7 = 111_2$ $7 = 111_2$

The binary number is 001111111_2.

Table 1.3: Octal equivalent of decimal numbers

Decimal	Octal	Decimal	Octal
0	0	16	20
1	1	17	21
2	2	18	22
3	3	19	23
4	4	20	24
5	5	21	25
6	6	22	26
7	7	23	27
8	10	24	30
9	11	25	31
10	12	26	32
11	13	27	33
12	14	28	34
13	15	29	35
14	16	30	36
15	17	31	37

Example 1.20

Convert octal number 75_8 into binary.

Solution 1.20

Write the binary equivalent of each octal digit:

$7 = 111_2$ $\quad$ $5 = 101_2$

The binary number is 111101_2.

1.15 Converting Binary Numbers into Octal

To convert a binary number into octal, arrange the number in groups of three and write the octal equivalent of each digit.

Example 1.21

Convert binary number 110111001_2 into octal.

Solution 1.21

Arranging in groups of three:

```
110111001 = 110 111 001
              6   7   1
```

The octal number is 671_8.

1.16 Negative Numbers

The most significant bit of a binary number is usually used as the sign bit. By convention, for positive numbers this bit is 0, and for negative numbers this bit is 1. Figure 1.5 shows the 4-bit positive and negative numbers. The largest positive and negative numbers are $+7$ and -8 respectively.

Binary number	Decimal equivalent
0111	+7
0110	+6
0101	+5
0100	+4
0011	+3
0010	+2
0001	+1
0000	0
1111	−1
1110	−2
1101	−3
1100	−4
1011	−5
1010	−6
1001	−7
1000	−8

Figure 1.5: 4-bit positive and negative numbers

To convert a positive number to negative, take the complement of the number and add 1. This process is also called the 2's complement of the number.

Example 1.22

Write decimal number −6 as a 4-bit number.

Solution 1.22

First, write the number as a positive number, then find the complement and add 1:

```
0110   +6
1001   complement
   1   add 1
____
1010   which is −6
```

Example 1.23

Write decimal number −25 as a 8-bit number.

Solution 1.23

First, write the number as a positive number, then find the complement and add 1:

```
00011001   +25
11100110   complement
       1   add 1
_____
11100111   which is −25
```

1.17   Adding Binary Numbers

The addition of binary numbers is similar to the addition of decimal numbers. Numbers in each column are added together with a possible carry from a previous column. The primitive addition operations are:

$0 + 0 = 0$
$0 + 1 = 1$
$1 + 0 = 1$
$1 + 1 = 10$ generate a carry bit
$1 + 1 + 1 = 11$ generate a carry bit

Some examples follow.

Example 1.24

Find the sum of binary numbers 011 and 110.

Solution 1.24

We can add these numbers as in the addition of decimal numbers:

$$
\begin{array}{ll}
011 & \text{First column:} \qquad 1 + 0 = 1 \\
+\,110 & \text{Second column:} \quad 1 + 1 = 0 \text{ and a carry bit} \\
\text{-----} & \text{Third column:} \qquad 1 + 1 = 10 \\
1001 &
\end{array}
$$

Example 1.25

Find the sum of binary numbers 01000011 and 00100010.

Solution 1.25

We can add these numbers as in the addition of decimal numbers:

$$
\begin{array}{ll}
01000011 & \text{First column:} \qquad 1 + 0 = 1 \\
+\,00100010 & \text{Second column:} \quad 1 + 1 = 10 \\
\text{----------} & \text{Third column:} \qquad 0 + \text{carry} = 1 \\
01100101 & \text{Fourth column:} \quad 0 + 0 = 0 \\
& \text{Fifth column:} \qquad 0 + 0 = 0 \\
& \text{Sixth column:} \qquad 0 + 1 = 1 \\
& \text{Seventh column:} \; 1 + 0 = 1 \\
& \text{Eighth column:} \quad 0 + 0 = 0
\end{array}
$$

1.18   Subtracting Binary Numbers

To subtract one binary number from another, convert the number to be subtracted into negative and then add the two numbers.

Example 1.26

Subtract binary number 0010 from 0110.

Solution 1.26

First, convert the number to be subtracted into negative:

```
0010   number to be subtracted
1101   complement
   1   add 1
————
1110
```

Now add the two numbers:

```
  0110
+ 1110
——————
  0100
```

Since we are using only 4 bits, we cannot show the carry bit.

1.19   Multiplication of Binary Numbers

Multiplication of two binary numbers is similar to the multiplication of two decimal numbers. The four possibilities are:

$0 \times 0 = 0$
$0 \times 1 = 0$
$1 \times 0 = 0$
$1 \times 1 = 1$

Some examples follow.

Example 1.27

Multiply the two binary numbers 0110 and 0010.

Solution 1.27

Multiplying the numbers:

```
     0110
     0010
     ----
     0000
    0110
   0000
  0000
  ------
  001100 or 1100
```

In this example 4 bits are needed to show the final result.

Example 1.28

Multiply binary numbers 1001 and 1010.

Solution 1.28

Multiplying the numbers:

```
     1001
     1010
     ----
     0000
    1001
   0000
  1001
  ------
  1011010
```

In this example 7 bits are required to show the final result.

1.20   Division of Binary Numbers

Division with binary numbers is similar to division with decimal numbers. An example follows.

Example 1.29

Divide binary number 1110 into binary number 10.

Solution 1.29

Dividing the numbers:

```
      111
10 | 1110
     10
    ----
     11
     10
    ----
      10
      10
    ----
      00
```

gives the result 111_2.

1.21   Floating Point Numbers

Floating point numbers are used to represent noninteger fractional numbers, for example, 3.256, 2.1, 0.0036, and so forth. Floating point numbers are used in most engineering and technical calculations. The most common floating point standard is the IEEE standard, according to which floating point numbers are represented with 32 bits (single precision) or 64 bits (double precision).

In this section we are looking at the format of 32-bit floating point numbers only and seeing how mathematical operations can be performed with such numbers.

According to the IEEE standard, 32-bit floating point numbers are represented as:

```
31 30           23  22                              0
X  XXXXXXXX  XXXXXXXXXXXXXXXXXXXXXXX
↑        ↑                          ↑
sign   exponent                  mantissa
```

The most significant bit indicates the sign of the number, where 0 indicates the number is positive, and 1 indicates it is negative.

The 8-bit exponent shows the power of the number. To make the calculations easy, the sign of the exponent is not shown; instead, the excess-128 numbering system is used. Thus, to find the real exponent we have to subtract 127 from the given exponent. For example, if the mantissa is "10000000," the real value of the mantissa is $128 - 127 = 1$.

The mantissa is 23 bits wide and represents the increasing negative powers of 2. For example, if we assume that the mantissa is "11100000000000000000000," the value of this mantissa is calculated as $2^{-1} + 2^{-2} + 2^{-3} = 7/8$.

The decimal equivalent of a floating point number can be calculated using the formula:

$$\text{Number} = (-1)^s \, 2^{e-127} \, 1.f$$

where

> $s = 0$ for positive numbers, 1 for negative numbers
> e = exponent (between 0 and 255)
> f = mantissa

As shown in this formula, there is a hidden 1 in front of the mantissa (i.e, the mantissa is shown as 1.f).

The largest number in 32-bit floating point format is:

```
0 11111110 11111111111111111111111
```

This number is $(2 - 2^{-23}) \, 2^{127}$ or decimal 3.403×10^{38}. The numbers keep their precision up to 6 digits after the decimal point.

The smallest number in 32-bit floating point format is:

0 00000001 00000000000000000000000

This number is 2^{-126} or decimal 1.175×10^{-38}.

1.22 Converting a Floating Point Number into Decimal

To convert a given floating point number into decimal, we have to find the mantissa and the exponent of the number and then convert into decimal as just shown.

Some examples are given here.

Example 1.30

Find the decimal equivalent of the floating point number: 0 10000001 10000000000000000000000

Solution 1.30

Here

```
sign = positive
exponent = 129 - 127 = 2
mantissa = 2⁻¹ = 0.5
```

The decimal equivalent of this number is $+1.5 \times 2^2 = +6.0$.

Example 1.31

Find the decimal equivalent of the floating point number: 0 10000010 11000000000000000000000

Solution 1.31

In this example,

```
sign = positive
exponent = 130 - 127 = 3
mantissa = 2⁻¹ + 2⁻² = 0.75
```

The decimal equivalent of the number is $+1.75 \times 2^3 = 14.0$.

1.22.1 Normalizing Floating Point Numbers

Floating point numbers are usually shown in normalized form. A normalized number has only one digit before the decimal point (a hidden number 1 is assumed before the decimal point).

To normalize a given floating point number, we have to move the decimal point repeatedly one digit to the left and increase the exponent after each move.

Some examples follow.

Example 1.32

Normalize the floating point number 123.56

Solution 1.32

If we write the number with a single digit before the decimal point we get:

```
1.2356 x 10²
```

Example 1.33

Normalize the binary number 1011.1_2

Solution 1.33

If we write the number with a single digit before the decimal point we get:

```
1.0111 x 2³
```

1.22.2 Converting a Decimal Number into Floating Point

To convert a given decimal number into floating point, carry out the following steps:

- Write the number in binary.

- Normalize the number.

- Find the mantissa and the exponent.

- Write the number as a floating point number.

Some examples follow:

Example 1.34

Convert decimal number 2.25_{10} into floating point.

Solution 1.34

Write the number in binary:

$$2.25_{10} = 10.01_2$$

Normalize the number:

$$10.01_2 = 1.001_2 \times 2^1$$

Here, s = 0, e – 127 = 1 or e = 128, and f = 00100000000000000000000.

(Remember that a number 1 is assumed on the left side, even though it is not shown in the calculation). The required floating point number can be written as:

```
s       e                    f
0   10000000   (1) 001 0000 0000 0000 0000 0000
```

or, the required 32-bit floating point number is:

```
01000000001000000000000000000000
```

Example 1.35

Convert the decimal number 134.0625_{10} into floating point.

Solution 1.35

Write the number in binary:

```
134.0625₁₀ = 10000110.0001
```

Normalize the number:

```
10000110.0001 = 1.00001100001 x 2⁷
```

Here, s = 0, e – 127 = 7 or e = 134, and f = 00001100001000000000000.

The required floating point number can be written as:

```
s       e                       f
0   10000110    (1) 00001100001000000000000
```

or, the required 32-bit floating point number is:

```
01000011000001100001000000000000
```

1.22.3 Multiplication and Division of Floating Point Numbers

Multiplication and division of floating point numbers are rather easy. Here are the steps:

- Add (or subtract) the exponents of the numbers.

- Multiply (or divide) the mantissa of the numbers.

- Correct the exponent.

- Normalize the number.

- The sign of the result is the EXOR of the signs of the two numbers.

Since the exponent is processed twice in the calculations, we have to subtract 127 from the exponent.

An example showing the multiplication of two floating point numbers follows.

Example 1.36

Show the decimal numbers 0.510 and 0.7510 in floating point and then calculate their multiplication.

Solution 1.36

Convert the numbers into floating point as:

$$0.5_{10} = 1.0000 \times 2^{-1}$$
```
here, s = 0, e - 127 = -1 or e = 126 and f = 0000
or,
```
$$0.5_{10} = 0\ 01110110\quad (1)\,000\ 0000\ 0000\ 0000\ 0000\ 0000$$
```
Similarly,
```
$$0.75_{10} = 1.1000 \times 2^{-1}$$

```
here, s = 0, e = 126 and f = 1000
or,
    0.75₁₀ = 0 01110110 (1)100 0000 0000 0000 0000 0000
```

Multiplying the mantissas results in "(1)100 0000 0000 0000 0000 0000." The sum of the exponents is $126 + 126 = 252$. Subtracting 127 from the mantissa, we obtain $252 - 127 = 125$. The EXOR of the signs of the numbers is 0. Thus, the result can be shown in floating point as:

```
0    01111101   (1)100 0000 0000 0000 0000 0000
```

This number is equivalent to decimal 0.375 ($0.5 \times 0.75 = 0.375$), which is the correct result.

1.22.4 Addition and Subtraction of Floating Point Numbers

The exponents of floating point numbers must be the same before they can be added or subtracted. The steps to add or subtract floating point numbers are:

- Shift the smaller number to the right until the exponents of both numbers are the same. Increment the exponent of the smaller number after each shift.

- Add (or subtract) the mantissa of each number as an integer calculation, without considering the decimal points.

- Normalize the result.

An example follows.

Example 1.37

Show decimal numbers 0.510 and 0.7510 in floating point and then calculate the sum of these numbers.

Solution 1.37

As shown in Example 1.36, we can convert the numbers into floating point as:

```
0.5₁₀ = 0 01110110   (1)000 0000 0000 0000 0000 0000
```

```
Similarly,
0.75₁₀ = 0 01110110 (1)100 0000 0000 0000 0000 0000
```

Since the exponents of both numbers are the same, there is no need to shift the smaller number. If we add the mantissa of the numbers without considering the decimal points, we get:

```
(1)000 0000 0000 0000 0000 0000
(1)100 0000 0000 0000 0000 0000
                              +
```
```
(10)100 0000 0000 0000 0000 0000
```

To normalize the number, shift it right by one digit and then increment its exponent. The resulting number is:

```
0 01111111   (1)010 0000 0000 0000 0000 0000
```

This floating point number is equal to decimal number 1.25, which is the sum of decimal numbers 0.5 and 0.75.

A program for converting floating point numbers into decimal, and decimal numbers into floating point, is available for free on the following web site:

```
http://babbage.cs.qc.edu/courses/cs341/IEEE-754.html
```

1.23   BCD Numbers

BCD (binary coded decimal) numbers are usually used in display systems such as LCDs and 7-segment displays to show numeric values. In BCD, each digit is a 4-bit number from 0 to 9. As an example, Table 1.4 shows the BCD numbers between 0 and 20.

Example 1.38

Write the decimal number 295 as a BCD number.

Solution 1.38

Write the 4-bit binary equivalent of each digit:

$2 = 0010_2$ $9 = 1001_2$ $5 = 0101_2$

The BCD number is $0010\ 1001\ 0101_2$.

Table 1.4: BCD numbers between 0 and 20

Decimal	BCD	Binary
0	0000	0000
1	0001	0001
2	0010	0010
3	0011	0011
4	0100	0100
5	0101	0101
6	0110	0110
7	0111	0111
8	1000	1000
9	1001	1001
10	0001 0000	1010
11	0001 0001	1011
12	0001 0010	1100
13	0001 0011	1101
14	0001 0100	1110
15	0001 0101	1111
16	0001 0110	1 0000
17	0001 0111	1 0001
18	0001 1000	1 0010
19	0001 1001	1 0011
20	0010 0000	1 0100

Example 1.39

Write the decimal equivalent of BCD number 1001 1001 0110 0001_2.

Solution 1.39

Writing the decimal equivalent of each group of 4-bit yields the decimal number:

9961

1.24   Summary

Chapter 1 has provided an introduction to the microprocessor and microcontroller systems. The basic building blocks of microcontrollers were described briefly. The chapter also provided an introduction to various number systems, and described how to convert a given number from one base into another. The important topics of floating point numbers and floating point arithmetic were also described with examples.

1.25   Exercises

1. What is a microcontroller? What is a microprocessor? Explain the main difference between a microprocessor and a microcontroller.

2. Identify some applications of microcontrollers around you.

3. Where would you use an EPROM memory?

4. Where would you use a RAM memory?

5. Explain the types of memory usually used in microcontrollers.

6. What is an input-output port?

7. What is an analog-to-digital converter? Give an example of how this converter is used.

8. Explain why a watchdog timer could be useful in a real-time system.

9. What is serial input-output? Where would you use serial communication?

10. Why is the current sink/source capability important in the specification of an output port pin?

11. What is an interrupt? Explain what happens when an interrupt is recognized by a microcontroller?

12. Why is brown-out detection important in real-time systems?

13. Explain the difference between an RISC-based microcontroller and a CISC-based microcontroller. What type of microcontroller is PIC?

14. Convert the following decimal numbers into binary:

    a) 23   b) 128   c) 255   d) 1023

    e) 120   f) 32000 g) 160   h) 250

15. Convert the following binary numbers into decimal:

    a) 1111   b) 0110   c) 11110000

    d) 00001111   e) 10101010   f) 10000000

16. Convert the following octal numbers into decimal:

    a) 177   b) 762   c) 777   d) 123

    e) 1777   f) 655   g) 177777 h) 207

17. Convert the following decimal numbers into octal:

    a) 255   b) 1024   c) 129   d) 2450

    e) 4096   f) 256   g) 180   h) 4096

18. Convert the following hexadecimal numbers into decimal:

    a) AA   b) EF   c) 1FF   d) FFFF

    e) 1AA   f) FEF   g) F0   h) CC

19. Convert the following binary numbers into hexadecimal:

    a) 0101   b) 11111111 c) 1111   d) 1010

    e) 1110   f) 10011111 g) 1001   h) 1100

20. Convert the following binary numbers into octal:

    a) 111000 b) 000111   c) 1111111 d) 010111

    e) 110001 f) 11111111 g) 1000001 h) 110000

21. Convert the following octal numbers into binary:

    a) 177   b) 7777   c) 555   d) 111

    e) 1777777   f) 55571   g) 171   h) 1777

22. Convert the following hexadecimal numbers into octal:

    a) AA   b) FF   c) FFFF   d) 1AC

    e) CC   f) EE   g) EEFF   h) AB

23. Convert the following octal numbers into hexadecimal:

    a) 177   b) 777   c) 123   d) 23

    e) 1111   f) 17777777   g) 349   h) 17

24. Convert the following decimal numbers into floating point:

    a) 23.45   b) 1.25   c) 45.86   d) 0.56

25. Convert the following decimal numbers into floating point and then calculate their sum:

    0.255 and 1.75

26. Convert the following decimal numbers into floating point and then calculate their product:

    2.125 and 3.75

27. Convert the following decimal numbers into BCD:

    a) 128   b) 970   c) 900   d) 125

CHAPTER 2

PIC18F Microcontroller Series

PIC16-series microcontrollers have been around for many years. Although these are excellent general purpose microcontrollers, they have certain limitations. For example, the program and data memory capacities are limited, the stack is small, and the interrupt structure is primitive, all interrupt sources sharing the same interrupt vector. PIC16-series microcontrollers also do not provide direct support for advanced peripheral interfaces such as USB, CAN bus, etc., and interfacing with such devices is not easy. The instruction set for these microcontrollers is also limited. For example, there are no multiplication or division instructions, and branching is rather simple, being a combination of *skip* and *goto* instructions.

Microchip Inc. has developed the PIC18 series of microcontrollers for use in high-pin-count, high-density, and complex applications. The PIC18F microcontrollers offer cost-efficient solutions for general purpose applications written in C that use a real-time operating system (RTOS) and require a complex communication protocol stack such as TCP/IP, CAN, USB, or ZigBee. PIC18F devices provide flash program memory in sizes from 8 to 128Kbytes and data memory from 256 to 4Kbytes, operating at a range of 2.0 to 5.0 volts, at speeds from DC to 40MHz.

The basic features of PIC18F-series microcontrollers are:

- 77 instructions
- PIC16 source code compatible
- Program memory addressing up to 2Mbytes
- Data memory addressing up to 4Kbytes

- DC to 40MHz operation

- 8 × 8 hardware multiplier

- Interrupt priority levels

- 16-bit-wide instructions, 8-bit-wide data path

- Up to two 8-bit timers/counters

- Up to three 16-bit timers/counters

- Up to four external interrupts

- High current (25mA) sink/source capability

- Up to five capture/compare/PWM modules

- Master synchronous serial port module (SPI and I^2C modes)

- Up to two USART modules

- Parallel slave port (PSP)

- Fast 10-bit analog-to-digital converter

- Programmable low-voltage detection (LVD) module

- Power-on reset (POR), power-up timer (PWRT), and oscillator start-up timer (OST)

- Watchdog timer (WDT) with on-chip RC oscillator

- In-circuit programming

In addition, some microcontrollers in the PIC18F family offer the following special features:

- Direct CAN 2.0 bus interface

- Direct USB 2.0 bus interface

- Direct LCD control interface

- TCP/IP interface

- ZigBee interface

- Direct motor control interface

Most devices in the PIC18F family are source compatible with each other. Table 2.1 gives the characteristics of some of the popular devices in this family. This chapter offers a detailed study of the PIC18FXX2 microcontrollers. The architectures of most of the other microcontrollers in the PIC18F family are similar.

The reader may be familiar with the programming and applications of the PIC16F series. Before going into the details of the PIC18F series, it is worthwhile to compare the features of the PIC18F series with those of the PIC16F series.

The following are similarities between PIC16F and PIC18F:

- Similar packages and pinouts

- Similar special function register (SFR) names and functions

- Similar peripheral devices

Table 2.1: The 18FXX2 microcontroller family

Feature	PIC18F242	PIC18F252	PIC18F442	PIC18F452
Program memory (Bytes)	16K	32K	16K	32K
Data memory (Bytes)	768	1536	768	1536
EEPROM (Bytes)	256	256	256	256
I/O Ports	A,B,C	A,B,C	A,B,C,D,E	A,B,C,D,E
Timers	4	4	4	4
Interrupt sources	17	17	18	18
Capture/compare/PWM	2	2	2	2
Serial communication	MSSP USART	MSSP USART	MSSP USART	MSSP USART
A/D converter (10-bit)	5 channels	5 channels	8 channels	8 channels
Low-voltage detect	yes	yes	yes	yes
Brown-out reset	yes	yes	yes	yes
Packages	28-pin DIP 28-pin SOIC	28-pin DIP 28-pin SOIC	40-pin DIP 44-pin PLCC 44-pin TQFP	40-pin DIP 44-pin PLCC 44-pin TQFP

- Subset of PIC18F instruction set

- Similar development tools

The following are new with the PIC18F series:

- Number of instructions doubled

- 16-bit instruction word

- Hardware 8×8 multiplier

- More external interrupts

- Priority-based interrupts

- Enhanced status register

- Increased program and data memory size

- Bigger stack

- Phase-locked loop (PLL) clock generator

- Enhanced input-output port architecture

- Set of configuration registers

- Higher speed of operation

- Lower power operation

2.1   PIC18FXX2 Architecture

As shown in Table 2.1, the PIC18FXX2 series consists of four devices. PIC18F2X2 microcontrollers are 28-pin devices, while PIC18F4X2 microcontrollers are 40-pin devices. The architectures of the two groups are almost identical except that the larger devices have more input-output ports and more A/D converter channels. In this section we shall be looking at the architecture of the PIC18F452 microcontroller in detail. The architectures of other standard PIC18F-series microcontrollers are similar, and the knowledge gained in this section should be enough to understand the operation of other PIC18F-series microcontrollers.

The pin configuration of the PIC18F452 microcontroller (DIP package) is shown in Figure 2.1. This is a 40-pin microcontroller housed in a DIL package, with a pin configuration similar to the popular PIC16F877.

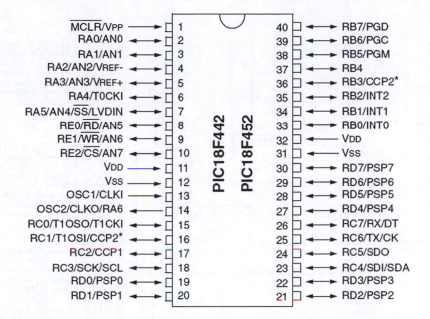

Figure 2.1: PIC18F452 microcontroller DIP pin configuration

Figure 2.2 shows the internal block diagram of the PIC18F452 microcontroller. The CPU is at the center of the diagram and consists of an 8-bit ALU, an 8-bit working accumulator register (WREG), and an 8×8 hardware multiplier. The higher byte and the lower byte of a multiplication are stored in two 8-bit registers called PRODH and PRODL respectively.

The program counter and program memory are shown in the upper left portion of the diagram. Program memory addresses consist of 21 bits, capable of accessing 2Mbytes of program memory locations. The PIC18F452 has only 32Kbytes of program memory, which requires only 15 bits. The remaining 6 address bits are redundant and not used. A table pointer provides access to tables and to the data stored in program memory. The program memory contains a 31-level stack which is normally used to store the interrupt and subroutine return addresses.

The data memory can be seen at the top center of the diagram. The data memory bus is 12 bits wide, capable of accessing 4Kbytes of data memory locations. As we shall see later, the data memory consists of special function registers (SFR) and general purpose registers, all organized in banks.

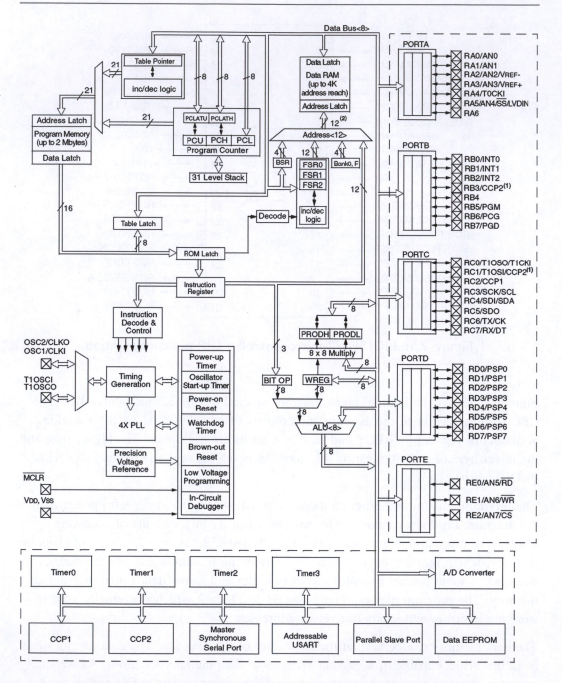

Figure 2.2: Block diagram of the PIC18F452 microcontroller

The bottom portion of the diagram shows the timers/counters, capture/compare/PWM registers, USART, A/D converter, and EEPROM data memory. The PIC18F452 consists of:

- 4 timers/counters

- 2 capture/compare/PWM modules

- 2 serial communication modules

- 8 10-bit A/D converter channels

- 256 bytes EEPROM

The oscillator circuit, located at the left side of the diagram, consists of:

- Power-up timer

- Oscillator start-up timer

- Power-on reset

- Watchdog timer

- Brown-out reset

- Low-voltage programming

- In-circuit debugger

- PLL circuit

- Timing generation circuit

The PLL circuit is new to the PIC18F series and provides the option of multiplying up the oscillator frequency to speed up the overall operation. The watchdog timer can be used to force a restart of the microcontroller in the event of a program crash. The in-circuit debugger is useful during program development and can be used to return diagnostic data, including the register values, as the microcontroller is executing a program.

The input-output ports are located at the right side of the diagram. The PIC18F452 has five parallel ports named PORTA, PORTB, PORTC, PORTD, and PORTE. Most port pins have multiple functions. For example, PORTA pins can be used as parallel inputs-outputs or analog inputs. PORTB pins can be used as parallel inputs-outputs or as interrupt inputs.

2.1.1   Program Memory Organization

The program memory map is shown in Figure 2.3. All PIC18F devices have a 21-bit program counter and hence are capable of addressing 2Mbytes of memory space. User memory space on the PIC18F452 microcontroller is 00000H to 7FFFH. Accessing a nonexistent memory location (8000H to 1FFFFFH) will cause a read of all 0s. The reset vector, where the program starts after a reset, is at address 0000. Addresses 0008H and

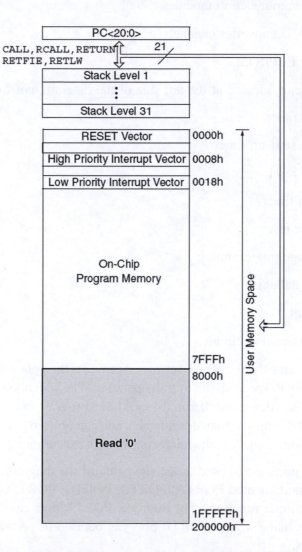

Figure 2.3: Program memory map of PIC18F452

0018H are reserved for the vectors of high-priority and low-priority interrupts respectively, and interrupt service routines must be written to start at one of these locations.

The PIC18F microcontroller has a 31-entry stack that is used to hold the return addresses for subroutine calls and interrupt processing. The stack is not part of the program or the data memory space. The stack is controlled by a 5-bit stack pointer which is initialized to 00000 after a reset. During a subroutine call (or interrupt) the stack pointer is first incremented, and the memory location it points to is written with the contents of the program counter. During the return from a subroutine call (or interrupt), the memory location the stack pointer has pointed to is decremented. The projects in this book are based on using the C language. Since subroutine and interrupt call/return operations are handled automatically by the C language compiler, their operation is not described here in more detail.

Program memory is addressed in bytes, and instructions are stored as two bytes or four bytes in program memory. The least significant byte of an instruction word is always stored in an even address of the program memory.

An instruction cycle consists of four cycles: A fetch cycle begins with the program counter incrementing in Q1. In the execution cycle, the fetched instruction is latched into the instruction register in cycle Q1. This instruction is decoded and executed during cycles Q2, Q3, and Q4. A data memory location is read during the Q2 cycle and written during the Q4 cycle.

2.1.2   Data Memory Organization

The data memory map of the PIC18F452 microcontroller is shown in Figure 2.4. The data memory address bus is 12 bits with the capability to address up to 4Mbytes. The memory in general consists of sixteen banks, each of 256 bytes, where only 6 banks are used. The PIC18F452 has 1536 bytes of data memory (6 banks × 256 bytes each) occupying the lower end of the data memory. Bank switching happens automatically when a high-level language compiler is used, and thus the user need not worry about selecting memory banks during programming.

The special function register (SFR) occupies the upper half of the top memory bank. SFR contains registers which control operations such as peripheral devices, timers/counters, A/D converter, interrupts, and USART. Figure 2.5 shows the SFR registers of the PIC18F452 microcontroller.

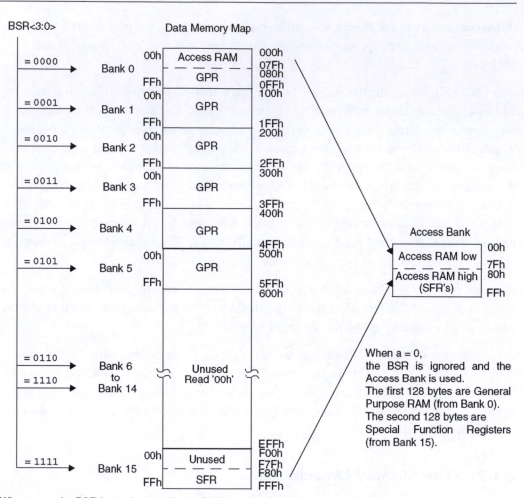

Figure 2.4: The PIC18F452 data memory map

2.1.3   The Configuration Registers

PIC18F452 microcontrollers have a set of configuration registers (PIC16-series microcontrollers had only one configuration register). Configuration registers are programmed during the programming of the flash program memory by the programming device. These registers are shown in Table 2.2. Descriptions of

Address	Name	Address	Name	Address	Name	Address	Name
FFFh	TOSU	FDFh	INDF2[3]	FBFh	CCPR1H	F9Fh	IPR1
FFEh	TOSH	FDEh	POSTINC2[3]	FBEh	CCPR1L	F9Eh	PIR1
FFDh	TOSL	FDDh	POSTDEC2[3]	FBDh	CCP1CON	F9Dh	PIE1
FFCh	STKPTR	FDCh	PREINC2[3]	FBCh	CCPR2H	F9Ch	—
FFBh	PCLATU	FDBh	PLUSW2[3]	FBBh	CCPR2L	F9Bh	—
FFAh	PCLATH	FDAh	FSR2H	FBAh	CCP2CON	F9Ah	—
FF9h	PCL	FD9h	FSR2L	FB9h	—	F99h	—
FF8h	TBLPTRU	FD8h	STATUS	FB8h	—	F98h	—
FF7h	TBLPTRH	FD7h	TMR0H	FB7h	—	F97h	—
FF6h	TBLPTRL	FD6h	TMR0L	FB6h	—	F96h	TRISE[2]
FF5h	TABLAT	FD5h	T0CON	FB5h	—	F95h	TRISD[2]
FF4h	PRODH	FD4h	—	FB4h	—	F94h	TRISC
FF3h	PRODL	FD3h	OSCCON	FB3h	TMR3H	F93h	TRISB
FF2h	INTCON	FD2h	LVDCON	FB2h	TMR3L	F92h	TRISA
FF1h	INTCON2	FD1h	WDTCON	FB1h	T3CON	F91h	—
FF0h	INTCON3	FD0h	RCON	FB0h	—	F90h	—
FEFh	INDF0[3]	FCFh	TMR1H	FAFh	SPBRG	F8Fh	—
FEEh	POSTINC0[3]	FCEh	TMR1L	FAEh	RCREG	F8Eh	—
FEDh	POSTDEC0[3]	FCDh	T1CON	FADh	TXREG	F8Dh	LATE[2]
FECh	PREINC0[3]	FCCh	TMR2	FACh	TXSTA	F8Ch	LATD[2]
FEBh	PLUSW0[3]	FCBh	PR2	FABh	RCSTA	F8Bh	LATC
FEAh	FSR0H	FCAh	T2CON	FAAh	—	F8Ah	LATB
FE9h	FSR0L	FC9h	SSPBUF	FA9h	EEADR	F89h	LATA
FE8h	WREG	FC8h	SSPADD	FA8h	EEDATA	F88h	—
FE7h	INDF1[3]	FC7h	SSPSTAT	FA7h	EECON2	F87h	—
FE6h	POSTINC1[3]	FC6h	SSPCON1	FA6h	EECON1	F86h	—
FE5h	POSTDEC1[3]	FC5h	SSPCON2	FA5h	—	F85h	—
FE4h	PREINC1[3]	FC4h	ADRESH	FA4h	—	F84h	PORTE[2]
FE3h	PLUSW1[3]	FC3h	ADRESL	FA3h	—	F83h	PORTD[2]
FE2h	FSR1H	FC2h	ADCON0	FA2h	IPR2	F82h	PORTC
FE1h	FSR1L	FC1h	ADCON1	FA1h	PIR2	F81h	PORTB
FE0h	BSR	FC0h	—	FA0h	PIE2	F80h	PORTA

Figure 2.5: The PIC18F452 SFR registers

these registers are given in Table 2.3. Some of the more important configuration registers are described in this section in detail.

CONFIG1H

The CONFIG1H configuration register is at address 300001H and is used to select the microcontroller clock sources. The bit patterns are shown in Figure 2.6.

Table 2.2: PIC18F452 configuration registers

	File Name	Bit 7	Bit 6	Bit 5	Bit 4	Bit 3	Bit 2	Bit 1	Bit 0	Default/ Unprogrammed Value
300001h	CONFIG1H	—	—	$\overline{OSCSEN}$	—	—	FOSC2	FOSC1	FOSC0	--1- -111
300002h	CONFIG2L	—	—	—	—	BORV1	BORV0	BOREN	$\overline{PWRTEN}$	---- 1111
300003h	CONFIG2H	—	—	—	—	WDTPS2	WDTPS1	WDTPS0	WDTEN	---- 1111
300005h	CONFIG3H	—	—	—	—	—	—	—	CCP2MX	---- ---1
300006h	CONFIG4L	$\overline{DEBUG}$	—	—	—	—	LVP	—	STVREN	1--- -1-1
300008h	CONFIG5L	—	—	—	—	CP3	CP2	CP1	CP0	---- 1111
300009h	CONFIG5H	CPD	CPB	—	—	—	—	—	—	11-- ----
30000Ah	CONFIG6L	—	—	—	—	WRT3	WRT2	WRT1	WRT0	---- 1111
30000Bh	CONFIG6H	WRTD	WRTB	WRTC	—	—	—	—	—	111- ----
30000Ch	CONFIG7L	—	—	—	—	EBTR3	EBTR2	EBTR1	EBTR0	---- 1111
30000Dh	CONFIG7H	—	EBTRB	—	—	—	—	—	—	-1-- ----
3FFFEh	DEVID1	DEV2	DEV1	DEV0	REV4	REV3	REV2	REV1	REV0	(1)
3FFFFh	DEVID2	DEV10	DEV9	DEV8	DEV7	DEV6	DEV5	DEV4	DEV3	0000 0100

Legend: x = unknown, u = unchanged, - = unimplemented, q = value depends on condition. Shaded cells are unimplemented, read as '0'.

Table 2.3: PIC18F452 configuration register descriptions

Configuration bits	Description
OSCSEN	Clock source switching enable
FOSC2:FOSC0	Oscillator modes
BORV1:BORV0	Brown-out reset voltage
BOREN	Brown-out reset enable
PWRTEN	Power-up timer enable
WDTPS2:WDTPS0	Watchdog timer postscale bits
WDTEN	Watchdog timer enable
CCP2MX	CCP2 multiplex
DEBUG	Debug enable
LVP	Low-voltage program enable
STVREN	Stack full/underflow reset enable
CP3:CP0	Code protection
CPD	EEPROM code protection
CPB	Boot block code protection
WRT3:WRT0	Program memory write protection
WRTD	EPROM write protection
WRTB	Boot block write protection
WRTC	Configuration register write protection
EBTR3:EBTR0	Table read protection
EBTRB	Boot block table read protection
DEV2:DEV0	Device ID bits (001 = 18F452)
REV4:REV0	Revision ID bits
DEV10:DEV3	Device ID bits

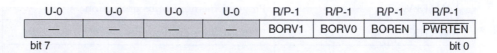

U-0	U-0	R/P-1	U-0	U-0	R/P-1	R/P-1	R/P-1
—	—	OSCSEN	—	—	FOSC2	FOSC1	FOSC0

bit 7   bit 0

bit 7-6   **Unimplemented:** Read as '0'

bit 5   **OSCSEN**: Oscillator System Clock Switch Enable bit

1 = Oscillator system clock switch option is disabled (main oscillator is source)
0 = Oscillator system clock switch option is enabled (oscillator switching is enabled)

bit 4-3   **Unimplemented:** Read as '0'

bit 2-0   **FOSC2:FOSC0**: Oscillator Selection bits

111 = RC oscillator w/ OSC2 configured as RA6
110 = HS oscillator with PLL enabled/Clock frequency = (4 x Fosc)
101 = EC oscillator w/ OSC2 configured as RA6
100 = EC oscillator w/ OSC2 configured as divide-by-4 clock output
011 = RC oscillator
010 = HS oscillator
001 = XT oscillator
000 = LP oscillator

Figure 2.6: CONFIG1H register bits

CONFIG2L

The CONFIG2L configuration register is at address 300002H and is used to select the brown-out voltage bits. The bit patterns are shown in Figure 2.7.

U-0	U-0	U-0	U-0	R/P-1	R/P-1	R/P-1	R/P-1
—	—	—	—	BORV1	BORV0	BOREN	PWRTEN

bit 7   bit 0

bit 7-4   **Unimplemented:** Read as '0'

bit 3-2   **BORV1:BORV0:** Brown-out Reset Voltage bits

11 = VBOR set to 2.5V
10 = VBOR set to 2.7V
01 = VBOR set to 4.2V
00 = VBOR set to 4.5V

bit 1   **BOREN:** Brown-out Reset Enable bit

1 = Brown-out Reset enabled
0 = Brown-out Reset disabled

bit 0   **PWRTEN:** Power-up Timer Enable bit

1 = PWRT disabled
0 = PWRT enabled

Figure 2.7: CONFIG2L register bits

CONFIG2H

The CONFIG2H configuration register is at address 300003H and is used to select the watchdog operations. The bit patterns are shown in Figure 2.8.

2.1.4   The Power Supply

The power supply requirements of the PIC18F452 microcontroller are shown in Figure 2.9. As shown in Figure 2.10, PIC18F452 can operate with a supply voltage of 4.2V to 5.5V at the full speed of 40MHz. The lower power version, PIC18LF452, can operate from 2.0 to 5.5 volts. At lower voltages the maximum clock frequency is 4MHz, which rises to 40MHz at 4.2V. The RAM data retention voltage is specified as 1.5V and will be lost if the power supply voltage is lowered below this value. In practice, most microcontroller-based systems are operated with a single +5V supply derived from a suitable voltage regulator.

2.1.5   The Reset

The reset action puts the microcontroller into a known state. Resetting a PIC18F microcontroller starts execution of the program from address 0000H of the

U-0	U-0	U-0	U-0	R/P-1	R/P-1	R/P-1	R/P-1
—	—	—	—	WDTPS2	WDTPS1	WDTPS0	WDTEN

bit 7   bit 0

bit 7-4   **Unimplemented:** Read as '0'

bit 3-1   **WDTPS2:WDTPS0:** Watchdog Timer Postscale Select bits

   111 = 1:128
   110 = 1:64
   101 = 1:32
   100 = 1:16
   011 = 1:8
   010 = 1:4
   001 = 1:2
   000 = 1:1

bit 0   **WDTEN:** Watchdog Timer Enable bit

   1 = WDT enabled
   0 = WDT disabled (control is placed on the SWDTEN bit)

Figure 2.8: CONFIG2H register bits

PIC18LFXX2 (Industrial)			Standard Operating Conditions (unless otherwise stated) Operating temperature -40°C ≤ TA ≤ +85°C for industrial				

PIC18FXX2
   (Industrial, Extended)
			Standard Operating Conditions (unless otherwise stated)
			Operating temperature -40°C ≤ TA ≤ +85°C for industrial
			                        -40°C ≤ TA ≤ +125°C for extended				

Param No.	Symbol	Characteristic	Min	Typ	Max	Units	Conditions
	V_{DD}	**Supply Voltage**					
D001		PIC18LFXX2	2.0	—	5.5	V	HS, XT, RC and LP Osc mode
D001		PIC18FXX2	4.2	—	5.5	V	
D002	V_{DR}	**RAM Data Retention Voltage**[1]	1.5	—	—	V	
D003	V_{POR}	**V_{DD} Start Voltage** to ensure internal Power-on Reset signal	—	—	0.7	V	See Section 3.1 (Power-on Reset) for details
D004	S_{VDD}	**V_{DD} Rise Rate** to ensure internal Power-on Reset signal	0.05	—	—	V/ms	See Section 3.1 (Power-on Reset) for details
	V_{BOR}	**Brown-out Reset Voltage**					
D005		PIC18LFXX2					
		BORV1:BORV0 = 11	1.98	—	2.14	V	85°C ≥ T ≥ 25°C
		BORV1:BORV0 = 10	2.67	—	2.89	V	
		BORV1:BORV0 = 01	4.16	—	4.5	V	
		BORV1:BORV0 = 00	4.45	—	4.83	V	
D005		PIC18FXX2					
		BORV1:BORV0 = 1x	N.A.	—	N.A.	V	Not in operating voltage range of device
		BORV1:BORV0 = 01	4.16	—	4.5	V	
		BORV1:BORV0 = 00	4.45	—	4.83	V	

Legend: Shading of rows is to assist in readability of the table.

Figure 2.9: The PIC8F452 power supply parameters

program memory. The microcontroller can be reset during one of the following operations:

- Power-on reset (POR)

- MCLR reset

- Watchdog timer (WDT) reset

- Brown-out reset (BOR)

- Reset instruction

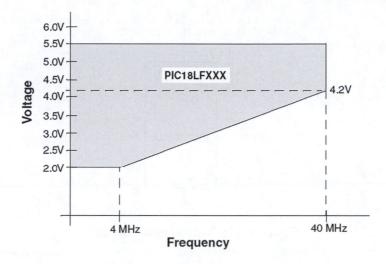

Figure 2.10: Operation of PIC18LF452 at different voltages

- Stack full reset
- Stack underflow reset

Two types of resets are commonly used: power-on reset and external reset using the MCLR pin.

Power-on Reset

The power-on reset is generated automatically when power supply voltage is applied to the chip. The MCLR pin should be tied to the supply voltage directly or, preferably, through a 10K resistor. Figure 2.11 shows a typical reset circuit.

For applications where the rise time of the voltage is slow, it is recommended to use a diode, a capacitor, and a series resistor as shown in Figure 2.12.

In some applications the microcontroller may have to be reset externally by pressing a button. Figure 2.13 shows the circuit that can be used to reset the microcontroller externally. Normally the MCLR input is at logic 1. When the RESET button is pressed, this pin goes to logic 0 and resets the microcontroller.

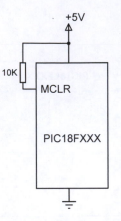

Figure 2.11: Typical reset circuit

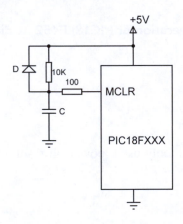

Figure 2.12: Reset circuit for slow-rising voltages

2.1.6 The Clock Sources

The PIC18F452 microcontroller can be operated from an external crystal or ceramic resonator connected to the microcontroller's OSC1 and OSC2 pins. In addition, an external resistor and capacitor, an external clock source, and in some models internal oscillators can be used to provide clock pulses to the microcontroller. There are eight clock sources on the PIC18F452 microcontroller, selected by the configuration register CONFIG1H. These are:

- Low-power crystal (LP)

- Crystal or ceramic resonator (XT)

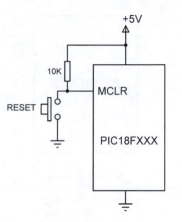

Figure 2.13: External reset circuit

- High-speed crystal or ceramic resonator (HS)

- High-speed crystal or ceramic resonator with PLL (HSPLL)

- External clock with $F_{OSC/4}$ on OSC2 (EC)

- External clock with I/O on OSC2 (port RA6) (ECIO)

- External resistor/capacitor with $F_{OSC/4}$ output on OSC2 (RC)

- External resistor/capacitor with I/O on OSC2 (port RA6) (RCIO)

Crystal or Ceramic Resonator Operation

The first several clock sources listed use an external crystal or ceramic resonator that is connected to the OSC1 and OSC2 pins. For applications where accuracy of timing is important, a crystal should be used. And if a crystal is used, a parallel resonant crystal must be chosen, since series resonant crystals do not oscillate when the system is first powered.

Figure 2.14 shows how a crystal is connected to the microcontroller. The capacitor values depend on the mode of the crystal and the selected frequency. Table 2.4 gives the recommended values. For example, for a 4MHz crystal frequency, use 15pF capacitors. Higher capacitance increases the oscillator stability but also increases the start-up time.

Resonators should be used in low-cost applications where high accuracy in timing is not required. Figure 2.15 shows how a resonator is connected to the microcontroller.

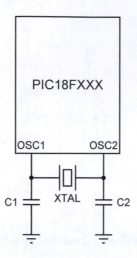

Figure 2.14: Using a crystal as the clock input

Table 2.4: Capacitor values

Mode	Frequency	C1,C2 (pF)
LP	32 KHz	33
	200 KHz	15
XT	200 KHz	22–68
	1.0 MHz	15
	4.0 MHz	15
HS	4.0 MHz	15
	8.0 MHz	15–33
	20.0 MHz	15–33
	25.0 MHz	15–33

The LP (low-power) oscillator mode is advised in applications to up to 200KHz clock. The XT mode is advised to up to 4MHz, and the HS (high-speed) mode is advised in applications where the clock frequency is between 4MHz to 25MHz.

An external clock source may also be connected to the OSC1 pin in the LP, XT, or HS modes as shown in Figure 2.16.

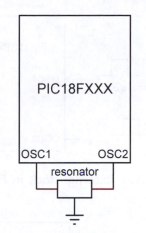

Figure 2.15: Using a resonator as the clock input

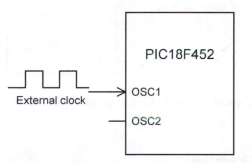

Figure 2.16: Connecting an external clock in LP, XT, or HS modes

External Clock Operation

An external clock source can be connected to the OSC1 input of the microcontroller in EC and ECIO modes. No oscillator start-up time is required after a power-on reset. Figure 2.17 shows the operation with the external clock in EC mode. Timing pulses at the frequency $F_{OSC/4}$ are available on the OSC2 pin. These pulses can be used for test purposes or to provide pulses to external devices.

The ECIO mode is similar to the EC mode, except that the OSC2 pin can be used as a general purpose digital I/O pin. As shown in Figure 2.18, this pin becomes bit 6 of PORTA (i.e., pin RA6).

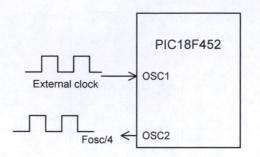

Figure 2.17: External clock in EC mode

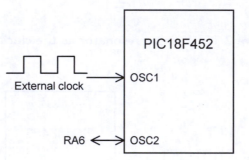

Figure 2.18: External clock in ECIO mode

Resistor/Capacitor Operation

In the many applications where accurate timing is not required we can use an external resistor and a capacitor to provide clock pulses. The clock frequency is a function of the resistor, the capacitor, the power supply voltage, and the temperature. The clock frequency is not accurate and can vary from unit to unit due to manufacturing and component tolerances. Table 2.5 gives the approximate clock frequency with various resistor and capacitor combinations. A close approximation of the clock frequency is 1/(4.2RC), where R should be between 3K and 100K and C should be greater than 20pF.

In RC mode, the oscillator frequency divided by 4 ($F_{OSC/4}$) is available on pin OSC2 of the microcontroller. Figure 2.19 shows the operation at a clock frequency of approximately 2MHz, where R = 3.9K and C = 30pF. In this application the clock frequency at the output of OSC2 is 2MHz/4 = 500KHz.

Table 2.5: Clock frequency with RC

C (pF)	R (K)	Frequency (MHz)
22	3.3	3.3
	4.7	2.3
	10	1.08
30	3.3	2.4
	4.7	1.7
	10	0.793

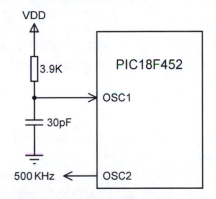

Figure 2.19: 2MHz clock in RC mode

RCIO mode is similar to RC mode, except that the OSC2 pin can be used as a general purpose digital I/O pin. As shown in Figure 2.20, this pin becomes bit 6 of PORTA (i.e., pin RA6).

Crystal or Resonator with PLL

One of the problems with using high-frequency crystals or resonators is electromagnetic interference. A Phase Locked Loop (PLL) circuit is provided that can be enabled to multiply the clock frequency by 4. Thus, for a crystal clock frequency of 10MHz, the

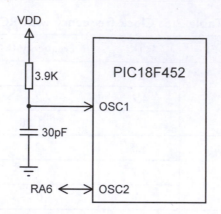

Figure 2.20: 2MHz clock in RCIO mode

internal operation frequency will be multiplied to 40MHz. The PLL mode is enabled when the oscillator configuration bits are programmed for HS mode.

Internal Clock

Some devices in the PIC18F family have internal clock modes (although the PIC18F452 does not). In this mode, OSC1 and OSC2 pins are available for general purpose I/O (RA6 and RA7) or as $F_{OSC/4}$ and RA7. An internal clock can be from 31KHz to 8MHz and is selected by registers OSCCON and OSCTUNE. Figure 2.21 shows the bits of internal clock control registers.

Clock Switching

It is possible to switch the clock from the main oscillator to a low-frequency clock source. For example, the clock can be allowed to run fast in periods of intense activity and slower when there is less activity. In the PIC18F452 microcontroller this is controlled by bit SCS of the OSCCON register. In microcontrollers of the PIC18F family that do support an internal clock, clock switching is controlled by bits SCS0 and SCS1 of OSCCON. It is important to ensure that during clock switching unwanted glitches do not occur in the clock signal. PIC18F microcontrollers contain circuitry to ensure error-free switching from one frequency to another.

OSCCON register

IDLEN	IRCF2	IRCF1	IRCF0	OSTS	IOFS	SCSI	SCS0

IDLEN	0	Run mode enabled
	1	Idle mode enabled

IRCF2:IRCF0	000	31 KHz
	001	125 KHz
	010	250 KHz
	011	500 KHz
	100	1 MHz
	101	2 MHz
	110	4 MHz
	111	8 MHz

OSTS	0	Oscillator start-up timer running
	1	Oscillator start-up timer expired

IOFS	0	Internal oscillator unstable
	1	Internal oscillator stable

SCSI:SCS0	00	Primary oscillator
	01	Timer 1 oscillator
	10	Internal oscillator
	11	Internal oscillator

OSCTUNE register

X	X	T5	T4	T3	T2	T1	T0

XX011111	Maximum frequency

XX000001	
XX000000	Center frequency
XX111111	

XX100000	Minimum frequency

Figure 2.21: Internal clock control registers

2.1.7 Watchdog Timer

In PIC18F-series microcontrollers family members the watchdog timer (WDT) is a free-running on-chip RC-based oscillator and does not require any external components. When the WDT times out, a device RESET is generated. If the device is in SLEEP mode, the WDT time-out will wake it up and continue with normal operation.

The watchdog is enabled/disabled by bit SWDTEN of register WDTCON. Setting SWDTEN = 1 enables the WDT, and clearing this bit turns off the WDT. On the PIC18F452 microcontroller an 8-bit postscaler is used to multiply the basic time-out period from 1 to 128 in powers of 2. This postscaler is controlled from configuration register CONFIG2H. The typical basic WDT time-out period is 18ms for a postscaler value of 1.

2.1.8   Parallel I/O Ports

The parallel ports in PIC18F microcontrollers are very similar to those of the PIC16 series. The number of I/O ports and port pins varies depending on which PIC18F microcontroller is used, but all of them have at least PORTA and PORTB. The pins of a port are labeled as RPn, where P is the port letter and n is the port bit number. For example, PORTA pins are labeled RA0 to RA7, PORTB pins are labeled RB0 to RB7, and so on.

When working with a port we may want to:

- Set port direction

- Set an output value

- Read an input value

- Set an output value and then read back the output value

The first three operations are the same in the PIC16 and the PIC18F series. In some applications we may want to send a value to the port and then read back the value just sent. The PIC16 series has a weakness in the port design such that the value read from a port may be different from the value just written to it. This is because the reading is the actual port bit pin value, and this value can be changed by external devices connected to the port pin. In the PIC18F series, a latch register (e.g., LATA for PORTA) is introduced to the I/O ports to hold the actual value sent to a port pin. Reading from the port reads the latched value, which is not affected by any external device.

In this section we shall be looking at the general structure of I/O ports.

PORTA

In the PIC18F452 microcontroller PORTA is 7 bits wide and port pins are shared with other functions. Table 2.6 shows the PORTA pin functions.

Table 2.6: PIC18F452 PORTA pin functions

Pin	Description
RA0/AN0	
RA0	Digital I/O
AN0	Analog input 0
RA1/AN1	
RA1	Digital I/O
AN1	Analog input 1
RA2/AN2/VREF−	
RA2	Digital I/O
AN2	Analog input 2
VREF−	A/D reference voltage (low) input
RA3/AN3/VREF+	
RA3	Digital I/O
AN3	Analog input 3
VREF+	A/D reference voltage (high) input
RA4/T0CKI	
RA4	Digital I/O
T0CKI	Timer 0 external clock input
RA5/AN4/SS/LVDIN	
RA5	Digital I/O
AN4	Analog input 4
SS	SPI Slave Select input
RA6	Digital I/O

The architecture of PORTA is shown in Figure 2.22. There are three registers associated with PORTA:

- Port data register—PORTA

- Port direction register—TRISA

- Port latch register—LATA

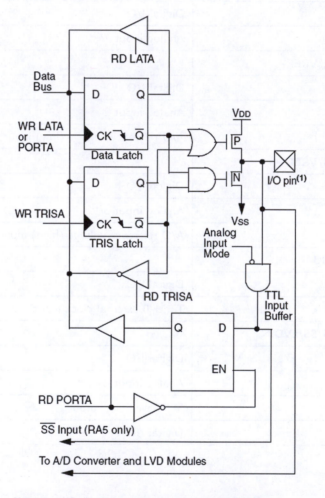

Note 1:   I/O pins have protection diodes to VDD and VSS.

Figure 2.22: PIC18F452 PORTA RA0–RA3 and RA5 pins

PORTA is the name of the port data register. The TRISA register defines the direction of PORTA pins, where a logic 1 in a bit position defines the pin as an input pin, and a 0 in a bit position defines it as an output pin. LATA is the output latch register which shares the same data latch as PORTA. Writing to one is equivalent to writing to the other. But reading from LATA activates the buffer at the top of the diagram, and the value held in the PORTA/LATA data latch is transferred to the data bus independent of the state of the actual output pin of the microcontroller.

Bits 0 through 3 and 5 of PORTA are also used as analog inputs. After a device reset, these pins are programmed as analog inputs and RA4 and RA6 are configured as digital inputs. To program the analog inputs as digital I/O, the ADCON1 register (A/D register) must be programmed accordingly. Writing 7 to ADCON1 configures all PORTA pins as digital I/O.

The RA4 pin is multiplexed with the Timer 0 clock input (T0CKI). This is a Schmitt trigger input and an open drain output.

RA6 can be used as a general purpose I/O pin, as the OSC2 clock input, or as a clock output providing $F_{OSC/4}$ clock pulses.

PORTB

In PIC18F452 microcontroller PORTB is an 8-bit bidirectional port shared with interrupt pins and serial device programming pins. Table 2.7 gives the PORTB bit functions.

PORTB is controlled by three registers:

- Port data register—PORTB

- Port direction register—TRISB

- Port latch register—LATB

The general operation of PORTB is similar to that of PORTA. Figure 2.23 shows the architecture of PORTB. Each port pin has a weak internal pull-up which can be enabled by clearing bit RBPU of register INTCON2. These pull-ups are disabled on a power-on reset and when the port pin is configured as an output. On a power-on reset, PORTB pins are configured as digital inputs. Internal pull-ups allow input devices such as switches to be connected to PORTB pins without the use of external pull-up resistors. This saves costs because the component count and wiring requirements are reduced.

Table 2.7: PIC18F452 PORTB pin functions

Pin	Description
RB0/INT0	
RB0	Digital I/O
INT0	External interrupt 0
RB1/INT1	
RB1	Digital I/O
INT1	External interrupt 1
RB2/INT2	
RB2	Digital I/O
INT2	External interrupt 2
RB3/ CCP2	
RB3	Digital I/O
CCP2	Capture 2 input, compare 2, and PWM2 output
RB4	Digital I/O, interrupt on change pin
RB5/PGM	
RB5	Digital I/O, interrupt on change pin
PGM	Low-voltage ICSP programming pin
RB6/PGC	
RB6	Digital I/O, interrupt on change pin
PGC	In-circuit debugger and ICSP programming pin
RB7/PGD	
RB7	Digital I/O, interrupt on change pin
PGD	In-circuit debugger and ICSP programming pin

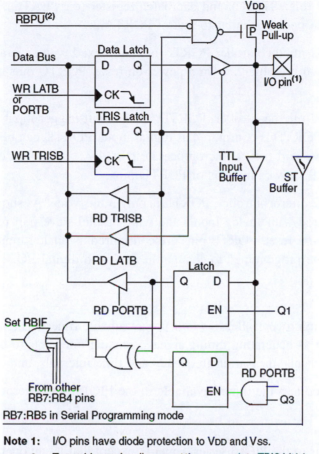

Figure 2.23: PIC18F452 PORTB RB4–RB7 pins

Port pins RB4–RB7 can be used as interrupt-on-change inputs, whereby a change on any of pins 4 through 7 causes an interrupt flag to be set. The interrupt enable and flag bits RBIE and RBIF are in register INTCON.

PORTC, PORTD, PORTE, and Beyond

In addition to PORTA and PORTB, the PIC18F452 has 8-bit bidirectional ports PORTC and PORTD, and 3-bit PORTE. Each port has its own data register (e.g., PORTC), data direction register (e.g., TRISC), and data latch register (e.g., LATC). The general operation of these ports is similar to that of PORTA.2.1.

In the PIC18F452 microcontroller PORTC is multiplexed with several peripheral functions as shown in Table 2.8. On a power-on reset, PORTC pins are configured as digital inputs.

In the PIC18F452 microcontroller, PORTD has Schmitt trigger input buffers. On a power-on reset, PORTD is configured as digital input. PORTD can be configured as an 8-bit parallel slave port (i.e., a microprocessor port) by setting bit 4 of the TRISE register. Table 2.9 shows functions of PORTD pins.

In the PIC18F452 microcontroller, PORTE is only 3 bits wide. As shown in Table 2.10, port pins are shared with analog inputs and with parallel slave port read/write control bits. On a power-on reset, PORTE pins are configured as analog inputs and register ADCON1 must be programmed to change these pins to digital I/O.

2.1.9   Timers

The PIC18F452 microcontroller has four programmable timers which can be used in many tasks, such as generating timing signals, causing interrupts to be generated at specific time intervals, measuring frequency and time intervals, and so on.

This section introduces the timers available in the PIC18F452 microcontroller.

Timer 0

Timer 0 is similar to the PIC16 series Timer 0, except that it can operate either in 8-bit or in 16-bit mode. Timer 0 has the following basic features:

- 8-bit or 16-bit operation
- 8-bit programmable prescaler
- External or internal clock source
- Interupt generation on overflow

Timer 0 control register is T0CON, shown in Figure 2.24. The lower 6 bits of this register have similar functions to the PIC16-series OPTION register. The top two bits are used to select the 8-bit or 16-bit mode of operation and to enable/disable the timer.

Table 2.8: PIC18F452 PORTC pin functions

Pin	Description
RC0/T1OSO/T1CKI	
RC0	Digital I/O
T1OSO	Timer 1 oscillator output
T1CKI	Timer 1/Timer 3 external clock input
RC1/T1OSI/CCP2	
RC1	Digital I/O
T1OSI	Timer 1 oscillator input
CCP2	Capture 2 input, Compare 2 and PWM2 output
RC2/CCP1	
RC2	Digital I/O
CCP1	Capture 1 input, Compare 1 and PWM1 output
RC3/SCK/SCL	
RC3	Digital I/O
SCK	Synchronous serial clock input/output for SPI
SCL	Synchronous serial clock input/output for I^2C
RC4/SDI/SDA	
RC4	Digital I/O
SDI	SPI data in
SDA	I^2C data I/O
RC5/SDO	
RC5	Digital I/O
SDO	SPI data output
RC6/TX/CK	
RC6	Digital I/O
TX	USART transmit pin
CK	USART synchronous clock pin
RC7/RX/DT	
RC7	Digital I/O
RX	USART receive pin
DT	USART synchronous data pin

Table 2.9: PIC18F452 PORTD pin functions

Pin	Description
RD0/PSP0	
RD0	Digital I/O
PSP0	Parallel slave port bit 0
RD1/PSP1	
RD1	Digital I/O
PSP1	Parallel slave port bit 1
RD2/PSP2	
RD2	Digital I/O
PSP2	Parallel slave port bit 2
RD3/PSP3	
RD3	Digital I/O
PSP3	Parallel slave port bit 3
RD4/PSP4	
RD4	Digital I/O
PSP4	Parallel slave port bit 4
RD5/PSP5	
RD5	Digital I/O
PSP5	Parallel slave port bit 5
RD6/PSP6	
RD6	Digital I/O
PSP6	Parallel slave port bit 6
RD7/PSP7	
RD7	Digital I/O
PSP7	Parallel slave port bit 7

Table 2.10: PIC18F452 PORTE pin functions

Pin	Description
RE0/RD/AN5	
RE0	Digital I/O
RD	Parallel slave port read control pin
AN5	Analog input 5
RE1/WR/ AN6	
RE1	Digital I/O
WR	Parallel slave port write control pin
AN6	Analog input 6
RE2/CS/AN7	
RE2	Digital I/O
CS	Parallel slave port CS
AN7	Analog input 7

Timer 0 can be operated either as a timer or as a counter. Timer mode is selected by clearing the T0CS bit, and in this mode the clock to the timer is derived from $F_{OSC/4}$. Counter mode is selected by setting the T0CS bit, and in this mode Timer 0 is incremented on the rising or falling edge of input RA4/T0CKI. Bit T0SE of T0CON selects the edge triggering mode.

An 8-bit prescaler can be used to change the timer clock rate by a factor of up to 256. The prescaler is selected by bits PSA and T0PS2:T0PS0 of register T0CON.

8-Bit Mode   Figure 2.25 shows Timer 0 in 8-bit mode. The following operations are normally carried out in a timer application:

- Clear T0CS to select clock $F_{OSC/4}$

- Use bits T0PS2:T0PS0 to select a suitable prescaler value

- Clear PSA to select the prescaler

R/W-1	R/W-1	R/W-1	R/W-1	R/W-1	R/W-1	R/W-1	R/W-1
TMR0ON	T08BIT	T0CS	T0SE	PSA	T0PS2	T0PS1	T0PS0

bit 7   bit 0

bit 7   **TMR0ON**: Timer0 On/Off Control bit

1 = Enables Timer0

0 = Stops Timer0

bit 6   **T08BIT**: Timer0 8-bit/16-bit Control bit

1 = Timer0 is configured as an 8-bit timer/counter

0 = Timer0 is configured as a 16-bit timer/counter

bit 5   **T0CS**: Timer0 Clock Source Select bit

1 = Transition on T0CKI pin

0 = Internal instruction cycle clock (CLKO)

bit 4   **T0SE**: Timer0 Source Edge Select bit

1 = Increment on high-to-low transition on T0CKI pin

0 = Increment on low-to-high transition on T0CKI pin

bit 3   **PSA**: Timer0 Prescaler Assignment bit

1 = TImer0 prescaler is NOT assigned. Timer0 clock input bypasses prescaler.

0 = Timer0 prescaler is assigned. Timer0 clock input comes from prescaler output.

bit 2-0   **T0PS2:T0PS0**: Timer0 Prescaler Select bits

111 = 1:256 prescale value

110 = 1:128 prescale value

101 = 1:64 prescale value

100 = 1:32 prescale value

011 = 1:16 prescale value

010 = 1:8   prescale value

001 = 1:4   prescale value

000 = 1:2   prescale value

Figure 2.24: Timer 0 control register, T0CON

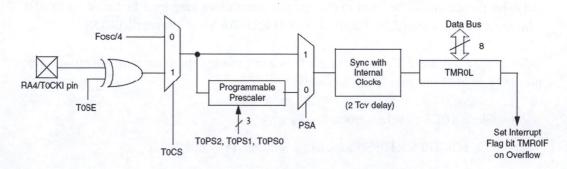

Figure 2.25: Timer 0 in 8-bit mode

- Load timer register TMR0L

- Optionally enable Timer 0 interrupts

- The timer counts up and an interrupt is generated when the timer value overflows from FFH to 00H in 8-bit mode (or from FFFFH to 0000H in 16-bit mode)

By loading a value into the TMR0 register we can control the count until an overflow occurs. The formula that follows can be used to calculate the time it will take for the timer to overflow (or to generate an interrupt) given the oscillator period, the value loaded into the timer, and the prescaler value:

$$\text{Overflow time} = 4 \times T_{OSC} \times \text{Prescaler} \times (256 - TMR0) \tag{2.1}$$

where

Overflow time is in μs

T_{OSC} is the oscillator period in μs

Prescaler is the prescaler value

TMR0 is the value loaded into TMR0 register

For example, assume that we are using a 4MHz crystal, and the prescaler is chosen as 1:8 by setting bits PS2:PS0 to 010. Also assume that the value loaded into the timer register TMR0 is decimal 100. The overflow time is then given by:

$$4\text{MHZ clock has a period}, T = 1/f = 0.25\mu s$$

using the above formula

$$\text{Overflow time} = 4 \times 0.25 \times 8 \times (256 - 100) = 1248\mu s$$

Thus, the timer will overflow after 1.248msec, and a timer interrupt will be generated if the timer interrupt and global interrupts are enabled.

What we normally want is to know what value to load into the TMR0 register for a required overflow time. This can be calculated by modifying Equation (2.1) as follows:

$$TMR0 = 256 - (\text{Overflow time})/(4 \times T_{OSC} \times \text{Prescaler}) \tag{2.2}$$

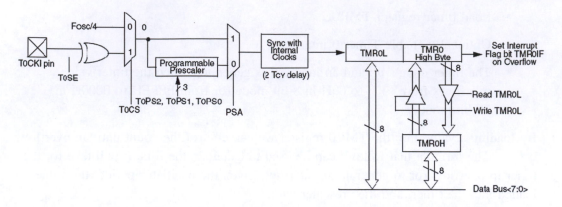

Figure 2.26: Timer 0 in 16-bit mode

For example, suppose we want an interrupt to be generated after 500μs and the clock and the prescaler values are as before. The value to be loaded into the TMR0 register can be calculated using Equation (2.2) as follows:

$$TMR0 = 256 - 500/(4 \times 0.25 \times 8) = 193.5$$

The closest number we can load into TMR0 register is 193.

16-Bit Mode The Timer 0 in 16-bit mode is shown in Figure 2.26. Here, two timer registers named TMR0L and TMR0 are used to store the 16-bit timer value. The low byte TMR0L is directly loadable from the data bus. The high byte TMR0 can be loaded through a buffer called TMR0H. During a read of TMR0L, the high byte of the timer (TMR0) is also loaded into TMR0H, and thus all 16 bits of the timer value can be read. To read the 16-bit timer value, first we have to read TMR0L, and then read TMR0H in a later instruction. Similarly, during a write to TMR0L, the high byte of the timer is also updated with the contents of TMR0H, allowing all 16 bits to be written to the timer. Thus, to write to the timer the program should first write the required high byte to TMR0H. When the low byte is written to TMR0L, then the value stored in TMR0H is automatically transferred to TMR0, thus causing all 16 bits to be written to the timer.

Timer 1

PIC18F452 Timer 1 is a 16-bit timer controlled by register T1CON, as shown in Figure 2.27. Figure 2.28 shows the internal structure of Timer 1.

R/W-0	U-0	R/W-0	R/W-0	R/W-0	R/W-0	R/W-0	R/W-0
RD16	—	T1CKPS1	T1CKPS0	T1OSCEN	$\overline{\text{T1SYNC}}$	TMR1CS	TMR1ON
bit 7							bit 0

bit 7   **RD16:** 16-bit Read/Write Mode Enable bit

                 1 = Enables register Read/Write of Timer1 in one 16-bit operation
                 0 = Enables register Read/Write of Timer1 in two 8-bit operations

bit 6   **Unimplemented:** Read as '0'

bit 5-4   **T1CKPS1:T1CKPS0**: Timer1 Input Clock Prescale Select bits

                 11 = 1:8 Prescale value
                 10 = 1:4 Prescale value
                 01 = 1:2 Prescale value
                 00 = 1:1 Prescale value

bit 3   **T1OSCEN:** Timer1 Oscillator Enable bit

                 1 = Timer1 Oscillator is enabled
                 0 = Timer1 Oscillator is shut-off
                     The oscillator inverter and feedback resistor are turned off to eliminate power drain.

bit 2   **T1SYNC:** Timer1 External Clock Input Synchronization Select bit

                 <u>When TMR1CS = 1:</u>
                 1 = Do not synchronize external clock input
                 0 = Synchronize external clock input
                 <u>When TMR1CS = 0:</u>
                 This bit is ignored. Timer1 uses the internal clock when TMR1CS = 0.

bit 1   **TMR1CS:** Timer1 Clock Source Select bit

                 1 = External clock from pin RC0/T1OSO/T13CKI (on the rising edge)
                 0 = Internal clock ($F_{OSC}/4$)

bit 0   **TMR1ON:** Timer1 On bit

                 1 = Enables Timer1
                 0 = Stops Timer1

Figure 2.27: Timer 1 control register, T1CON

Timer 1 can be operated as either a timer or a counter. When bit TMR1CS of register T1CON is low, clock $F_{OSC/4}$ is selected for the timer. When TMR1CS is high, the module operates as a counter clocked from input T1OSI. A crystal oscillator circuit, enabled from bit T1OSCEN of T1CON, is built between pins T1OSI and T1OSO where a crystal up to 200KHz can be connected between these pins. This oscillator is primarily intended for a 32KHz crystal operation in real-time clock applications. A prescaler is used in Timer 1 that can change the timing rate as a factor of 1, 2, 4, or 8.

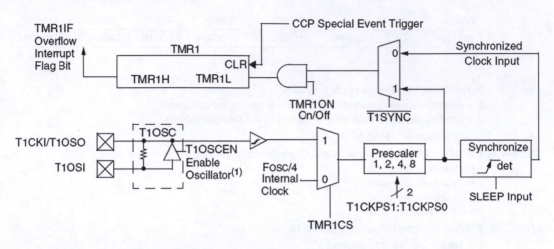

Figure 2.28: Internal structure of Timer 1

Timer 1 can be configured so that read/write can be performed either in 16-bit mode or in two 8-bit modes. Bit RD16 of register T1CON controls the mode. When RD16 is low, timer read and write operations are performed as two 8-bit operations. When RD16 is high, the timer read and write operations are as in Timer 0 16-bit mode (i.e., a buffer is used between the timer register and the data bus) (see Figure 2.29).

If the Timer 1 interrupts are enabled, an interrupt will be generated when the timer value rolls over from FFFFH to 0000H.

Timer 2

Timer 2 is an 8-bit timer with the following features:

- 8-bit timer (TMR2)

- 8-bit period register (PR2)

- Programmable prescaler

- Programmable postscaler

- Interrupt when TM2 matches PR2

Timer 2 is controlled from register T2CON, as shown in Figure 2.30. Bits T2CKPS1: T2CKPS0 set the prescaler for a scaling of 1, 4, and 16. Bits TOUTPS3:TOUTPS0 set

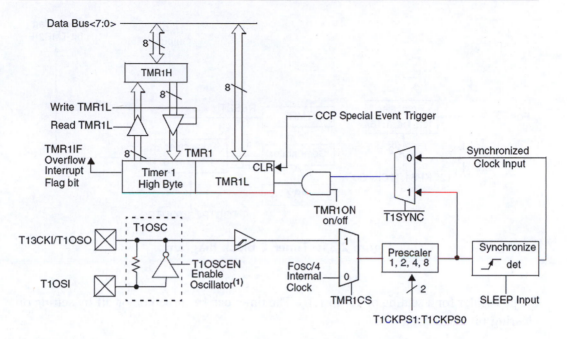

Figure 2.29: Timer 1 in 16-bit mode

U-0	R/W-0	R/W-0	R/W-0	R/W-0	R/W-0	R/W-0	R/W-0
—	TOUTPS3	TOUTPS2	TOUTPS1	TOUTPS0	TMR2ON	T2CKPS1	T2CKPS0

bit 7　　　　　　　　　　　　　　　　　　　　　　　　　　　　　bit 0

bit 7　　　**Unimplemented:** Read as '0'

bit 6-3　　**TOUTPS3:TOUTPS0**: Timer2 Output Postscale Select bits

　　　　　0000 = 1:1 Postscale
　　　　　0001 = 1:2 Postscale
　　　　　•
　　　　　•
　　　　　•
　　　　　1111 = 1:16 Postscale

bit 2　　　**TMR2ON**: Timer2 On bit

　　　　　1 = Timer2 is on
　　　　　0 = Timer2 is off

bit 1-0　　**T2CKPS1:T2CKPS0**: Timer2 Clock Prescale Select bits

　　　　　00 = Prescaler is 1
　　　　　01 = Prescaler is 4
　　　　　1x = Prescaler is 16

Figure 2.30: Timer 2 control register, T2CON

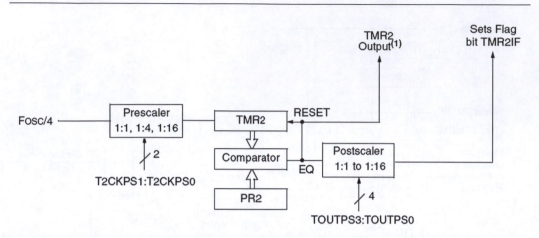

Figure 2.31: Timer 2 block diagram

the postscaler for a scaling of 1:1 to 1:16. The timer can be turned on or off by setting or clearing bit TMR2ON.

The block diagram of Timer 2 is shown in Figure 2.31. Timer 2 can be used for the PWM mode of the CCP module. The output of Timer 2 can be software selected by the SSP module as a baud clock. Timer 2 increments from 00H until it matches PR2 and sets the interrupt flag. It then resets to 00H on the next cycle.

Timer 3

The structure and operation of Timer 3 is the same as for Timer 1, having registers TMR3H and TMR3L. This timer is controlled from register T3CON as shown in Figure 2.32.

The block diagram of Timer 3 is shown in Figure 2.33.

2.1.10   Capture/Compare/PWM Modules (CCP)

The PIC18F452 microcontroller has two capture/compare/PWM (CCP) modules, and they work with Timers 1, 2, and 3 to provide capture, compare, and pulse width modulation (PWM) operations. Each module has two 8-bit registers. Module 1 registers are CCPR1L and CCPR1H, and module 2 registers are CCPR2L and CCPR2H. Together, each register pair forms a 16-bit register and can be used to capture, compare, or generate waveforms with a specified duty cycle. Module 1 is controlled by register

R/W-0	R/W-0	R/W-0	R/W-0	R/W-0	R/W-0	R/W-0	R/W-0
RD16	T3CCP2	T3CKPS1	T3CKPS0	T3CCP1	T3SYNC	TMR3CS	TMR3ON
bit 7							bit 0

bit 7   **RD16:** 16-bit Read/Write Mode Enable bit

1 = Enables register Read/Write of Timer3 in one 16-bit operation
0 = Enables register Read/Write of Timer3 in two 8-bit operations

bit 6-3   **T3CCP2:T3CCP1:** Timer3 and Timer1 to CCPx Enable bits

1x = Timer3 is the clock source for compare/capture CCP modules
01 = Timer3 is the clock source for compare/capture of CCP2,
      Timer1 is the clock source for compare/capture of CCP1
00 = Timer1 is the clock source for compare/capture CCP modules

bit 5-4   **T3CKPS1:T3CKPS0:** Timer3 Input Clock Prescale Select bits

11 = 1:8 Prescale value
10 = 1:4 Prescale value
01 = 1:2 Prescale value
00 = 1:1 Prescale value

bit 2   **T3SYNC:** Timer3 External Clock Input Synchronization Control bit
(Not usable if the system clock comes from Timer1/Timer3)
<u>When TMR3CS = 1:</u>
1 = Do not synchronize external clock input
0 = Synchronize external clock input
<u>When TMR3CS = 0:</u>
This bit is ignored. Timer3 uses the internal clock when TMR3CS = 0.

bit 1   **TMR3CS:** Timer3 Clock Source Select bit

1 = External clock input from Timer1 oscillator or T1CKI
      (on the rising edge after the first falling edge)
0 = Internal clock (Fosc/4)

bit 0   **TMR3ON:** Timer3 On bit

1 = Enables Timer3
0 = Stops Timer3

Figure 2.32: Timer 3 control register, T3CON

CCP1CON, and module 2 is controlled by CCP2CON. Figure 2.34 shows the bit allocations of the CCP control registers.

Capture Mode

In capture mode, the registers operate like a stopwatch. When an event occurs, the time of the event is recorded, although the clock continues running (a stopwatch, on the other hand, stops when the event time is recorded).

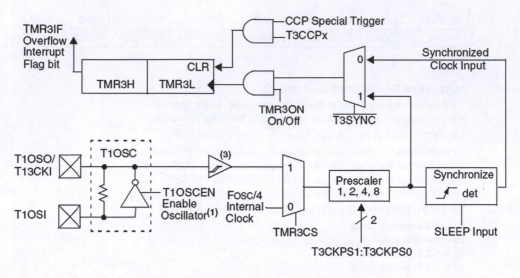

Figure 2.33: Block diagram of Timer 3

Figure 2.35 shows the capture mode of operation. Here, CCP1 will be considered, but the operation of CCP2 is identical with the register and port names changed accordingly. In this mode CCPR1H:CCPR1L captures the 16-bit value of the TMR1 or TMR3 registers when an event occurs on pin RC2/CCP1 (pin RC2/CCP1 must be configured as an input pin using TRISC). An external signal can be prescaled by 4 or 16. The event is selected by control bits CCP1M3:CCP1M0, and any of the following events can be selected:

- Every falling edge

- Every rising edge

- Every fourth rising edge

- Every sixteenth rising edge

If the capture interrupt is enabled, the occurrence of an event causes an interrupt to be generated in software. If another capture occurs before the value in register CCPR1 is read, the old captured value is overwritten by the new captured value.

Either Timer 1 or Timer 3 can be used in capture mode. They must be running in timer mode, or in synchronized counter mode, selected by register T3CON.

U-0	U-0	R/W-0	R/W-0	R/W-0	R/W-0	R/W-0	R/W-0
—	—	DCxB1	DCxB0	CCPxM3	CCPxM2	CCPxM1	CCPxM0
bit 7							bit 0

bit 7-6   **Unimplemented:** Read as '0'

bit 5-4   **DCxB1:DCxB0:** PWM Duty Cycle bit1 and bit0

<u>Capture mode:</u>
Unused

<u>Compare mode:</u>
Unused

<u>PWM mode:</u>
These bits are the two LSbs (bit1 and bit0) of the 10-bit PWM duty cycle. The upper eight bits (DCx9:DCx2) of the duty cycle are found in CCPRxL.

bit 3-0   **CCPxM3:CCPxM0:** CCPx Mode Select bits

0000 = Capture/Compare/PWM disabled (resets CCPx module)
0001 = Reserved
0010 = Compare mode, toggle output on match (CCPxIF bit is set)
0011 = Reserved
0100 = Capture mode, every falling edge
0101 = Capture mode, every rising edge
0110 = Capture mode, every 4th rising edge
0111 = Capture mode, every 16th rising edge
1000 = Compare mode,
        Initialize CCP pin Low, on compare match force CCP pin High (CCPIF bit is set)
1001 = Compare mode,
        Initialize CCP pin High, on compare match force CCP pin Low (CCPIF bit is set)
1010 = Compare mode,
        Generate software interrupt on compare match (CCPIF bit is set, CCP pin is unaffected)
1011 = Compare mode,
        Trigger special event (CCPIF bit is set)
11xx = PWM mode

Figure 2.34: CCPxCON register bit allocations

Compare Mode

In compare mode, a digital comparator is used to compare the value of Timer 1 or Timer 3 to the value in a 16-bit register pair. When a match occurs, the output state of a pin is changed. Figure 2.36 shows the block diagram of compare mode in operation.

Here only module CCP1 is considered, but the operation of module CCP2 is identical.

The value of the 16-bit register pair CCPR1H:CCPR1L is continuously compared against the Timer 1 or Timer 3 value. When a match occurs, the state of the RC2/CCP1

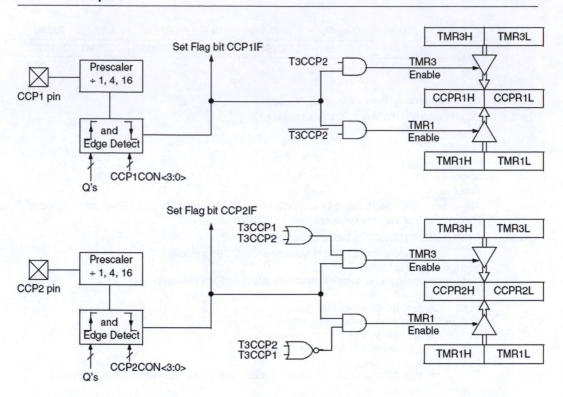

Figure 2.35: Capture mode of operation

pin is changed depending on the programming of bits CCP1M2:CCP1M0 of register CCP1CON. The following changes can be programmed:

- Force RC2/CCP1 high

- Force RC2/CCP1 low

- Toggle RC2/CCP1 pin (low to high or high to low)

- Generate interrupt when a match occurs

- No change

Timer 1 or Timer 3 must be running in timer mode or in synchronized counter mode, selected by register T3CON.

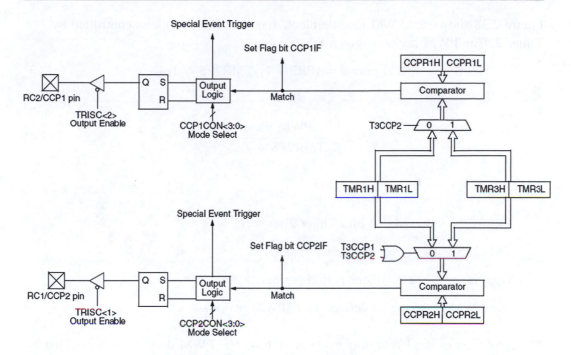

Figure 2.36: Compare mode of operation

PWM Module

The pulse width modulation (PWM) mode produces a PWM output at 10-bit resolution. A PWM output is basically a square waveform with a specified period and duty cycle. Figure 2.37 shows a typical PWM waveform.

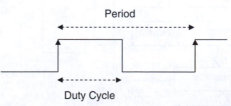

Figure 2.37: Typical PWM waveform

Figure 2.38 shows the PWM module block diagram. The module is controlled by Timer 2. The PWM period is given by:

$$\text{PWM period} = (PR2 + 1)^*\text{TMR2PS}^*4^*T_{OSC} \qquad (2.3)$$

or

$$PR2 = \frac{\text{PWM period}}{\text{TMR2PS}^*4^*T_{OSC}} - 1 \qquad (2.4)$$

where

PR2 is the value loaded into Timer 2 register

TMR2PS is the Timer 2 prescaler value

T_{OSC} is the clock oscillator period (seconds)

The PWM frequency is defined as 1/(PWM period).

The resolution of the PWM duty cycle is 10 bits. The PWM duty cycle is selected by writing the eight most significant bits into the CCPR1L register and the two least

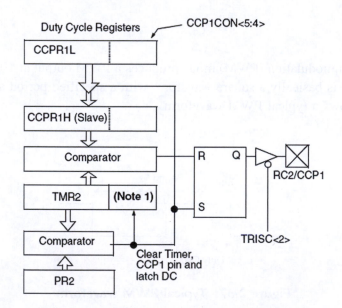

Figure 2.38: PWM module block diagram

significant bits into bits 4 and 5 of CCP1CON register. The duty cycle (in seconds) is given by:

$$\text{PWM duty cycle} = (\text{CCPR1L}:\text{CCP1CON} < 5:4 >)^*\text{TMR2PS}^*\text{T}_{OSC} \qquad (2.5)$$

or

$$\text{CCPR1L}:\text{CCP1CON} < 5:4 > = \frac{\text{PWM duty cycle}}{\text{TMR2PS}^*\text{T}_{OSC}} \qquad (2.6)$$

The steps to configure the PWM are as follows:

- Specify the required period and duty cycle.

- Choose a value for the Timer 2 prescaler (TMR2PS).

- Calculate the value to be written into the PR2 register using Equation (2.2).

- Calculate the value to be loaded into the CCPR1L and CCP1CON registers using Equation (2.6).

- Clear bit 2 of TRISC to make CCP1 pin an output pin.

- Configure the CCP1 module for PWM operation using register CCP1CON.

The following example shows how the PWM can be set up.

Example 2.1

PWM pulses must be generated from pin CCP1 of a PIC18F452 microcontroller. The required pulse period is 44μs and the required duty cycle is 50%. Assuming that the microcontroller operates with a 4MHz crystal, calculate the values to be loaded into the various registers.

Solution 2.1

$$\text{Using a 4MHz crystal}, \text{T}_{OSC} = 1/4 = 0.25 \times 10^{-6}$$

The required PWM duty cycle is 44/2 = 22μs.

From Equation (2.4), assuming a timer prescaler factor of 4, we have:

$$\text{PR2} = \frac{\text{PWM period}}{\text{TMR2PS}^*4^*\text{T}_{OSC}} - 1$$

or

$$PR2 = \frac{44 \times 10^{-6}}{4 * 4 * 0.25 \times 10^{-6}} - 1 = 10 \qquad \text{i.e., 0AH}$$

and from Equation (2.6)

$$CCPR1L : CCP1CON < 5 : 4 > = \frac{\text{PWM duty cycle}}{\text{TMR2PS} * T_{OSC}}$$

or

$$CCPR1L : CCP1CON < 5 : 4 > = \frac{22 \times 10^{-6}}{4 * 0.25 \times 10^{-6}} = 22$$

But the equivalent of number 22 in 10-bit binary is:

"00 00010110"

Therefore, the value to be loaded into bits 4 and 5 of CCP1CON is "00." Bits 2 and 3 of CCP1CON must be set to high for PWM operation. Therefore, CCP1CON must be set to bit pattern ("X" is "don't care"):

XX001100

Taking the don't-care entries as 0, we can set CCP1CON to hexadecimal 0CH.

The value to be loaded into CCPR1L is "00010110" (i.e., hexadecimal number 16H).

The required steps are summarized as follows:

- Load Timer 2 with prescaler of 4 (i.e., load T2CON) with 00000101 (i.e., 05H).

- Load 0AH into PR2.

- Load 16H into CCPR1L.

- Load 0 into TRISC (make CCP1 pin output).

- Load 0CH into CCP1CON.

One period of the generated PWM waveform is shown in Figure 2.39.

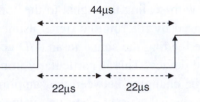

Figure 2.39: Generated PWM waveform

2.1.11   Analog-to-Digital Converter (A/D) Module

An analog-to-digital converter (A/D) is another important peripheral component of a microcontroller. The A/D converts an analog input voltage into a digital number so it can be processed by a microcontroller or any other digital system. There are many analog-to-digital converter chips available on the market, and an embedded systems designer should understand the characteristics of such chips so they can be used efficiently.

As far as the input and output voltage are concerned A/D converters can be classified as either unipolar and bipolar. Unipolar A/D converters accept unipolar input voltages in the range 0 to $+0V$, and bipolar A/D converters accept bipolar input voltages in the range $\pm V$. Bipolar converters are frequently used in signal processing applications, where the signals by nature are bipolar. Unipolar converters are usually cheaper, and they are used in many control and instrumentation applications.

Figure 2.40 shows the typical steps involved in reading and converting an analog signal into digital form, a process also known as signal conditioning. Signals received from sensors usually need to be processed before being fed to an A/D converter. This

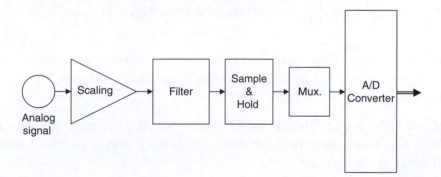

Figure 2.40: Signal conditioning and A/D conversion process

processing usually begins with scaling the signal to the correct value. Unwanted signal components are then removed by filtering the signal using classical filters (e.g., a low-pass filter). Finally, before feeding the signal to an A/D converter, the signal is passed through a sample-and-hold device. This is particularly important with fast real-time signals whose value may be changing between the sampling instants. A sample-and-hold device ensures that the signal stays at a constant value during the actual conversion process. Many applications required more than one A/D, which normally involves using an analog multiplexer at the input of the A/D. The multiplexer selects only one signal at any time and presents this signal to the A/D converter. An A/D converter usually has a single analog input and a digital parallel output. The conversion process is as follows:

- Apply the processed signal to the A/D input

- Start the conversion

- Wait until conversion is complete

- Read the converted digital data

The A/D conversion starts by triggering the converter. Depending on the speed of the converter, the conversion process itself can take several microseconds. At the end of the conversion, the converter either raises a flag or generates an interrupt to indicate that the conversion is complete. The converted parallel output data can then be read by the digital device connected to the A/D converter.

Most members of the PIC18F family contain a 10-bit A/D converter. If the chosen voltage reference is +5V, the voltage step value is:

$$\left(\frac{5V}{1023}\right) = 0.00489V \text{ or } 4.89mV$$

Therefore, for example, if the input voltage is 1.0V, the converter will generate a digital output of 1.0/0.00489 = 205 decimal. Similarly, if the input voltage is 3.0V, the converter will generate 3.0/0.00489 = 613.

The A/D converter used by the PIC18F452 microcontroller has eight channels, named AN0–AN7, which are shared by the PORTA and PORTE pins. Figure 2.41 shows the block diagram of the A/D converter.

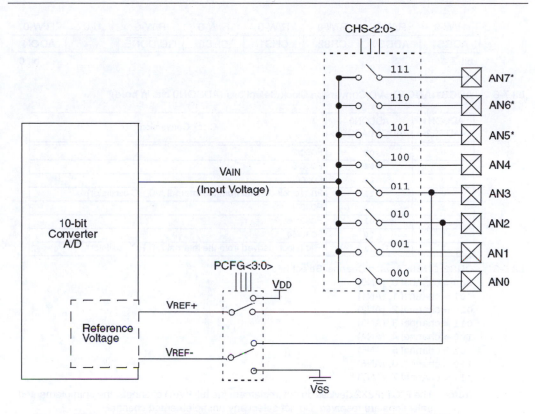

Figure 2.41: Block diagram of the PIC18F452 A/D converter

The A/D converter has four registers. Registers ADRESH and ADRESL store the higher and lower results of the conversion respectively. Register ADCON0, shown in Figure 2.42, controls the operation of the A/D module, such as selecting the conversion clock together with register ADCON1, selecting an input channel, starting a conversion, and powering up and shutting down the A/D converter.

Register ADCON1 (see Figure 2.43) is used for selecting the conversion format, configuring the A/D channels for analog input, selecting the reference voltage, and selecting the conversion clock together with register ADCON0.

A/D conversion starts by setting the GO/DONE bit of ADCON0. When the conversion is complete, the 2 bits of the converted data is written into register ADRESH, and the remaining 8 bits are written into register ADRESL. At the same time the GO/DONE bit is cleared to indicate the end of conversion. If required, interrupts can be enabled so that a software interrupt is generated when the conversion is complete.

R/W-0	R/W-0	R/W-0	R/W-0	R/W-0	R/W-0	U-0	R/W-0
ADCS1	ADCS0	CHS2	CHS1	CHS0	GO/$\overline{\text{DONE}}$	—	ADON

bit 7   bit 0

bit 7-6   **ADCS1:ADCS0:** A/D Conversion Clock Select bits (ADCON0 bits in **bold**)

ADCON1 <ADCS2>	ADCON0 <ADCS1:ADCS0>	Clock Conversion
0	00	Fosc/2
0	01	Fosc/8
0	10	Fosc/32
0	11	F_RC (clock derived from the internal A/D RC oscillator)
1	00	Fosc/4
1	01	Fosc/16
1	10	Fosc/64
1	11	F_RC (clock derived from the internal A/D RC oscillator)

bit 5-3   **CHS2:CHS0:** Analog Channel Select bits
000 = channel 0, (AN0)
001 = channel 1, (AN1)
010 = channel 2, (AN2)
011 = channel 3, (AN3)
100 = channel 4, (AN4)
101 = channel 5, (AN5)
110 = channel 6, (AN6)
111 = channel 7, (AN7)

Note: The PIC18F2X2 devices do not implement the full 8 A/D channels; the unimplemented selections are reserved. Do not select any unimplemented channel.

bit 2   **GO/$\overline{\text{DONE}}$:** A/D Conversion Status bit
<u>When ADON = 1:</u>
1 = A/D conversion in progress (setting this bit starts the A/D conversion which is automatically cleared by hardware when the A/D conversion is complete)
0 = A/D conversion not in progress

bit 1   **Unimplemented:** Read as '0'

bit 0   **ADON:** A/D On bit
1 = A/D converter module is powered up
0 = A/D converter module is shut-off and consumes no operating current

Figure 2.42: ADCON0 register

R/W-0	R/W-0	U-0	U-0	R/W-0	R/W-0	R/W-0	R/W-0
ADFM	ADCS2	—	—	PCFG3	PCFG2	PCFG1	PCFG0

bit 7   bit 0

bit 7   **ADFM:** A/D Result Format Select bit
           1 = Right justified. Six (6) Most Significant bits of ADRESH are read as '0'.
           0 = Left justified. Six (6) Least Significant bits of ADRESL are read as '0'.

bit 6   **ADCS2:** A/D Conversion Clock Select bit (ADCON1 bits in **bold**)

ADCON1 <ADCS2>	ADCON0 <ADCS1:ADCS0>	Clock Conversion
0	00	Fosc/2
0	01	Fosc/8
0	10	Fosc/32
0	11	FRC (clock derived from the internal A/D RC oscillator)
1	00	Fosc/4
1	01	Fosc/16
1	10	Fosc/64
1	11	FRC (clock derived from the internal A/D RC oscillator)

bit 5-4   **Unimplemented:** Read as '0'

bit 3-0   **PCFG3:PCFG0:** A/D Port Configuration Control bits

PCFG <3:0>	AN7	AN6	AN5	AN4	AN3	AN2	AN1	AN0	VREF+	VREF-	C / R
0000	A	A	A	A	A	A	A	A	VDD	VSS	8 / 0
0001	A	A	A	A	VREF+	A	A	A	AN3	VSS	7 / 1
0010	D	D	D	A	A	A	A	A	VDD	VSS	5 / 0
0011	D	D	D	A	VREF+	A	A	A	AN3	VSS	4 / 1
0100	D	D	D	D	A	D	A	A	VDD	VSS	3 / 0
0101	D	D	D	D	VREF+	D	A	A	AN3	VSS	2 / 1
011x	D	D	D	D	D	D	D	D	—	—	0 / 0
1000	A	A	A	A	VREF+	VREF-	A	A	AN3	AN2	6 / 2
1001	D	D	A	A	A	A	A	A	VDD	VSS	6 / 0
1010	D	D	A	A	VREF+	A	A	A	AN3	VSS	5 / 1
1011	D	D	A	A	VREF+	VREF-	A	A	AN3	AN2	4 / 2
1100	D	D	D	A	VREF+	VREF-	A	A	AN3	AN2	3 / 2
1101	D	D	D	D	VREF+	VREF-	A	A	AN3	AN2	2 / 2
1110	D	D	D	D	D	D	D	A	VDD	VSS	1 / 0
1111	D	D	D	D	VREF+	VREF-	D	A	AN3	AN2	1 / 2

A = Analog input   D = Digital I/O

Figure 2.43: ADCON1 register

The steps in carrying out an A/D conversion are as follows:

- Use ADCON1 to configure required channels as analog and configure the reference voltage.

- Set the TRISA or TRISE bits so the required channel is an input port.

- Use ADCON0 to select the required analog input channel.

- Use ADCON0 and ADCON1 to select the conversion clock.

- Use ADCON0 to turn on the A/D module.

- Configure the A/D interrupt (if desired).

- Set the GO/DONE bit to start conversion.

- Wait until the GO/DONE bit is cleared, or until a conversion complete interrupt is generated.

- Read the converted data from ADRESH and ADRESL.

- Repeat these steps as required.

For correct A/D conversion, the A/D conversion clock must be selected to ensure a minimum bit conversion time of 1.6µs. Table 2.11 gives the recommended A/D clock sources for various microcontroller operating frequencies. For example, if the

Table 2.11: A/D conversion clock selection

A/D clock source		
Operation	ADCS2:ADCS0	Maximum microcontroller frequency
2 T_{OSC}	000	1.25 MHz
4 T_{OSC}	100	2.50 MHz
8 T_{OSC}	001	5.0 MHz
16 T_{OSC}	101	10.0 MHz
32 T_{OSC}	010	20.0 MHz
64 T_{OSC}	110	40.0 MHz
RC	011	–

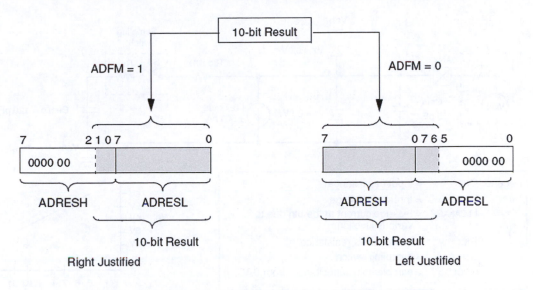

Figure 2.44: Formatting the A/D conversion result

microcontroller is operated from a 10MHz clock, the A/D clock source should be $F_{OSC/16}$ or higher (e.g., $F_{OSC/32}$).

Bit ADFM of register ADCON1 controls the format of a conversion. When ADFM is cleared, the 10-bit result is left justified (see Figure 2.44) and lower 6 bits of ADRESL are cleared to 0. When ADFM is set to 1 the result is right justified and the upper 6 bits of ADRESH are cleared to 0. This is the mode most commonly used, in which ADRESL contains the lower 8 bits, and bits 0 and 1 of ADRESH contain the upper 2 bits of the 10-bit result.

Analog Input Model and Acquisition Time

An understanding of the A/D analog input model is necessary to interface the A/D to external devices. Figure 2.45 shows the analog input model of the A/D. The analog input voltage V_{AIN} and the source resistance R_S are shown on the left side of the diagram. It is recommended that the source resistance be no greater than 2.5K. The analog signal is applied to the pin labeled ANx. There is a small capacitance (5pF) and a leakage current to the ground of approximately 500nA. R_{IC} is the interconnect resistance, which has a value of less than 1K. The sampling process is shown with switch SS having a resistance R_{SS} whose value depends on the voltage as shown in the

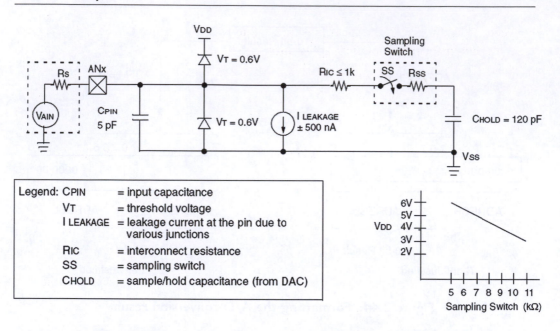

Figure 2.45: Analog input model of the A/D converter

small graph at the bottom of Figure 2.45. The value of R_{SS} is approximately 7K at 5V supply voltage.

The A/D converter is based on a switched capacitor principle, and capacitor C_{HOLD} shown in Figure 2.45 must be charged fully before the start of a conversion. This is a 120pF capacitor which is disconnected from the input pin once the conversion is started.

The acquisition time can be calculated by using Equation (2.7), provided by Microchip Inc:

$$T_{ACQ} = \text{Amplifier settling time} + \text{Holding capacitor charging time} \\ + \text{temperature coefficient} \qquad (2.7)$$

The amplifier settling time is specified as a fixed 2μs. The temperature coefficient, which is only applicable if the temperature is above 25°C, is specified as:

$$\text{Temperature coefficient} = (\text{Temperature} - 25°\text{C})(0.05μ\text{s}/°\text{C}) \qquad (2.8)$$

Equation (2.8) shows that the effect of the temperature is very small, creating about 0.5μs delay for every 10°C above 25°C. Thus, assuming a working environment between 25°C and 35°C, the maximum delay due to temperature will be 0.5µs, which can be ignored for most practical applications.

The holding capacitor charging time as specified by Microchip Inc is:

$$\text{Holding capacitor charging time} = -(120\text{pF})(1\text{K} + R_{SS} + R_S)\text{Ln}(1/2048) \qquad (2.9)$$

Assuming that $R_{SS} = 7\text{K}$, $R_S = 2.5\text{K}$, Equation (2.9) gives the holding capacitor charging time as 9.6µs.

The acquisition time is then calculated as:

$$T_{ACQ} = 2 + 9.6 + 0.5 = 12.1\,\mu s$$

A full 10-bit conversion takes 12 A/D cycles, and each A/D cycle is specified at a minimum of 1.6µs. Thus, the fastest conversion time is 19.2µs. Adding this to the best possible acquisition time gives a total time to complete a conversion of $19.2 + 12.1 = 31.3$µs.

When a conversion is complete, it is specified that the converter should wait for two conversion periods before starting a new conversion. This corresponds to $2 \times 1.6 = 3.2$µs. Adding this to the best possible conversion time of 31.3µs gives a complete conversion time of 34.5µs. Assuming the A/D converter is used successively, and ignoring the software overheads, this implies a maximum sampling frequency of about 29KHz.

2.1.12   Interrupts

An interrupt is an event that requires the CPU to stop normal program execution and then execute a program code related to the event causing the interrupt. Interrupts can be generated internally (by some event inside the chip) or externally (by some external event). An example of an internal interrupt is a timer overflowing or the A/D completing a conversion. An example of an external interrupt is an I/O pin changing state.

Interrupts can be useful in many applications such as:

- *Time critical applications.* Applications which require the immediate attention of the CPU can use interrupts. For example, in an emergency such as a power failure or fire in a plant the CPU may have to shut down the system immediately in an orderly manner. In such applications an external interrupt can force the CPU to stop whatever it is doing and take immediate action.

- *Performing routine tasks.* Many applications require the CPU to perform routine work at precise times, such as checking the state of a peripheral device exactly every millisecond. A timer interrupt scheduled with the required timing can divert the CPU from normal program execution to accomplish the task at the precise time required.

- *Task switching in multi-tasking applications.* In multi-tasking applications, each task may have a finite time to execute its code. Interrupt mechanisms can be used to stop a task should it consume more than its allocated time.

- *To service peripheral devices quickly.* Some applications may need to know when a task, such as an A/D conversion, is completed. This can be accomplished by continuously checking the completion flag of the A/D converter. A more elegant solution would be to enable the A/D completion interrupt so the CPU is forced to read the converted data as soon as it becomes available.

The PIC18F452 microcontroller has both core and peripheral interrupt sources. The core interrupt sources are:

- External edge-triggered interrupt on INT0, INT1, and INT2 pins.

- PORTB pins change interrupts (any one of the RB4–RB7 pins changing state)

- Timer 0 overflow interrupt

The peripheral interrupt sources are:

- Parallel slave port read/write interrupt

- A/D conversion complete interrupt

- USART receive interrupt

- USART transmit interrupt

- Synchronous serial port interrupt

- CCP1 interrupt

- TMR1 overflow interrupt

- TMR2 overflow interrupt

- Comparator interrupt

- EEPROM/FLASH write interrupt

- Bus collision interrupt

- Low-voltage detect interrupt

- Timer 3 overflow interrupt

- CCP2 interrupt

Interrupts in the PIC18F family can be divided into two groups: high priority and low priority. Applications that require more attention can be placed in the higher priority group. A high-priority interrupt can stop a low-priority interrupt that is in progress and gain access to the CPU. However, high-priority interrupts cannot be stopped by low-priority interrupts. If the application does not need to set priorities for interrupts, the user can choose to disable the priority scheme so all interrupts are at the same priority level. High-priority interrupts are vectored to address 00008H and low-priority ones to address 000018H of the program memory. Normally, a user program code (interrupt service routine, ISR) should be at the interrupt vector address to service the interrupting device.

In the PIC18F452 microcontroller there are ten registers that control interrupt operations. These are:

- RCON

- INTCON

- INTCON2

- INTCON3

- PIR1, PIR2

- PIE1, PIE2

- IPR1, IPR2

Every interrupt source (except INT0) has three bits to control its operation. These bits are:

- A flag bit to indicate whether an interrupt has occurred. This bit has a name ending in ...**IF**

- An interrupt enable bit to enable or disable the interrupt source. This bit has the name ending in ...**IE**

- A priority bit to select high or low priority. This bit has a name ending in ...**IP**

RCON Register

The top bit of the RCON register, called IPEN, is used to enable the interrupt priority scheme. When IPEN = 0, interrupt priority levels are disabled and the microcontroller interrupt structure is similar to that of the PIC16 series. When IPEN = 1, interrupt priority levels are enabled. Figure 2.46 shows the bits of register RCON.

Enabling/Disabling Interrupts—No Priority Structure

When the IPEN bit is cleared, the priority feature is disabled. All interrupts branch to address 00008H of the program memory. In this mode, bit PEIE of register INTCON enables/disables all peripheral interrupt sources. Similarly, bit GIE of INTCON enables/disables all interrupt sources. Figure 2.47 shows the bits of register INTCON.

R/W-0	U-0	U-0	R/W-1	R-1	R-1	R/W-0	R/W-0
IPEN	—	—	$\overline{RI}$	$\overline{TO}$	$\overline{PD}$	POR	BOR

bit 7   bit 0

bit 7   **IPEN:** Interrupt Priority Enable bit
        1 = Enable priority levels on interrupts
        0 = Disable priority levels on interrupts (16CXXX Compatibility mode)

bit 6-5 **Unimplemented:** Read as '0'

bit 4   **$\overline{RI}$:** RESET Instruction Flag bit
        For details of bit operation, see Register 4-3

bit 3   **$\overline{TO}$:** Watchdog Time-out Flag bit
        For details of bit operation, see Register 4-3

bit 2   **$\overline{PD}$:** Power-down Detection Flag bit
        For details of bit operation, see Register 4-3

bit 1   **POR:** Power-on Reset Status bit
        For details of bit operation, see Register 4-3

bit 0   **$\overline{BOR}$:** Brown-out Reset Status bit
        For details of bit operation, see Register 4-3

Figure 2.46: RCON register bits

R/W-0	R/W-0	R/W-0	R/W-0	R/W-0	R/W-0	R/W-0	R/W-x
GIE/GIEH	PEIE/GIEL	TMR0IE	INT0IE	RBIE	TMR0IF	INT0IF	RBIF

bit 7   bit 0

bit 7   **GIE/GIEH:** Global Interrupt Enable bit

<u>When IPEN = 0:</u>
1 = Enables all unmasked interrupts
0 = Disables all interrupts

<u>When IPEN = 1:</u>
1 = Enables all high priority interrupts
0 = Disables all interrupts

bit 6   **PEIE/GIEL:** Peripheral Interrupt Enable bit

<u>When IPEN = 0:</u>
1 = Enables all unmasked peripheral interrupts
0 = Disables all peripheral interrupts

<u>When IPEN = 1:</u>
1 = Enables all low priority peripheral interrupts
0 = Disables all low priority peripheral interrupts

bit 5   **TMR0IE:** TMR0 Overflow Interrupt Enable bit
1 = Enables the TMR0 overflow interrupt
0 = Disables the TMR0 overflow interrupt

bit 4   **INT0IE:** INT0 External Interrupt Enable bit
1 = Enables the INT0 external interrupt
0 = Disables the INT0 external interrupt

bit 3   **RBIE:** RB Port Change Interrupt Enable bit
1 = Enables the RB port change interrupt
0 = Disables the RB port change interrupt

bit 2   **TMR0IF:** TMR0 Overflow Interrupt Flag bit
1 = TMR0 register has overflowed (must be cleared in software)
0 = TMR0 register did not overflow

bit 1   **INT0IF:** INT0 External Interrupt Flag bit
1 = The INT0 external interrupt occurred (must be cleared in software)
0 = The INT0 external interrupt did not occur

bit 0   **RBIF:** RB Port Change Interrupt Flag bit
1 = At least one of the RB7:RB4 pins changed state (must be cleared in software)
0 = None of the RB7:RB4 pins have changed state

Note: A mismatch condition will continue to set this bit. Reading PORTB will end the mismatch condition and allow the bit to be cleared.

Figure 2.47: INTCON register bits

For an interrupt to be accepted by the CPU the following conditions must be satisfied:

- The interrupt enable bit of the interrupt source must be enabled. For example, if the interrupt source is external interrupt pin INT0, then bit INT0IE of register INTCON must be set to 1.

- The interrupt flag of the interrupt source must be cleared. For example, if the interrupt source is external interrupt pin INT0, then bit INT0IF of register INTCON must be cleared to 0.

- The peripheral interrupt enable/disable bit PEIE of INTCON must be set to 1 if the interrupt source is a peripheral.

- The global interrupt enable/disable bit GIE of INTCON must be set to 1.

With an external interrupt source we normally have to define whether the interrupt should occur on the low-to-high or high-to-low transition of the interrupt source. With INT0 interrupts, for example, this is done by setting/clearing bit INTEDG0 of register INTCON2.

When an interrupt occurs, the CPU stops its normal flow of execution, pushes the return address onto the stack, and jumps to address 00008H in the program memory where the user interrupt service routine program resides. Once the CPU is in the interrupt service routine, the global interrupt enable bit (GIE) is cleared to disable further interrupts. When multiple interrupt sources are enabled, the source of the interrupt can be determined by polling the interrupt flag bits. The interrupt flag bits must be cleared in the software before reenabling interrupts to avoid recursive interrupts. When the CPU has returned from the interrupt service routine, the global interrupt bit GIE is automatically set by the software.

Enabling/Disabling Interrupts—Priority Structure

When the IPEN bit is set to 1, the priority feature is enabled and the interrupts are grouped into two: low priority and high priority. Low-priority interrupts branch to address 00008H and high-priority interrupts branch to address 000018H of the program memory. Setting the priority bit makes the interrupt source a high-priority interrupt, and clearing this bit makes the interrupt source a low-priority interrupt.

Setting the GIEH bit of INTCON enables all high-priority interrupts that have the priority bit set. Similarly, setting the GIEL bit of INTCON enables all low-priority interrupts (the priority is bit cleared).

For a high-priority interrupt to be accepted by the CPU, the following conditions must be satisfied:

- The interrupt enable bit of the interrupt source must be enabled. For example, if the interrupt source is external interrupt pin INT1, then bit INT1IE of register INTCON3 must be set to 1.

- The interrupt flag of the interrupt source must be cleared. For example, if the interrupt source is external interrupt pin INT1, then bit INT1IF of register INTCON3 must be cleared to 0.

- The priority bit must be set to 1. For example, if the interrupt source is external interrupt INT1, then bit INT1P of register INTCON3 must be set to 1.

- The global interrupt enable/disable bit GIEH of INTCON must be set to 1.

For a low-priority interrupt to be accepted by the CPU, the following conditions must be satisfied:

- The interrupt enable bit of the interrupt source must be enabled. For example, if the interrupt source is external interrupt pin INT1, then bit INT1IE of register INTCON3 must be set to 1.

- The interrupt flag of the interrupt source must be cleared. For example, if the interrupt source is external interrupt pin INT1, then bit INT1IF of register INTCON3 must be cleared to 0.

- The priority bit must be cleared to 0. For example, if the interrupt source is external interrupt INT1, then bit INT1P of register INTCON3 must be cleared to 0.

- Low-priority interrupts must be enabled by setting bit GIEL of INTCON to 1.

- The global interrupt enable/disable bit GIEH of INTCON must be set to 1.

Table 2.12 gives a listing of the PIC18F452 microcontroller interrupt bit names and register names for every interrupt source.

Table 2.12: PIC18F452 interrupt bits and registers

Interrupt source	Flag bit	Enable bit	Priority bit
INT0 external	INT0IF	INT0IE	–
INT1 external	INT1IF	INT1IE	INT1IP
INT2 external	INT2IF	INT2IE	INT2IP
RB port change	RBIF	RBIE	RBIP
TMR0 overflow	TMR0IF	TMR0IE	TMR0IP
TMR1overflow	TMR1IF	TMR1IE	TMR1IP
TMR2 match PR2	TMR2IF	TMR2IE	TMR2IP
TMR3 overflow	TMR3IF	TMR3IE	TMR3IP
A/D complete	ADIF	ADIE	ADIP
CCP1	CCP1IF	CCP1IE	CCP1IP
CCP2	CCP2IF	CCP2IE	CCP2IP
USART RCV	RCIF	RCIE	RCIP
USART TX	TXIF	TXIE	TXIP
Parallel slave port	PSPIF	PSPIE	PSPIP
Sync serial port	SSPIF	SSPIE	SSPIP
Low-voltage detect	LVDIF	LVDIE	LVDIP
Bus collision	BCLIF	BCLIE	BCLIP
EEPROM/FLASH write	EEIF	EEIE	EEIP

Figures 2.48 to 2.55 show the bit definitions of interrupt registers INTCON2, INTCON3, PIR1, PIR2, PIE1, PIE2, IPR1, and IPR2.

Examples are given in this section to illustrate how the CPU can be programmed for an interrupt.

Example 2.2

Set up INT1 as a falling-edge triggered interrupt input having low priority.

Solution 2.2

The following bits should be set up before the INT1 falling-edge triggered interrupts can be accepted by the CPU in low-priority mode:

R/W-1	R/W-1	R/W-1	R/W-1	U-0	R/W-1	U-0	R/W-1
RBPU	INTEDG0	INTEDG1	INTEDG2	—	TMR0IP	—	RBIP
bit 7							bit 0

bit 7   **RBPU:** PORTB Pull-up Enable bit
         1 = All PORTB pull-ups are disabled
         0 = PORTB pull-ups are enabled by individual port latch values

bit 6   **INTEDG0:**External Interrupt0 Edge Select bit
         1 = Interrupt on rising edge
         0 = Interrupt on falling edge

bit 5   **INTEDG1:** External Interrupt1 Edge Select bit
         1 = Interrupt on rising edge
         0 = Interrupt on falling edge

bit 4   **INTEDG2:** External Interrupt2 Edge Select bit
         1 = Interrupt on rising edge
         0 = Interrupt on falling edge

bit 3   **Unimplemented:** Read as '0'

bit 2   **TMR0IP:** TMR0 Overflow Interrupt Priority bit
         1 = High priority
         0 = Low priority

bit 1   **Unimplemented:** Read as '0'

bit 0   **RBIP:** RB Port Change Interrupt Priority bit
         1 = High priority
         0 = Low priority

Figure 2.48: INTCON2 bit definitions

R/W-1	R/W-1	U-0	R/W-0	R/W-0	U-0	R/W-0	R/W-0
INT2IP	INT1IP	—	INT2IE	INT1IE	—	INT2IF	INT1IF

bit 7   bit 0

bit 7   **INT2IP:** INT2 External Interrupt Priority bit
1 = High priority
0 = Low priority

bit 6   **INT1IP:** INT1 External Interrupt Priority bit
1 = High priority
0 = Low priority

bit 5   **Unimplemented:** Read as '0'

bit 4   **INT2IE:** INT2 External Interrupt Enable bit
1 = Enables the INT2 external interrupt
0 = Disables the INT2 external interrupt

bit 3   **INT1IE:** INT1 External Interrupt Enable bit
1 = Enables the INT1 external interrupt
0 = Disables the INT1 external interrupt

bit 2   **Unimplemented:** Read as '0'

bit 1   **INT2IF:** INT2 External Interrupt Flag bit
1 = The INT2 external interrupt occurred (must be cleared in software)
0 = The INT2 external interrupt did not occur

bit 0   **INT1IF:** INT1 External Interrupt Flag bit
1 = The INT1 external interrupt occurred (must be cleared in software)
0 = The INT1 external interrupt did not occur

Figure 2.49: INTCON3 bit definitions

- Enable the priority structure. Set IPEN = 1

- Make INT1 an input pin. Set TRISB = 1

- Set INT1 interrupts for falling edge. SET INTEDG1 = 0

- Enable INT1 interrupts. Set INT1IE = 1

- Enable low priority. Set INT1IP = 0

- Clear INT1 flag. Set INT1IF = 0

- Enable low-priority interrupts. Set GIEL = 1

- Enable all interrupts. Set GIEH = 1

R/W-0	R/W-0	R-0	R-0	R/W-0	R/W-0	R/W-0	R/W-0
PSPIF[(1)]	ADIF	RCIF	TXIF	SSPIF	CCP1IF	TMR2IF	TMR1IF

bit 7   bit 0

bit 7   **PSPIF**[(1)]: Parallel Slave Port Read/Write Interrupt Flag bit
        1 = A read or a write operation has taken place (must be cleared in software)
        0 = No read or write has occurred

bit 6   **ADIF**: A/D Converter Interrupt Flag bit
        1 = An A/D conversion completed (must be cleared in software)
        0 = The A/D conversion is not complete

bit 5   **RCIF**: USART Receive Interrupt Flag bit
        1 = The USART receive buffer, RCREG, is full (cleared when RCREG is read)
        0 = The USART receive buffer is empty

bit 4   **TXIF**: USART Transmit Interrupt Flag bit (see Section 16.0 for details on TXIF functionality)
        1 = The USART transmit buffer, TXREG, is empty (cleared when TXREG is written)
        0 = The USART transmit buffer is full

bit 3   **SSPIF**: Master Synchronous Serial Port Interrupt Flag bit
        1 = The transmission/reception is complete (must be cleared in software)
        0 = Waiting to transmit/receive

bit 2   **CCP1IF**: CCP1 Interrupt Flag bit
        Capture mode:
        1 = A TMR1 register capture occurred (must be cleared in software)
        0 = No TMR1 register capture occurred
        Compare mode:
        1 = A TMR1 register compare match occurred (must be cleared in software)
        0 = No TMR1 register compare match occurred
        PWM mode:
        Unused in this mode

bit 1   **TMR2IF**: TMR2 to PR2 Match Interrupt Flag bit
        1 = TMR2 to PR2 match occurred (must be cleared in software)
        0 = No TMR2 to PR2 match occurred

bit 0   **TMR1IF**: TMR1 Overflow Interrupt Flag bit
        1 = TMR1 register overflowed (must be cleared in software)
        0 = MR1 register did not overflow

Figure 2.50: PIR1 bit definitions

U-0	U-0	U-0	R/W-0	R/W-0	R/W-0	R/W-0	R/W-0
—	—	—	EEIF	BCLIF	LVDIF	TMR3IF	CCP2IF

bit 7   bit 0

bit 7-5   **Unimplemented:** Read as '0'

bit 4   **EEIF:** Data EEPROM/FLASH Write Operation Interrupt Flag bit
                 1 = The Write operation is complete (must be cleared in software)
                 0 = The Write operation is not complete, or has not been started

bit 3   **BCLIF:** Bus Collision Interrupt Flag bit
                 1 = A bus collision occurred (must be cleared in software)
                 0 = No bus collision occurred

bit 2   **LVDIF:** Low Voltage Detect Interrupt Flag bit
                 1 = A low voltage condition occurred (must be cleared in software)
                 0 = The device voltage is above the Low Voltage Detect trip point

bit 1   **TMR3IF:** TMR3 Overflow Interrupt Flag bit
                 1 = TMR3 register overflowed (must be cleared in software)
                 0 = TMR3 register did not overflow

bit 0   **CCP2IF:** CCPx Interrupt Flag bit
                 <u>Capture mode:</u>
                 1 = A TMR1 register capture occurred (must be cleared in software)
                 0 = No TMR1 register capture occurred

                 <u>Compare mode:</u>
                 1 = A TMR1 register compare match occurred (must be cleared in software)
                 0 = No TMR1 register compare match occurred

                 <u>PWM mode:</u>
                 Unused in this mode

Figure 2.51: PIR2 bit definitions

When an interrupt occurs, the CPU jumps to address 00008H in the program memory to execute the user program at the interrupt service routine.

Example 2.3

Set up INT1 as a rising-edge triggered interrupt input having high priority.

Solution 2.3

The following bits should be set up before the INT1 rising-edge triggered interrupts can be accepted by the CPU in high-priority mode:

- Enable the priority structure. Set IPEN = 1

- Make INT1 an input pin. Set TRISB = 1

R/W-0	R/W-0	R/W-0	R/W-0	R/W-0	R/W-0	R/W-0	R/W-0
PSPIE[(1)]	ADIE	RCIE	TXIE	SSPIE	CCP1IE	TMR2IE	TMR1IE
bit 7							bit 0

bit 7   **PSPIE[(1)]:** Parallel Slave Port Read/Write Interrupt Enable bit

1 = Enables the PSP read/write interrupt
0 = Disables the PSP read/write interrupt

bit 6   **ADIE:** A/D Converter Interrupt Enable bit

1 = Enables the A/D interrupt
0 = Disables the A/D interrupt

bit 5   **RCIE:** USART Receive Interrupt Enable bit

1 = Enables the USART receive interrupt
0 = Disables the USART receive interrupt

bit 4   **TXIE:** USART Transmit Interrupt Enable bit

1 = Enables the USART transmit interrupt
0 = Disables the USART transmit interrupt

bit 3   **SSPIE:** Master Synchronous Serial Port Interrupt Enable bit

1 = Enables the MSSP interrupt
0 = Disables the MSSP interrupt

bit 2   **CCP1IE:** CCP1 Interrupt Enable bit

1 = Enables the CCP1 interrupt
0 = Disables the CCP1 interrupt

bit 1   **TMR2IE:** TMR2 to PR2 Match Interrupt Enable bit

1 = Enables the TMR2 to PR2 match interrupt
0 = Disables the TMR2 to PR2 match interrupt

bit 0   **TMR1IE:** TMR1 Overflow Interrupt Enable bit

1 = Enables the TMR1 overflow interrupt
0 = Disables the TMR1 overflow interrupt

Figure 2.52: PIE1 bit definitions

- Set INT1 interrupts for rising edge. SET INTEDG1 = 1

- Enable INT1 interrupts. Set INT1IE = 1

- Enable high priority. Set INT1IP = 1

- Clear INT1 flag. Set INT1IF = 0

- Enable all interrupts. Set GIEH = 1

When an interrupt occurs, the CPU jumps to address 000018H of the program memory to execute the user program at the interrupt service routine.

U-0	U-0	U-0	R/W-0	R/W-0	R/W-0	R/W-0	R/W-0
—	—	—	EEIE	BCLIE	LVDIE	TMR3IE	CCP2IE
bit 7							bit 0

bit 7-5   **Unimplemented:** Read as '0'

bit 4   **EEIE:** Data EEPROM/FLASH Write Operation Interrupt Enable bit
1 = Enabled
0 = Disabled

bit 3   **BCLIE:** Bus Collision Interrupt Enable bit
1 = Enabled
0 = Disabled

bit 2   **LVDIE:** Low Voltage Detect Interrupt Enable bit
1 = Enabled
0 = Disabled

bit 1   **TMR3IE:** TMR3 Overflow Interrupt Enable bit
1 = Enables the TMR3 overflow interrupt
0 = Disables the TMR3 overflow interrupt

bit 0   **CCP2IE:** CCP2 Interrupt Enable bit
1 = Enables the CCP2 interrupt
0 = Disables the CCP2 interrupt

Figure 2.53: PIE2 bit definitions

R/W-1	R/W-1	R/W-1	R/W-1	R/W-1	R/W-1	R/W-1	R/W-1
PSPIP[1]	ADIP	RCIP	TXIP	SSPIP	CCP1IP	TMR2IP	TMR1IP
bit 7							bit 0

bit 7   **PSPIP[1]:** Parallel Slave Port Read/Write Interrupt Priority bit
1 = High priority
0 = Low priority

bit 6   **ADIP:** A/D Converter Interrupt Priority bit
1 = High priority
0 = Low priority

bit 5   **RCIP:** USART Receive Interrupt Priority bit
1 = High priority
0 = Low priority

bit 4   **TXIP:** USART Transmit Interrupt Priority bit
1 = High priority
0 = Low priority

bit 3   **SSPIP:** Master Synchronous Serial Port Interrupt Priority bit
1 = High priority
0 = Low priority

bit 2   **CCP1IP:** CCP1 Interrupt Priority bit
1 = High priority
0 = Low priority

bit 1   **TMR2IP:** TMR2 to PR2 Match Interrupt Priority bit
1 = High priority
0 = Low priority

bit 0   **TMR1IP:** TMR1 Overflow Interrupt Priority bit
1 = High priority
0 = Low priority

Figure 2.54: IPR1 bit definitions

U-0	U-0	U-0	R/W-1	R/W-1	R/W-1	R/W-1	R/W-1
—	—	—	EEIP	BCLIP	LVDIP	TMR3IP	CCP2IP

bit 7 .... bit 0

bit 7-5   **Unimplemented:** Read as '0'

bit 4   **EEIP:** Data EEPROM/FLASH Write Operation Interrupt Priority bit
1 = High priority
0 = Low priority

bit 3   **BCLIP:** Bus Collision Interrupt Priority bit
1 = High priority
0 = Low priority

bit 2   **LVDIP:** Low Voltage Detect Interrupt Priority bit
1 = High priority
0 = Low priority

bit 1   **TMR3IP:** TMR3 Overflow Interrupt Priority bit
1 = High priority
0 = Low priority

bit 0   **CCP2IP:** CCP2 Interrupt Priority bit
1 = High priority
0 = Low priority

Figure 2.55: IPR2 bit definitions

2.2   Summary

This chapter has described the architecture of the PIC18F family of microcontrollers. The PIC18F452 was used as a typical sample microcontroller in this family. Other members of the same family, such as the PIC18F242, have smaller pin counts and less functionality. And some, such as the PIC18F6680, have larger pin counts and more functionality.

Important parts and peripheral circuits of the PIC18F series have been described, including data memory, program memory, clock circuits, reset circuits, watchdog timer, general purpose timers, capture and compare module, PWM module, A/D converter, and the interrupt structure.

2.3   Exercises

1. Describe the data memory structure of the PIC18F452 microcontroller. What is a bank? How many banks are there?

2. Explain the differences between a general purpose register (GPR) and a special function register (SFR).

3. Explain the various ways the PIC18F microcontroller can be reset. Draw a circuit diagram to show how an external push-button switch can be used to reset the microcontroller.

4. Describe the various clock sources that can be used to provide a clock to a PIC18F452 microcontroller. Draw a circuit diagram to show how a 10MHz crystal can be connected to the microcontroller.

5. Draw a circuit diagram to show how a resonator can be connected to a PIC18F microcontroller.

6. In a non-time-critical application a clock must be provided for a PIC18F452 microcontroller using an external resistor and a capacitor. Draw a circuit diagram to show how this can be done and find the component values for a required clock frequency of 5MHz.

7. Explain how an external clock can provide clock pulses to a PIC18F microcontroller.

8. What are the registers of PORTA? Explain the operation of the port by drawing the port block diagram.

9. The watchdog timer must be set to provide an automatic reset every 0.5 seconds. Describe how to do this, including the appropriate register bits.

10. PWM pulses must be generated from pin CCP1 of a PIC18F452 microcontroller. The required pulse period is 100μs, and the required duty cycle is 50%. Assuming the microcontroller is operating with a 4MHz crystal, calculate the values to be loaded into the various registers.

11. Again, with regard to PWM pulses generated from pin CCP1 of a PIC18F452 microcontroller: If the required pulse frequency is 40KHz, and the required duty cycle is 50%, and assuming the microcontroller is operating with a 4MHz crystal, calculate the values to be loaded into the various registers.

12. An LM35DZ-type analog temperature sensor is connected to analog port AN0 of a PIC18F452 microcontroller. The sensor provides an analog output voltage proportional to the temperature (i.e., V0 = 10 mV/°C). Show the steps required to read the temperature.

13. Explain the difference between a priority interrupt and a nonpriority interrupt.

14. Show the steps required to set up INT2 as a falling-edge triggered interrupt input having low priority. What is the interrupt vector address?

15. Show the steps required to set up both INT1 and INT2 as falling-edge triggered interrupt inputs having low priority.

16. Show the steps required to set up INT1 as falling-edge triggered and INT2 as rising-edge triggered interrupt inputs having high priorities. Explain how to find the source of the interrupt when an interrupt occurs.

17. Show the steps required to set up Timer 0 to generate interrupts every millisecond with a high priority. What is the interrupt vector address?

18. In an application the CPU registers have been configured to accept interrupts from external sources INT0, INT1, and INT2. An interrupt has been detected. Explain how to find the source of the interrupt.

CHAPTER 3

C Programming Language

There are several C compilers on the market for the PIC18 series of microcontrollers. These compilers have many similar features, and they can all be used to develop C-based high-level programs for PIC18 microcontrollers.

Some of the C compilers used most often in commercial, industrial, and educational PIC18 microcontroller applications are:

- mikroC
- PICC18
- C18
- CCS

The popular and powerful mikroC, developed by MikroElektronika (web site: www.microe.com), is easy to learn and comes with rich resources, such as a large number of library functions and an integrated development environment with a built-in simulator and an in-circuit debugger (e.g., mikroICD). A demo version of the compiler with a 2K program limit is available from MikroElektronika.

PICC18, another popular C compiler, was developed by Hi-Tech Software (web site: www.htsoft.com) and is available in two versions: standard and professional. A powerful simulator and an integrated development environment (Hi-Tide) are provided by the company. PICC18 is supported by the PROTEUS simulator (www.labcenter.co.uk) which can be used to simulate PIC microcontroller–based systems. A limited-period demo version of this compiler is available on the developer's web site.

C18 is a product of Microchip Inc. (web site: www.microchip.com). A limited-period demo version, as well as a limited functionality version of C18 with no time limit, are available from the Microchip web site. C18 includes a simulator and supports hardware and software development tools such as in-circuit emulators (e.g., ICE2000) and in-circuit debuggers (e.g., ICD2).

CCS has been developed by the Custom Computer Systems Inc. (web site: www. ccsinfo.com). The company offers a limited-period demo version of their compiler. CCS provides a large number of built-in functions and supports an in-circuit debugger (e.g., ICD-U40) which are very helpful in the development of PIC18 microcontroller–based systems.

In this book we are mainly concentrating on the use of the mikroC compiler, and most of the projects are based on this compiler.

3.1   Structure of a mikroC Program

Figure 3.1 shows the simplest structure of a mikroC program. This program flashes an LED connected to port RB0 (bit 0 of PORTB) of a PIC microcontroller in one-second

```
/**********************************************************************

                          LED FLASHING PROGRAM
                      *********************************

This program flashes an LED connected to port pin RB0 of PORTB with one
second intervals.

    Programmer   : D. Ibrahim
    File         : LED.C
    Date         : May, 2007
    Micro        : PIC18F452
**********************************************************************/

void main()
{
    for(;;)                          // Endless loop
    {
        TRISB = 0;                   // Configure PORTB as output
        PORTB.0 = 0;                 // RB0 = 0
        Delay_Ms(1000);              // Wait 1 second
        PORTB.0 = 1;                 // RB0 = 1
        Delay_Ms(1000);              // Wait 1 second
    }                                // End of loop
}
```

Figure 3.1: Structure of a simple C program

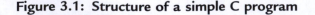

intervals. Do not worry if you don't understand the operation of the program at this stage, as all will come clear as this chapter progresses. Some of the programming elements in Figure 3.1 are described in detail here.

3.1.1   Comments

Comments are used to clarify the operation of the program or a programming statement. Comment lines are ignored and not compiled by the compiler. In mikroC programs comments can be of two types: long comments, extending several lines, and short comments, occupying only a single line. Comment lines at the beginning of a program can describe briefly the program's operation and provide the author's name, the program filename, the date the program was written, and a list of version numbers, together with the modifications in each version. As shown in Figure 3.1, comments can also be added after statements to describe the operations that the statements perform. Clear and succinct comment lines are important for the maintenance and thus the lifetime of a program, as a program with good comments is easier to modify and/or update.

As shown in Figure 3.1, long comments start with the character "/*" and terminate with the character "*/". Similarly, short comments start with the character "//" and do not need a terminating character.

3.1.2   Beginning and Ending of a Program

In C language, a program begins with the keywords:

```
void main ()
```

After this, a curly opening bracket is used to indicate the beginning of the program body. The program is terminated with a closing curly bracket. Thus, as shown in Figure 3.1, the program has the following structure:

```
void main()
{
        program body
}
```

3.1.3   Terminating Program Statements

In C language, all program statements must be terminated with the semicolon (";") character; otherwise a compiler error will be generated:

```
j = 5;            // correct
j = 5             // error
```

3.1.4 White Spaces

White spaces are spaces, blanks, tabs, and newline characters. The C compiler ignores
all white spaces. Thus, the following three sequences are identical:

```
int i;    char j;
or
int i;
char j;
or
int i;
          char j;
```

Similarly, the following sequences are identical:

```
i = j + 2;
or
i = j
      + 2;
```

3.1.5 Case Sensitivity

In general, C language is case sensitive and variables with lowercase names are
different from those with uppercase names. Currently, however, mikroC variables are
not case sensitive (although future releases of mikroC may offer case sensitivity) so
the following variables are equivalent:

```
total    TOTAL    Total    ToTal    total    totaL
```

The only exception is the identifiers *main* and *interrupt*, which must be written in
lowercase in mikroC. In this book we are assuming that the variables are case sensitive,
for the sake of compatibility with other C compilers, and variables with the same name
but different cases are not used.

3.1.6 Variable Names

In C language, variable names can begin with an alphabetical character or with the underscore character. In essence, variable names can include any of the characters a to z and A to Z, the digits 0 to 9, and the underscore character "_". Each variable name should be unique within the first 31 characters of its name. Variable names can contain uppercase and lowercase characters (see Section 3.1.5), and numeric characters can be used inside a variable name. Examples of valid variable names are:

```
Sum    count    sum100    counter    i1    UserName
_myName
```

Some names are reserved for the compiler itself and cannot be used as variable names in a program. Table 3.1 gives a list of these reserved names.

3.1.7 Variable Types

The mikroC language supports the variable types shown in Table 3.2. Examples of variables are given in this section.

Table 3.1: mikroC reserved names

asm	enum	signed
auto	extern	sizeof
break	float	static
case	for	struct
char	goto	switch
const	if	typedef
continue	int	union
default	long	unsigned
do	register	void
double	return	volatile
else	short	while

Table 3.2: mikroC variable types

Type	Size (bits)	Range
unsigned char	8	0 to 255
unsigned short int	8	0 to 255
unsigned int	16	0 to 65535
unsigned long int	32	0 to 4294967295
signed char	8	−128 to 127
signed short int	8	−128 to 127
signed int	16	−32768 to 32767
signed long int	32	−2147483648 to 2147483647
float	32	±1.17549435082E-38 to ±6.80564774407E38
double	32	±1.17549435082E-38 to ±6.80564774407E38
long double	32	±1.17549435082E-38 to ±6.80564774407E38

(unsigned) char or unsigned short (int)

The variables *(unsigned) char*, or *unsigned short (int)*, are 8-bit unsigned variables with a range of 0 to 255. In the following example two 8-bit variables named *total* and *sum* are created, and *sum* is assigned decimal value 150:

```
unsigned char total, sum;
sum = 150;
```

or

```
char total, sum;
sum = 150;
```

Variables can be assigned values during their declaration. Thus, the above statements can also be written as:

```
char total, sum = 150;
```

signed char or (signed) short (int)

The variables *signed char*, or *(signed) short (int)*, are 8-bit signed character variables with a range of −128 to +127. In the following example a signed 8-bit variable named *counter* is created with a value of −50:

```
signed char counter = −50;
```

or

```
short counter = −50;
```

or

```
short int counter = −50;
```

(signed) int

Variables called *(signed) int* are 16-bit variables with a range −32768 to +32767. In the following example a signed integer named *Big* is created:

```
int Big;
```

unsigned (int)

Variables called *(unsigned) int* are 16-bit unsigned variables with a range 0 to 65535. In the following example an unsigned 16-bit variable named *count* is created and is assigned value 12000:

```
unsigned int count = 12000;
```

(signed) long (int)

Variables called *(signed) long (int)* are 32 bits long with a range −2147483648 to +2147483647. An example is:

```
signed long LargeNumber;
```

unsigned long (int)

Variables called *(unsigned) long (int)* are 32-bit unsigned variables having the range 0 to 4294967295. An example is:

```
unsigned long VeryLargeNumber;
```

float or double or long double

The variables called *float* or *double* or *long double*, are floating point variables implemented in mikroC using the Microchip AN575 32-bit format, which is IEEE 754 compliant. Floating point numbers range from $\pm 1.17549435082\text{E-}38$ to $\pm 6.80564774407\text{E}38$. In the following example, a floating point variable named *area* is created and assigned the value 12.235:

```
float area;
area = 12.235;
```

To avoid confusion during program development, specifying the sign of the variable (signed or unsigned) as well as the type of variable is recommended. For example, use *unsigned char* instead of *char* only, and *unsigned int* instead of *unsigned* only.

In this book we are using the following mikroC data types, which are easy to remember and also compatible with most other C compilers:

unsigned char	0 to 255
signed char	−128 to 127
unsigned int	0 to 65535
signed int	−32768 to 32767
unsigned long	0 to 4294967295
signed long	−2147483648 to 2147483647
float	$\pm 1.17549435082\text{E-}38$ to $\pm 6.80564774407\text{E}38$

3.1.8   Constants

Constants represent fixed values (numeric or character) in programs that cannot be changed. Constants are stored in the flash program memory of the PIC microcontroller, thus not wasting valuable and limited RAM memory. In mikroC, constants can be integers, floating points, characters, strings, or enumerated types.

Integer Constants

Integer constants can be decimal, hexadecimal, octal, or binary. The data type of a constant is derived by the compiler from its value. But suffixes can be used to change the type of a constant.

In Table 3.2 we saw that decimal constants can have values from -2147483648 to $+4294967295$. For example, constant number 210 is stored as an *unsigned char* (or *unsigned short int*). Similarly, constant number -200 is stored as a *signed int*.

Using the suffix u or U forces the constant to be *unsigned*. Using the suffix L or l forces the constant to be *long*. Using both U (or u) and L (or l) forces the constant to be *unsigned long*.

Constants are declared using the keyword *const* and are stored in the flash program memory of the PIC microcontroller, thus not wasting valuable RAM space. In the following example, constant *MAX* is declared as 100 and is stored in the flash program memory of the PIC microcontroller:

```
const MAX = 100;
```

Hexadecimal constants start with characters 0x or 0X and may contain numeric data 0 to 9 and hexadecimal characters A to F. In the following example, constant *TOTAL* is given the hexadecimal value FF:

```
const TOTAL = 0xFF;
```

Octal constants have a zero at the beginning of the number and may contain numeric data 0 to 7. In the following example, constant CNT is given octal value 17:

```
const CNT = 017;
```

Binary constant numbers start with 0b or 0B and may contain only 0 or 1. In the following example a constant named *Min* is declared as having the binary value 11110000:

```
const Min = 0b11110000
```

Floating Point Constants

Floating point constant numbers have integer parts, a dot, a fractional part, and an optional e or E followed by a signed integer exponent. In the following example, a constant named *TEMP* is declared as having the fractional value 37.50:

```
const TEMP = 37.50
```

or

```
const TEMP = 3.750E1
```

Character Constants

A character constant is a character enclosed within single quote marks. In the following example, a constant named *First_Alpha* is declared as having the character value "A":

```
const First_Alpha = 'A';
```

String Constants

String constants are fixed sequences of characters stored in the flash memory of the microcontroller. The string must both begin and terminate with a double quote character ("). The compiler automatically inserts a null character as a terminator. An example string constant is:

```
"This is an example string constant"
```

A string constant can be extended across a line boundary by using a backslash character ("\"):

```
"This is first part of the string \
and this is the continuation of the string"
```

This string constant declaration is the same as:

```
"This is first part of the string and this is the continuation
of the string"
```

Enumerated Constants

Enumerated constants are integer type and are used to make a program easier to follow. In the following example, constant *colors* stores the names of colors. The first element is given the value 0:

```
enum colors {black, brown, red, orange, yellow, green, blue, gray,
white};
```

3.1.9   Escape Sequences

Escape sequences are used to represent nonprintable ASCII characters. Table 3.3 shows some commonly used escape sequences and their representation in C language. For example, the character combination "\n" represents the newline character.

Table 3.3: Some commonly used escape sequences

Escape sequence	Hex value	Character
\a	0×07	BEL (bell)
\b	0×08	BS (backspace)
\t	0×09	HT (horizontal tab)
\n	0×0A	LF (linefeed)
\v	0×0B	VT (vertical feed)
\f	0×0C	FF (formfeed)
\r	0×0D	CR (carriage return)
\xH		String of hex digits

An ASCII character can also be represented by specifying its hexadecimal code after a backslash. For example, the newline character can also be represented as "\x0A".

3.1.10   Static Variables

Static variables are local variables used in functions (see Chapter 4) when the last value of a variable between successive calls to the function must be preserved. As the following example shows, static variables are declared using the keyword *static*:

```
static unsigned int count;
```

3.1.11   External Variables

Using the keyword *extern* before a variable name declares that variable as external. It tells the compiler that the variable is declared elsewhere in a separate source code module. In the following example, variables sum1 and sum2 are declared as external unsigned integers:

```
extern int sum1, sum2;
```

3.1.12 Volatile Variables

Volatile variables are especially important in interrupt-based programs and input-output routines. Using the keyword *volatile* indicates that the value of the variable may change during the lifetime of the program independent of the normal flow of the program. Variables declared as volatile are not optimized by the compiler, since their values can change unexpectedly. In the following example, variable *Led* is declared as a volatile unsigned char:

```
volatile unsigned char Led;
```

3.1.13 Enumerated Variables

Enumerated variables are used to make a program more readable. In an enumerated variable, a list of items is specified and the value of the first item is set to 0, the next item is set to 1, and so on. In the following example, type *Week* is declared as an enumerated list and MON = 0, TUE = 1, WED = 2, and so on):

```
enum Week {MON, TUE, WED, THU, FRI, SAT, SUN};
```

It is possible to imply the values of the elements in an enumerated list. In the following example, black = 2, blue = 3, red = 4, and so on.

```
enum colors {black = 2, blue, red, white, gray};
```

Similarly, in the following example, black = 2, blue = 3, red = 8, and gray = 9:

```
enum colors {black = 2, blue, red = 8, gray};
```

Variables of type enumeration can be declared by specifying them after the list of items. For example, to declare variable My_Week of enumerated type Week, use the following statement:

```
enum Week {MON, TUE, WED, THU, FRI, SAT, SUN} My_Week;
```

Now we can use variable My_Week in a program:

```
My_Week = WED    // assign 2 to My_Week
```

or

```
My_Week = 2    // same as above
```

After defining the enumerated type *Week*, we can declare variables This_Week and Next_Week of type Week as:

```
enum Week This_Week, Next_Week;
```

3.1.14   Arrays

Arrays are used to store related items in the same block of memory and under a specified name. An array is declared by specifying its type, name, and the number of elements it will store. For example:

```
unsigned int Total[5];
```

This array of type unsigned int has the name *Total* and has five elements. The first element of an array is indexed with 0. Thus, in this example, Total[0] refers to the first element of the array and Total[4] refers to the last element. The array Total is stored in memory in five consecutive locations as follows:

| Total[0] |
| Total[1] |
| Total[2] |
| Total[3] |
| Total[4] |

Data can be stored in the array by specifying the array name and index. For example, to store 25 in the second element of the array we have to write:

```
Total[1] = 25;
```

Similarly, the contents of an array can be read by specifying the array name and its index. For example, to copy the third array element to a variable called *Temp* we have to write:

```
Temp = Total[2];
```

The contents of an array can be initialized during the declaration of the array by assigning a sequence of comma-delimited values to the array. An example follows where array *months* has twelve elements and months[0] = 31, months[1] = 28, and so on:

```
unsigned char months[12] = {31,28,31,30,31,30,31,31,30,31,30,31};
```

The same array can also be declared without specifying its size:

```
unsigned char months[ ] = {31,28,31,30,31,30,31,31,30,31,30,31};
```

Character arrays can be declared similarly. In the following example, a character array named *Hex_Letters* is declared with 6 elements:

```
unsigned char Hex_Letters[ ] = {'A', 'B', 'C', 'D', 'E', 'F'};
```

Strings are character arrays with a null terminator. Strings can be declared either by enclosing the string in double quotes, or by specifying each character of the array within single quotes and then terminating the string with a null character. The two string declarations in the following example are identical, and both occupy five locations in memory:

```
unsigned char Mystring[ ] = "COMP";
```

and

```
unsigned char Mystring[ ] = {'C', 'O', 'M', 'P', '\0'};
```

In C programming language, we can also declare arrays with multiple dimensions. One-dimensional arrays are usually called vectors, and two-dimensional arrays are called matrices. A two-dimensional array is declared by specifying the data type of the array, the array name, and the size of each dimension. In the following example, a two-dimensional array named P is created having three rows and four columns. Altogether, the array has twelve elements. The first element of the array is P[0][0], and the last element is P[2][3]. The structure of this array is shown below:

P[0][0]	P[0][1]	P[0][2]	P[0][3]
P[1][0]	P[1][1]	P[1][2]	P[1][3]
P[2][0]	P[2][1]	P[2][2]	P[2][3]

Elements of a multidimensional array can be specified during the declaration of the array. In the following example, two-dimensional array Q has two rows and two columns, its diagonal elements are set to 1, and its nondiagonal elements are cleared to 0:

```
unsigned char Q[2][2] = { {1,0}, {0,1} };
```

3.1.15 Pointers

Pointers are an important part of the C language, as they hold the memory addresses of variables. Pointers are declared in the same way as other variables, but with the character ("*") in front of the variable name. In general, pointers can be created to point to (or hold the addresses of) character variables, integer variables, long variables, floating point variables, or functions (although mikroC currently does not support pointers to functions).

In the following example, an unsigned character pointer named *pnt* is declared:

```
unsigned char *pnt;
```

When a new pointer is created, its content is initially unspecified and it does not hold the address of any variable. We can assign the address of a variable to a pointer using the ("&") character:

```
pnt = &Count;
```

Now *pnt* holds the address of variable *Count*. Variable Count can be set to a value by using the character ("*") in front of its pointer. For example, Count can be set to 10 using its pointer:

```
*pnt = 10;      // Count = 10
```

which is the same as

```
Count = 10;     // Count = 10
```

Or, the value of Count can be copied to variable Cnt using its pointer:

```
Cnt = *pnt;     // Cnt = Count
```

Array Pointers

In C language the name of an array is also a pointer to the array. Thus, for the array:

```
unsigned int Total[10];
```

The name *Total* is also a pointer to this array, and it holds the address of the first element of the array. Thus the following two statements are equal:

```
Total[2] = 0;
```

and

```
*(Total + 2) = 0;
```

Also, the following statement is true:

```
&Total[j] = Total + j
```

In C language we can perform pointer arithmetic which may involve:

- Comparing two pointers

- Adding or subtracting a pointer and an integer value

- Subtracting two pointers

- Assigning one pointer to another

- Comparing a pointer to null

For example, let's assume that pointer P is set to hold the address of array element Z[2]:

```
P = &Z[2];
```

We can now clear elements 2 and 3 of array Z, as in the two examples that follow. The two examples are identical except that in the first example pointer P holds the address of Z[3] at the end of the statements, and it holds the address of Z[2] at the end of the second set of statements:

```
*P = 0;              // Z[2] = 0
P = P + 1;           // P now points to element 3 of Z
*P = 0;              // Z[3] = 0
```

or

```
*P = 0;              // Z[2] = 0
*(P + 1) = 0;        // Z[3] = 0
```

A pointer can be assigned to another pointer. In the following example, variables *Cnt* and *Tot* are both set to 10 using two different pointers:

```
unsigned int *i, *j;      // declare 2 pointers
unsigned int Cnt, Tot;    // declare two variables
```

```
i = &Cnt;       // i points to Cnt
*i = 10;        // Cnt = 10
j = i;          // copy pointer i to pointer j
Tot = *j;       // Tot = 10
```

3.1.16   Structures

A structure can be used to collect related items that are then treated as a single object. Unlike an array, a structure can contain a mixture of data types. For example, a structure can store the personal details (name, surname, age, date of birth, etc.) of a student.

A structure is created by using the keyword *struct*, followed by a structure name and a list of member declarations. Optionally, variables of the same type as the structure can be declared at the end of the structure.

The following example declares a structure named *Person*:

```
struct Person
{
        unsigned char name[20];
        unsigned char surname[20];
        unsigned char nationality[20];
        unsigned char age;
}
```

Declaring a structure does not occupy any space in memory; rather, the compiler creates a template describing the names and types of the data objects or member elements that will eventually be stored within such a structure variable. Only when variables of the same type as the structure are created do these variables occupy space in memory. We can declare variables of the same type as the structure by giving the name of the structure and the name of the variable. For example, two variables *Me* and *You* of type Person can be created by the statement:

```
struct Person Me, You;
```

Variables of type Person can also be created during the declaration of the structure as follows:

```
struct Person
{
      unsigned char name[20];
      unsigned char surname[20];
      unsigned char nationality[20];
      unsigned char age;
} Me, You;
```

We can assign values to members of a structure by specifying the name of the structure, followed by a dot (".") and the name of the member. In the following example, the *age* of structure variable *Me* is set to 25, and variable *M* is assigned to the value of *age* in structure variable *You*:

```
Me.age = 25;
M = You.age;
```

Structure members can be initialized during the declaration of the structure. In the following example, the radius and height of structure *Cylinder* are initialized to 1.2 and 2.5 respectively:

```
struct Cylinder
{
      float radius;
      float height;
} MyCylinder = {1.2, 2.5};
```

Values can also be set to members of a structure using pointers by defining the variable types as pointers. For example, if *TheCylinder* is defined as a pointer to structure Cylinder, then we can write:

```
struct Cylinder
{
      float radius;
      float height;
} *TheCylinder;

TheCylinder -> radius = 1.2;
TheCylinder -> height = 2.5;
```

The size of a structure is the number of bytes contained within the structure. We can use the *sizeof* operator to get the size of a structure. Considering the above example,

```
sizeof(MyCylinder)
```

returns 8, since each float variable occupies 4 bytes in memory.

Bit fields can be defined using structures. With bit fields we can assign identifiers to bits of a variable. For example, to identify bits 0, 1, 2, and 3 of a variable as *LowNibble* and to identify the remaining 4 bits as *HighNibble* we can write:

```
struct
{
      LowNibble    : 4;
      HighNibble   : 4;
} MyVariable;
```

We can then access the nibbles of variable *MyVariable* as:

```
MyVariable.LowNibble  = 12;
MyVariable.HighNibble = 8;
```

In C language we can use the typedef statements to create new types of variables. For example, a new structure data type named *Reg* can be created as follows:

```
typedef struct
{
      unsigned char name[20];
      unsigned char surname[20];
      unsigned age;
} Reg;
```

Variables of type *Reg* can then be created in the same way other types of variables are created. In the following example, variables *MyReg*, *Reg1*, and *Reg2* are created from data type *Reg*:

```
Reg MyReg, Reg1, Reg2;
```

The contents of one structure can be copied to another structure, provided that both structures are derived from the same template. In the following example, structure variables of the same type, P1 and P2, are created, and P2 is copied to P1:

```
struct Person
{
      unsigned char name[20];
      unsigned char surname[20];
      unsigned int age;
      unsigned int height;
      unsigned weight;
}
struct Person P1, P2;
```

```
.....................
.....................
P2 = P1;
```

3.1.17   Unions

Unions are used to overlay variables. A union is similar to a structure and is even
defined in a similar manner. Both are based on templates, and the members of both are
accessed using the "." or "->" operators. A union differs from a structure in that all
variables in a union occupy the same memory area, that is, they share the same storage.
An example of a union declaration is:

```
union flags
{
      unsigned char x;
      unsigned int y;
} P;
```

In this example, variables *x* and *y* occupy the same memory area, and the size of this
union is 2 bytes long, which is the size of the biggest member of the union. When
variable y is loaded with a 2-byte value, variable x will have the same value as the low
byte of y. In the following example, y is loaded with 16-bit hexadecimal value
0xAEFA, and x is loaded with 0xFA:

```
P.y = 0xAEFA;
```

The size of a union is the size (number of bytes) of its largest member. Thus, the
statement:

```
sizeof(P)
```

returns 2.

This union can also be declared as:

```
union flags
{
      unsigned char x;
      unsigned int y;
}
union flags P;
```

3.1.18 Operators in C

Operators are applied to variables and other objects in expressions to cause certain conditions or computations to occur.

mikroC language supports the following operators:

- Arithmetic operators
- Relational operators
- Logical operators
- Bitwise operators
- Assignment operators
- Conditional operators
- Preprocessor operators

Arithmetic Operators

Arithmetic operators are used in arithmetic computations. Arithmetic operators associate from left to right, and they return numerical results. The mikroC arithmetic operators are listed in Table 3.4.

Table 3.4: mikroC arithmetic operators

Operator	Operation
+	Addition
−	Subtraction
*	Multiplication
/	Division
%	Remainder (integer division)
++	Auto increment
−−	Auto decrement

The following example illustrates the use of arithmetic operators:

```
/* Adding two integers */
5 + 12                                      // equals 17

/* Subtracting two integers */
120 - 5                                     // equals 115
10 - 15                                     // equals -5

/* Dividing two integers */
5 / 3                                       // equals 1
12 / 3                                      // equals 4

/* Multiplying two integers */
3 * 12                                      // equals 36

/* Adding two floating point numbers */
3.1 + 2.4                                   // equals 5.5

/* Multiplying two floating point numbers */
2.5 * 5.0                                   // equals 12.5

/* Dividing two floating point numbers */
25.0 / 4.0                                  // equals 6.25

/* Remainder (not for float) */
7 % 3                                       // equals 1

/* Post-increment operator */
j = 4;
k = j++;                                    // k = 4, j = 5

/* Pre-increment operator */
j = 4;
k = ++j;                                    // k = 5, j = 5

/* Post-decrement operator */
j = 12;
k = j--;                                    // k = 12, j = 11
```

```
/* Pre-decrement operator */

j = 12;
k = --j;                                    // k = 11, j = 11
```

Relational Operators

Relational operators are used in comparisons. If the expression evaluates to TRUE, a 1 is returned; otherwise a 0 is returned.

All relational operators associate from left to right. A list of mikroC relational operators is given in Table 3.5.

The following example illustrates the use of relational operators:

```
x = 10
x > 8       // returns 1
x == 10     // returns 1
x < 100     // returns 1
x > 20      // returns 0
x != 10     // returns 0
x >= 10     // returns 1
x <= 10     // returns 1
```

Logical Operators

Logical operators are used in logical and arithmetic comparisons, and they return TRUE (i.e., logical 1) if the expression evaluates to nonzero, and FALSE (i.e., logical 0) if the

Table 3.5: mikroC relational operators

Operator	Operation
==	Equal to
!=	Not equal to
>	Greater than
<	Less than
>=	Greater than or equal to
<=	Less than or equal to

Table 3.6: mikroC logical operators

Operator	Operation
&&	AND
||	OR
!	NOT

expression evaluates to zero. If more than one logical operator is used in a statement, and if the first condition evaluates to FALSE, the second expression is not evaluated.

The mikroC logical operators are listed in Table 3.6.

The following example illustrates the use of logical operators:

```
/* Logical AND */
x = 7;

x > 0 && x < 10       // returns 1
x > 0 || x < 10       // returns 1
x >=0 && x <=10       // returns 1
x >=0 && x < 5        // returns 0

a = 10; b = 20; c = 30; d = 40;
a > b && c > d        // returns 0
b > a && d > c        // returns 1
a > b || d > c        // returns 1
```

Bitwise Operators

Bitwise operators are used to modify the bits of a variable. The mikroC bitwise operators are listed in Table 3.7.

Bitwise AND returns 1 if both bits are 1, otherwise it returns 0.

Bitwise OR returns 0 if both bits are 0, otherwise it returns 1.

Bitwise XOR returns 1 if both bits are complementary, otherwise it returns 0.

Bitwise complement inverts each bit.

Bitwise shift left and shift right move the bits to the left or right respectively.

Table 3.7: mikroC bitwise operators

Operator	Operation
&	Bitwise AND
|	Bitwise OR
^	Bitwise EXOR
~	Bitwise complement
<<	Shift left
>>	Shift right

The following example illustrates the use of bitwise operators:

```
 i.  0xFA & 0xEE returns 0xEA
     0xFA:     1111 1010
     0xEE:     1110 1110
     - - - - - - - - - - -
     0xEA:     1110 1010


 ii. 0x01 | 0xFE returns 0xFF
     0x08:     0000 0001
     0xFE:     1111 1110
     - - - - - - - - - - -
     0xFE:     1111 1111


iii. 0xAA ^ 0x1F returns
     0xAA:     1010 1010
     0x1F:     0001 1111
     - - - - - - - - - - -
     0xB5:     1011 0101


 iv. ~0xAA returns 0x
     0xAA:     1010 1010
     ~ :       0101 0101
     - - - - - - - - - - -
     0x55:     0101 0101
```

    v.   0x14 >> 1 returns 0x08 (shift 0x14 right by 1 digit)
        0x14:   0001 0100
        >>1 :   0000 1010
        - - - - - - - - - - -
        0x0A: 0000 1010

   vi.   0x14 >> 2 returns 0x05 (shift 0x14 right by 2 digits)
        0x14:   0001 0100
        >> 2:   0000 0101
        - - - - - - - - - - -
        0x05:   0000 0101

  vii.   0x235A << 1 returns 0x46B4 (shift left 0x235A left by 1 digit)
        0x235A: 0010 0011 0101 1010
        <<1 :   0100 0110 1011 0100
        - - - - - - - - - - - - - - - - - -
        0x46B4 : 0100 0110 1011 0100

 viii.   0x1A << 3 returns 0xD0 (shift left 0x1A by 3 digits)
        0x1A: 0001 1010
        <<3 :   1101 0000
        - - - - - - - - - - -
        0xD0:   1101 0000

Assignment Operators

In C language there are two types of assignments: simple and compound. In simple assignments an expression is simply assigned to another expression, or an operation is performed using an expression and the result is assigned to another expression:

Expression1 = Expression2

or

Result = Expression1 operation Expression2

Examples of simple assignments are:

Temp = 10;
Cnt = Cnt + Temp;

Compound assignments have the general format:

Result operation = Expression1

Here the specified operation is performed on *Expression1* and the result is stored in *Result*. For example:

```
j += k;     is same as:    j = j + k;
```

also

```
p * = m;    is same as    p = p * m;
```

The following compound operators can be used in mikroC programs:

```
+=      -=      * =     /=     %=
&=      |=      ^=      >>=    <<=
```

Conditional Operators

The syntax of a conditional operator is:

```
Result = Expression1 ? Expression2 : Expression3
```

Expression1 is evaluated first, and if its value is true, *Expression2* is assigned to *Result*, otherwise *Expression3* is assigned to Result. In the following example, the maximum of *x* and *y* is found where x is compared with y and if x > y then max = x, otherwise max = y:

$$max = (x > y) ? x : y;$$

In the following example, lowercase characters are converted to uppercase. If the character is lowercase (between a and z), then by subtracting 32 from the character we obtain the equivalent uppercase character:

$$c = (c > = a \&\& c < = z) ? (c - 32) : c;$$

Preprocessor Operators

The preprocessor allows a programmer to:

- Compile a program conditionally, such that parts of the code are not compiled
- Replace symbols with other symbols or values
- Insert text files into a program

The preprocessor operator is the ("#") character, and any line of code leading with a ("#") is assumed to be a preprocessor command. The semicolon character (";") is not needed to terminate a preprocessor command.

mikroC compiler supports the following preprocessor commands:

```
#define    #undef
#if        #elif      #endif
#ifdef     #ifndef
#error
#line
```

#define, #undef, #ifdef, #ifndef   The *#define* preprocessor command provides macro expansion where every occurrence of an identifier in the program is replaced with the value of that identifier. For example, to replace every occurrence of *MAX* with value 100 we can write:

#define MAX 100

An identifier that has already been defined cannot be defined again unless both definitions have the same value. One way to get around this problem is to remove the macro definition:

#undef MAX

Alternatively, the existence of a macro definition can be checked. In the following example, *if* MAX has *not* already been defined, it is given value 100, otherwise the #define line is skipped:

#ifndef MAX
      #define MAX 100
#endif

Note that the #define preprocessor command does not occupy any space in memory.

We can pass parameters to a macro definition by specifying the parameters in a parenthesis after the macro name. For example, consider the macro definition:

#define ADD(a, b)   (a + b)

When this macro is used in a program, (a, b) will be replaced with (a + b) as shown:

p = ADD(x, y)   will be transformed into p = (x + y)

Similarly, we can define a macro to calculate the square of two numbers:

#define SQUARE(a)   (a * a)

We can now use this macro in a program:

p = SQUARE(x)   will be transformed into p = (x * x)

#include   The preprocessor directive *#include* is used to include a source file in our program. Usually header files with extension ".h" are used with #include. There are two formats for using #include:

#include <file>

and

#include "file"

In first option the file is searched in the mikroC installation directory first and then in user search paths. In second option the specified file is searched in the mikroC project folder, then in the mikroC installation folder, and then in user search paths. It is also possible to specify a complete directory path as:

#include "C:\temp\last.h"

The file is then searched only in the specified directory path.

#if, #elif, #else, #endif   The preprocessor commands *#if*, *#elif*, *#else*, and *#endif* are used for conditional compilations, where parts of the source code can be compiled only if certain conditions are met. In the following example, the code section where variables A and B are cleared to zero is compiled if M has a nonzero value, otherwise the code section where A and B are both set to 1 is compiled. Notice that the #if must be terminated with #endif:

```
#if M
        A = 0;
        B = 0;
#else
        A = 1;
        B = 1;
#endif
```

We can also use the #elif condition, which tests for a new condition if the previous condition was false:

```
#if M
        A = 0;
        B = 0;
#elif N
        A = 1;
        B = 1;
#else
        A = 2;
        B = 2;

#endif
```

In the above example, if M has a nonzero value code section, A = 0; B = 0; are compiled. Otherwise, if N has a nonzero value, then code section A = 1; B = 1; is compiled. Finally, if both M and N are zero, then code section A = 2; B = 2; is compiled. Notice that only one code section is compiled between #if and #endif and that a code section can contain any number of statements.

3.1.19 Modifying the Flow of Control

Statements are normally executed sequentially from the beginning to the end of a program. We can use control statements to modify this normal sequential flow in a C program. The following control statements are available in mikroC programs:

- Selection statements
- Unconditional modifications of flow
- Iteration statements

Selection Statements

There are two selection statements: *if* and *switch*.

if *Statement*   The general format of the *if* statement is:

```
if(expression)

        Statement1;
```

```
else
        Statement2;
```

or

```
if(expression)Statement1; else Statement2;
```

If the *expression* evaluates to TRUE, *Statement1* is executed, otherwise *Statement2* is executed. The *else* keyword is optional and may be omitted. In the following example, if the value of x is greater than *MAX* then variable P is incremented by 1, otherwise it is decremented by 1:

```
if(x > MAX)
        P++;
else
        P--;
```

We can have more than one statement by enclosing the statements within curly brackets. For example:

```
if(x > MAX)
{
        P++;
        Cnt = P;
        Sum = Sum + Cnt;
}
else
        P--;
```

In this example, if x is greater than *MAX* then the three statements within the curly brackets are executed, otherwise the statement $P--$ is executed.

Another example using the *if* statement is:

```
if(x > 0 && x < 10)
{
        Total += Sum;
        Sum++;
}
else
{
        Total = 0;
        Sum = 0;
}
```

switch *Statement*   The *switch* statement is used when a number of conditions and different operations are performed if a condition is true. The syntax of the switch statement is:

```
switch (condition)
{
        case condition1:
                  Statements;
                  break;
        case condition2:
                  Statements;
                  break;
        ....................
        ....................
        case condition:
                  Statements;
                  break;
        default:
                  Statements;
}
```

The switch statement functions as follows: First the *condition* is evaluated. The condition is then compared to *condition1* and if a match is found, statements in that case block are evaluated and control jumps outside the switch statement when the *break* keyword is encountered. If a match is not found, condition is compared to *condition2* and if a match is found, statements in that case block are evaluated and control jumps outside the switch statements, and so on. The *default* is optional, and statements following default are evaluated if the condition does not match any of the conditions specified after the *case* keywords.

In the following example, the value of variable *Cnt* is evaluated. If Cnt = 1, A is set to 1. If Cnt = 10, B is set to 1, and if Cnt = 100, C is set to 1. If Cnt is not equal to 1, 10, or 100 then D is set to 1:

```
switch (Cnt)
{
        case 1:
                  A = 1;
                  break;
        case 10:
                  B = 1;
                  break;
```

```
        case 100:
                C = 1;
                break;
        default:
                D = 1;
}
```

Because white spaces are ignored in C language we can also write the preceding code as:

```
switch (Cnt)
{
        case 1:        A = 1;    break;
        case 10:       B = 1;    break;
        case 100:      C = 1;    break;
        default:       D = 1;
}
```

Example 3.1

In an experiment the relationship between X and Y values are found to be:

X	Y
1	3.2
2	2.5
3	8.9
4	1.2
5	12.9

Write a switch statement that will return the Y value, given the X value.

Solution 3.1

The required switch statement is:

```
switch (X)
{
        case 1:
                Y = 3.2;
                break;
        case 2:
                Y = 2.5;
                break;
```

```
        case 3:
                  Y = 8.9;
                  break;
        case 4:
                  Y = 1.2;
                  break;
        case 5:
                  Y = 12.9;
}
```

Iteration Statements

Iteration statements enable us to perform loops in a program, where part of a code must be repeated a number of times. In mikroC iteration can be performed in four ways. We will look at each one with examples:

- Using *for* statement

- Using *while* statement

- Using *do* statement

- Using *goto* statement

for *Statement*   The syntax of a *for* statement is:

```
for(initial expression; condition expression; increment expression)
{
      Statements;
}
```

The *initial expression* sets the starting variable of the loop, and this variable is compared against the *condition expression* before entry into the loop. Statements inside the loop are executed repeatedly, and after each iteration the value of the *increment expression* is incremented. The iteration continues until the condition expression becomes false. An endless loop is formed if the condition expression is always true.

The following example shows how a loop can be set up to execute 10 times. In this example, variable i starts from 0 and increments by 1 at the end of each iteration. The loop terminates when $i = 10$, in which case the condition $i < 10$ becomes false. On exit from the loop, the value of i is 10:

```
for(i = 0; i < 10; i ++)
{
      statements;
}
```

This loop could also be started by an initial expression with a nonzero value. Here, *i* starts with 1 and the loop terminates when *i* = 11. Thus, on exit from the loop, the value of *i* is 11:

```
for(i = 1; i <= 10; i++)
{
      Statements;
}
```

The parameters of a *for* loop are all optional and can be omitted. If the condition expression is left out, it is assumed to be true. In the following example, an endless loop is formed where the condition expression is always true and the value of *i* starts with 0 and is incremented after each iteration:

```
/* Endless loop with incrementing i */
for(i=0; ; i++)
{
      Statements;
}
```

In the following example of an endless loop all the parameters are omitted:

```
/* Example of endless loop */
for(; ;)
{
      Statements;
}
```

In the following endless loop, *i* starts with 1 and is not incremented inside the loop:

```
/* Endless loop with i = 1 */
for(i=1; ;)
{
      Statements;
}
```

If there is only one statement inside the *for* loop, he curly brackets can be omitted as shown in the following example:

```
for(k = 0; k < 10; k++) Total = Total + Sum;
```

Nested *for* loops can also be used. In a nested *for* loop, the inner loop is executed for each iteration of the outer loop. In the following example the inner loop is executed five times and the outer loop is executed ten times. The total iteration count is fifty:

```
/* Example of nested for loops */
for(i = 0; i < 10; i++)
{
        for(j = 0; j < 5; j++)
        {
                Statements;
        }
}
```

In the following example, the sum of all the elements of a 3 × 4 matrix M is calculated and stored in a variable called *Sum*:

```
/* Add all elements of a 3x4 matrix */
Sum = 0;
for(i = 0; i < 3; i++)
{
        for(j = 0; j < 4; j++)
        {
            Sum = Sum + M[i][j];
        }
}
```

Since there is only one statement to be executed, the preceding example could also be written as:

```
/* Add all elements of a 3x4 matrix */
Sum = 0;
for(i = 0; i < 3; i++)
{
        for(j = 0; j < 4; j++) Sum = Sum + M[i][j];
}
```

while *Statement* The syntax of a *while* statement is:

```
while (condition)
{
      Statements;
}
```

Here, the statements are executed repeatedly until the *condition* becomes false, or the statements are executed repeatedly as long as the condition is true. If the condition is false on entry to the loop, then the loop will not be executed and the program will continue from the end of the *while* loop. It is important that the condition is changed inside the loop, otherwise an endless loop will be formed.

The following code shows how to set up a loop to execute 10 times, using the *while* statement:

```
/* A loop that executes 10 times */
k = 0;
while (k < 10)
{
      Statements;
      k++;
}
```

At the beginning of the code, variable *k* is 0. Since k is less than 10, the *while* loop starts. Inside the loop the value of k is incremented by 1 after each iteration. The loop repeats as long as k < 10 and is terminated when k = 10. At the end of the loop the value of k is 10.

Notice that an endless loop will be formed if k is not incremented inside the loop:

```
/* An endless loop */
k = 0;
while (k < 10)
{
      Statements;
}
```

An endless loop can also be formed by setting the condition to be always true:

```
/* An endless loop */
while (k = k)
{
      Statements;
}
```

Here is an example of calculating the sum of numbers from 1 to 10 and storing the result in a variable called *sum*:

```
/* Calculate the sum of numbers from 1 to 10 */
unsigned int k, sum;
k = 1;
sum = 0;
while (k <= 10)
{
        sum = sum + k;
        k++;
}
```

It is possible to have a *while* statement with no body. Such a statement is useful, for example, if we are waiting for an input port to change its value. An example follows where the program will wait as long as bit 0 of PORTB (PORTB.0) is at logic 0. The program will continue when the port pin changes to logic 1:

```
while (PORTB.0 == 0);    // Wait until PORTB.0 becomes 1
```

or

```
while (PORTB.0);
```

It is also possible to have nested **while** statements.

do *Statement*   A *do* statement is similar to a *while* statement except that the loop executes until the *condition* becomes false, or, the loop executes as long as the condition is true. The condition is tested at the end of the loop. The syntax of a *do* statement is:

```
do
{
        Statements;
} while (condition);
```

The first iteration is always performed whether the condition is true or false. This is the main difference between a *while* statement and a *do* statement.

The following code shows how to set up a loop to execute 10 times using the *do* statement:

```
/* Execute 10 times */
k = 0;
do
{
      Statements;
      k++;
} while (k < 10);
```

The loop starts with k = 0, and the value of *k* is incremented inside the loop after each iteration. At the end of the loop k is tested, and if k is not less than 10, the loop terminates. In this example because k = 0 is at the beginning of the loop, the value of k is 10 at the end of the loop.

An endless loop will be formed if the condition is not modified inside the loop, as shown in the following example. Here k is always less than 10:

```
/* An endless loop */
k = 0;
do
{
      Statements;
} while (k < 10);
```

An endless loop can also be created if the condition is set to be true all the time:

```
/* An endless loop */
do
{
      Statements;
} while (k = k);
```

It is also possible to have nested *do* statements.

Unconditional Modifications of Flow

goto *Statement*   A *goto* statement can be used to alter the normal flow of control in a program. It causes the program to jump to a specified label. A label can be any alphanumeric character set starting with a letter and terminating with the colon (":") character.

Although not recommended, a *goto* statement can be used together with an *if* statement to create iterations in a program. The following example shows how to set up a loop to execute 10 times using *goto* and *if* statements:

```
/* Execute 10 times */
k = 0;
Loop:
        Statements;
        k++;
if (k < 10) goto Loop;
```

The loop starts with label *Loop* and variable k = 0 at the beginning of the loop. Inside the loop the statements are executed and k is incremented by 1. The value of k is then compared with 10 and the program jumps back to label *Loop* if k < 10. Thus, the loop is executed 10 times until the condition at the end becomes false. At the end of the loop the value of k is 10.

continue *and* break *Statements   continue* and *break* statements can be used inside iterations to modify the flow of control. A *continue* statement is usually used with an *if* statement and causes the loop to skip an iteration. An example follows that calculates the sum of numbers from 1 to 10 except number 5:

```
/* Calculate sum of numbers 1,2,3,4,6,7,8,9,10 */
Sum = 0;
i = 1;
for (i = 1; i <= 10; i++)
{
        if (i == 5) continue;      // Skip number 5
        Sum = Sum + i;
}
```

Similarly, a *break* statement can be used to terminate a loop from inside the loop. In the following example, the sum of numbers from 1 to 5 is calculated even though the loop parameters are set to iterate 10 times:

```
/* Calculate sum of numbers 1,2,3,4,5 */
Sum = 0;
i = 1;
for (i = 1; i <= 10; i++)
{
        if (i > 5) break;      // Stop loop if i > 5
        Sum = Sum + i;
}
```

3.1.20   Mixing mikroC with Assembly Language Statements

It sometimes becomes necessary to mix PIC microcontroller assembly language statements with the mikroC language. For example, very accurate program delays can be generated by using assembly language statements. The topic of assembly language is beyond the scope of this book, but techniques for including assembly language instructions in mikroC programs are discussed in this section for readers who are familiar with the PIC microcontroller assembly languages.

Assembly language instructions can be included in a mikroC program by using the keyword *asm* (or _*asm*, or __*asm*). A group of assembly instructions or a single such instruction can be included within a pair of curly brackets. The syntax is:

```
asm
{
        assembly instructions
}
```

Assembly language style comments (a line starting with a semicolon character) are not allowed, but mikroC does allow both types of C style comments to be used with assembly language programs:

```
asm
{
        /* This assembly code introduces delay to the program*/
        MOVLW 6                 // Load W with 6
        ...............
        ...............
}
```

User-declared C variables can be used in assembly language routines, but they must be declared and initialized before use. For example, C variable *Temp* can be initialized and then loaded to the W register as:

```
unsigned char Temp = 10;
asm
{
        MOVLW Temp              // W = Temp = 10
        .................
        .................
}
```

Global symbols such as predefined port names and register names can be used in assembly language routines without having to initialize them:

```
asm
{
        MOVWF PORTB
        ....................
        ....................
}
```

3.2   PIC Microcontroller Input-Output Port Programming

Depending on the type of microcontroller used, PIC microcontroller input-output ports are named as PORTA, PORTB, PORTC, and so on. Port pins can be in analog or digital mode. In analog mode, ports are input only and a built-in analog-to-digital converter and multiplexer circuits are used. In digital mode, a port pin can be configured as either input or output. The TRIS registers control the port directions, and there are TRIS registers for each port, named as TRISA, TRISB, TRISC, and so on. Clearing a TRIS register bit to 0 sets the corresponding port bit to output mode. Similarly, setting a TRIS register bit to 1 sets the corresponding port bit to input mode.

Ports can be accessed as a single 8-bit register, or individual bits of a port can be accessed. In the following example, PORTB is configured as an output port and all its bits are set to a 1:

```
TRISB = 0;                      // Set PORTB as output
PORTB = 0xFF;                   // Set PORTB bits to 1
```

Similarly, the following example shows how the 4 upper bits of PORTC can be set as input and the 4 lower bits of PORTC can be set as output:

```
TRISC = 0xF0;
```

Bits of an input-output port can be accessed by specifying the required bit number. In the following example, variable P2 is loaded with bit 2 of PORTB:

```
P2 = PORTB.2;
```

All the bits of a port can be complemented by the statement:

```
PORTB = ~PORTB;
```

3.3   Programming Examples

In this section, some simple programming examples are given to familiarize the reader with programming in C.

Example 3.2

Write a program to set all eight port pins of PORTB to logic 1.

Solution 3.2

PORTB is configured as an output port, and then all port pins are set to logic 1 by sending hexadecimal number 0xFF:

```
void main()
{
      TRISB = 0;              // Configure PORTB as output
      PORTB = 0xFF;           // Set all port pins to logic a
}
```

Example 3.3

Write a program to set the odd-numbered PORTB pins (bits 1, 3, 5, and 7) to logic 1.

Solution 3.3

Odd-numbered port pins can be set to logic 1 by sending the bit pattern 10101010 to the port. This bit pattern is the hexadecimal number 0xAA and the required program is:

```
void main()
{
      TRISB = 0;              // Configure PORTB as output
      PORTB = 0xAA;           // Turn on odd numbered port pins
}
```

Example 3.4

Write a program to continuously count up in binary and send this data to PORTB. Thus PORTB requires the binary data:

```
00000000
00000001
00000010
00000011
...........
...........
11111110
11111111
00000000
...........
```

Solution 3.4

A *for* loop can be used to create an endless loop, and inside this loop the value of a variable can be incremented and then sent to PORTB:

```
void main()
{
        unsigned char Cnt = 0;
        for(;;)                        // Endless loop
        {
                PORTB = Cnt;           // Send Cnt to PORTB
                Cnt++;                 // Increment Cnt
        }
}
```

Example 3.5

Write a program to set all bits of PORTB to logic 1 and then to logic 0, and to repeat this process ten times.

Solution 3.5

A *for* statement can be used to create a loop to repeat the required operation ten times:

```
void main()
{
        unsigned char j;
        for(j = 0; j < 10; j++)    // Repeat 10 times
        {
        PORTB = 0xFF;              // Set PORTB pins to 1

        PORTB = 0;                // Clear PORTB pins
        }
}
```

Example 3.6

The radius and height of a cylinder are 2.5cm and 10cm respectively. Write a program to calculate the volume of this cylinder.

Solution 3.6

The required program is:

```c
void main()
{
      float Radius = 2.5, Height = 10;
      float Volume;
      Volume = PI * Radius* Radius* Height;
}
```

Example 3.7

Write a program to find the largest element of an integer array having ten elements.

Solution 3.7

At the beginning, variable *m* is set to the first element of the array. A loop is then formed and the largest element of the array is found:

```c
void main()
{
      unsigned char j;
      int m, A[10];
      m = A[0];      // First element of array
      for(j = 1; j < 10; j++)
      {
            if(A[j] > m) m = A[j];
      }
}
```

Example 3.8

Write a program using a *while* statement to clear all ten elements of an integer array M.

Solution 3.8

As shown in the program that follows, *NUM* is defined as 10 and variable *j* is used as the loop counter:

```
#define NUM 10
void main()
{
      int M[NUM] ;
      unsigned char j = 0;
      while (j < NUM)
      {
            M[j] = 0;
            j++;
      }
}
```

Example 3.9

Write a program to convert the temperature from °C to °F starting from 0°C, in steps of 1°C up to and including 100°C, and store the results in an array called F.

Solution 3.9

Given the temperature in °C, the equivalent in °F is calculated using the formula:

$$F = (C - 32.0)/1.8$$

A *for* loop is used to calculate the temperature in °F and store in array F:

```
void main()
{
      float F[100] ;
      unsigned char C;
      for(C = 0; C <= 100; C++)
      {
            F[C] = (C - 32.0) / 1.8;
      }
}
```

3.4   Summary

There are many assembly and high-level languages for the PIC18 series of microcontrollers. This book focuses on the mikroC compiler, since it is easy to learn and a free demo version is available that allows users to develop programs as large as 2K in size.

This chapter presented an introduction to the mikroC language. A C program may contain a number of functions and variables plus a main program. The beginning of the main program is indicated by the statement *void main( )*.

A variable stores a value used during the computation. All variables in C must be declared before they are used. A variable can be an 8-bit character, a 16-bit integer, a 32-bit long, or a floating point number. Constants are stored in the flash program memory of PIC microcontrollers, so using them avoids using valuable and limited RAM memory.

Various flow control and iteration statements such as *if*, *switch*, *while*, *do*, *break,* and so on have been described in the chapter, with examples.

Pointers are used to store the addresses of variables. As we shall see in the next chapter, pointers can be used to pass information back and forth between a function and its calling point. For example, pointers can be used to pass variables between a main program and a function.

3.5   Exercises

1. Write a C program to set bits 0 and 7 of PORTC to logic 1.

2. Write a C program to count down continuously and send the count to PORTB.

3. Write a C program to multiply each element of a ten element array by 2.

4. Write a C program to add two matrices P and Q. Assume that the dimension of each matrix is 3×3 and store the result in another matrix called W.

5. Repeat Exercise 4 but this time multiply matrices P and Q and store the product in matrix R.

6. What do the terms *variable* and *constant* mean?

7. What does *program repetition* mean? Describe the operation of *while*, *do-while*, and *for* loops in C.

8. What is an array? Write example statements to define the following arrays:

    a) An array of ten integers

    b) An array of thirty floats

    c) A two-dimensional array having six rows and ten columns

9. Trace the operation of the following loops. What will be the value of variable *z* at the end of each loop?

    a) ```
        unsigned char j = 0, z = 0;
        while (j < 10)
           {
               z++;
                j++;
           }
        ```

    b)  ```
        unsigned char z = 10;
        for (j = 0; j < 10; j++) z--;
        ```

10. Given the following variable definitions, list the outcome of the following conditional tests in terms of "true" or "false":

    ```
    unsigned int a = 10, b = 2;
    if (a > 10)
    if (b >= 2)
    if (a == 10)
    if (a > 0)
    ```

11. Write a program to calculate whether a number is odd or even.

12. Determine the value of the following bitwise operations using AND, OR, and EXOR operations:

    ```
    Operand 1:    00010001
    Operand 2:    11110001
    ```

13. How many times does each of the following loops iterate, and what is the final value of the variable *j* in each case?

    a) ```
        for (j = 0;   j < 5; j++)
        ```
    b)  ```
        for (j = 1;   j < 10; j++)
        ```
    c) ```
        for (j = 0;   j <= 10; j++)
        ```
    d)  ```
        for (j = 0;   j <= 10; j += 2)
        ```
    e) ```
        for (j = 10; j > 0; j -= 2)
        ```
        ```

14. Write a program to calculate the sum of all positive integer numbers from 1 to 100.

15. Write a program to evaluate factorial n, where 0! and 1! evaluate to 1 and n! = n × (n − 1)!

16. Write a program to calculate the average value of the numbers stored in an array. Assume that the array is called M and has twenty elements.

17. Modify the program in Exercise 16 to find the smallest and largest values of the array. Store the smallest value in a variable called *Sml* and the largest value in a variable called *Lrg*.

18. Derive equivalent if-else statements for the following tests:

    a)   (a > b) ? 0 : 1
    b)   (x < y) ? (a > b) : (c > d)

19. Given that f1 and f2 are both floating point variables, explain why the following test expression controlling the *while* loop may not be safe:

```
do
{
        . . . . . . . . . . . . . .
        . . . . . . . . . . . . . .
} while (f1 != f2);
```

Why would the problem not occur if both f1 and f2 were integers? How would you correct this *while* loop?

20. What can you say about the following *while* loop?

```
k = 0;
Total = 0;
while (k < 10)
{
        Sum++;
        Total += Sum;
}
```

21. What can you say about the following *for* loop?

```
Cnt = 0;
for (; ;)
{
        Cnt++;
}
```

Functions and Libraries in mikroC

4.1 mikroC Functions

A function is a self-contained section of code written to perform a specifically defined action. Functions are usually created when a single operation must be performed in different parts of the main program. It is, moreover, good programming practice to divide a large program into a number of smaller, independent functions. The statements within a function are executed by calling (or invoking) the function.

The general syntax of a function definition is shown in Figure 4.1. The data type indicates the type of data returned by the function. This is followed by the name of the function and then a set of parentheses, within which the arguments, separated by commas, are declared. The body of the function, which includes the function's operational code, is written inside a set of curly brackets.

In the sample function definition that follows, the function, named *Mult*, receives two integer arguments, a and b, and returns their product. Note that using parentheses in a return statement is optional:

```
int Mult(int a, int b)
{
        return (a*b);
}
```

When a function is called, it generally expects to be given the number of arguments expressed in the function's argument list. For example, the preceding function can be called as:

```
z = Mult(x, y);
```

```
type name (parameter1, parameter2,.....)
{
        .............
        function body
        .............
}
```

Figure 4.1: General syntax of a function definition

where variable z has the data type *int*. Note that the arguments declared in the function definition and the arguments passed when the function is called are independent of each other, even if they have the same name. In the preceding example, when the function is called, variable x is copied to a and variable y is copied to b on entry into function *Mult*.

Some functions do not return any data. The data type of such functions must be declared as *void*. For example:

```
void LED(unsigned char D)
{
        PORTB = D;
}
```

void functions can be called without any assignment statements, but the parentheses are needed to tell the compiler that a function call is made:

```
LED( );
```

Also, some functions do not have any arguments. In the following example, the function, named *Compl*, complements PORTC of the microcontroller. It returns no data and has no arguments:

```
void Compl( )
{
        PORTC = ~PORTC;
}
```

This function can be called as:

```
Compl( );
```

Functions are normally defined before the start of the main program.

Some function definitions and their use in main programs are illustrated in the following examples:

Example 4.1

Write a function called *Circle_Area* to calculate the area of a circle where the radius is to be used as an argument. Use this function in a main program to calculate the area of a circle whose radius is 2.5cm. Store the area in a variable called *Circ*.

Solution 4.1

The data type of the function is declared as *float*. The area of a circle is calculated by the formula:

```
Area = πr²
```

where r is the radius of the circle. The area is calculated and stored in a local variable called *s*, which is then returned from the function:

```
float Circle_Area(float radius)

{

    float s;

    s = PI * radius * radius;
    return s;

}
```

Figure 4.2 shows how the function Circle_Area can be used in a main program to calculate the area of a circle whose radius is 2.5cm. The function is defined before the main program. Inside the main program the function is called to calculate and store the area in variable Circ.

Example 4.2

Write a function called *Area* and a function called *Volume* to calculate the area and volume of a cylinder respectively. Then write a main program to calculate the area and the volume of cylinder whose radius is 2.0cm and height is 5.0cm. Store the area in variable *cyl_area* and the volume in variable *cyl_volume*.

Solution 4.2

The area of a cylinder is calculated by the formula:

```
Area = 2πrh
```

```
/*****************************************************************

                         AREA OF A CIRCLE
                         ================

This program calls to function Circle_Area to calculate the area of a circle.


Programmer:   Dogan Ibrahim
File:         CIRCLE.C
Date:         May, 2007


*****************************************************************/

/* This function calculates the area of a circle given the radius */
float Circle_Area(float radius)
{
        float s;

        s = PI * radius * radius;
        return s;
}


/* Start of main program. Calculate the area of a circle where radius = 2.5 */
void main()
{
        float r, Circ;


        r = 2.5;
        Circ = Circle_Area(r);
}
```

Figure 4.2: Program to calculate the area of a circle

where r and h are the radius and height of the cylinder. The volume of a cylinder is calculated by the formula:

$$\text{Volume} = \pi r^2 h$$

Figure 4.3 shows the functions that calculate the area and volume of a cylinder.

The main program that calculates the area and volume of a cylinder whose radius = 2.0cm and height = 5.0cm is shown in Figure 4.4.

Example 4.3

Write a function called *LowerToUpper* to convert a lowercase character to uppercase.

```
float Area(float radius, float height)
{
        float s;

        s = 2.0*PI * radius*height;
        return s;
}


float Volume(float radius, float height)
{
        float s;

        s = PI *radius*radius*height;
        return s;
}
```

Figure 4.3: Functions to calculate cylinder area and volume

Solution 4.3

The ASCII value of the first uppercase character ('A') is 0×41. Similarly, the ASCII value of the first lowercase character ('a') is 0×61. An uppercase character can be converted to its equivalent lowercase by subtracting 0×20 from the character. The required function listing is shown in Figure 4.5.

Example 4.4

Use the function you created in Example 4.3 in a main program to convert letter 'r' to uppercase.

Solution 4.4

The required program is shown in Figure 4.6. Function *LowerToUpper* is called to convert the lowercase character in variable *Lc* to uppercase and store in *Uc*.

4.1.1   Function Prototypes

If a function is not defined before it is called, the compiler will generate an error message. One way around this problem is to create a function prototype. A function prototype is easily constructed by making a copy of the function's header and appending a semicolon to it. If the function has parameters, giving names to these

```
/*************************************************************************

                    AREA AND VOLUME OF A CYLINDER
                    ================================

This program calculates the area and volume of a cylinder whose radius is 2.0cm
and height is 5.0cm.

Programmer:   Dogan Ibrahim
File:         CYLINDER.C
Date:         May, 2007
*************************************************************************/

/* Function to calculate the area of a cylinder */
float Area(float radius, float height)
{
        float s;

        s = 2.0*PI * radius*height;
        return s;
}


/* Function to calculate the volume of a cylinder */
float Volume(float radius, float height)
{
        float s;

        s = PI *radius*radius*height;
        return s;
}

/* Start of the main program */
void main()
{
        float r = 2.0, h = 5.0;
        float cyl_area, cyl_volume;

        cyl_area = Area(r, h);
        cyl_volume(r, h);
}
```

Figure 4.4: Program that calculates the area and volume of a cylinder

parameters is not compulsory, but the data type of the parameters must be defined. An example follows in which a function prototype called *Area* is declared and the function is expected to have a floating point type parameter:

```
float Area(float radius);
```

```
unsigned char LowerToUpper(unsigned char c)
{
        if(c >= 'a' && c <= 'z')
                return (c – 0x20);
        else
                return c;
}
```

Figure 4.5: Function to convert lowercase to uppercase

```
/*******************************************************************

                    LOWERCASE TO UPPERCASE
                    ==========================

This program converts the lowercase character in variable Lc to uppercase
and stores in variable Uc.


Programmer:  Dogan Ibrahim
File:        LTOUPPER.C
Date:        May, 2007
*******************************************************************/

/* Function to convert a lower case character to upper case */
unsigned char LowerToUpper(unsigned char c)
{
        if(c >= 'a' && c <= 'z')
                return (c – 0x20);
        else
                return c;
}


/* Start of main program */
void main()
{
        unsigned char Lc, Uc;

        Lc = 'r';
        Uc = LowerToUpper(Lc);
}
```

Figure 4.6: Program calling function LowerToUpper

This function prototype could also be declared as:

```
float Area(float);
```

Function prototypes should be declared at the beginning of a program. Function definitions and function calls can then be made at any point in the program.

Example 4.5

Repeat Example 4.4 but declare *LowerToUpper* as a function prototype.

Solution 4.5

Figure 4.7 shows the program where function LowerToUpper is declared as a function prototype at the beginning of the program. In this example, the actual function definition is written after the main program.

One important advantage of using function prototypes is that if the function prototype does not match the actual function definition, mikroC will detect this and modify the

```
/************************************************************************

                        LOWERCASE TO UPPERCASE
                        ==========================

This program converts the lowercase character in variable Lc to uppercase
and stores in variable Uc.


Programmer:  Dogan Ibrahim
File:        LTOUPPER2.C
Date:        May, 2007
************************************************************************/

unsigned char LowerToUpper(unsigned char);


/* Start of main program */
void main()
{
        unsigned char Lc, Uc;

        Lc = 'r';
        Uc = LowerToUpper(Lc);
}


/* Function to convert a lower case character to upper case */
unsigned char LowerToUpper(unsigned char c)
{
        if(c >= 'a' && c <= 'z')
                return (c – 0x20);
        else
                return c;
}
```

Figure 4.7: Program using function prototype

data types in the function call to match the data types declared in the function prototype. Suppose we have the following code:

```
unsigned char c = 'A';
unsigned int x = 100;
long Tmp;
long MyFunc(long a, long b);      // function prototype

void main()

{
      ..............
      ..............
      Tmp = MyFunc(c, x);
      ..............
      ..............
}
```

In this example, because the function prototype declares the two arguments as *long*, variables *c* and *x* are converted to *long* before they are used inside function *MyFunc*.

4.1.2 Passing Arrays to Functions

There are many applications where we may want to pass arrays to functions. Passing a single array element is straightforward, as we simply specify the index of the array element to be passed, as in the following function call which passes the second element (index = 1) of array A to function *Calc*. It is important to realize that an individual array element is passed *by value* (i.e., a copy of the array element is passed to the function):

```
x = Calc(A[1]);
```

In some applications we may want to pass complete arrays to functions. An array name can be used as an argument to a function, thus permitting the entire array to be passed. To pass a complete array to a function, the array name must appear by itself within the brackets. The size of the array is not specified within the formal argument declaration. In the function header the array name must be specified with a pair of empty brackets. It is important to realize that when a complete array is passed to a function, what is actually passed is not a copy of the array but the address of the first element of the array (i.e., the array elements are passed *by reference*, which means that the original array elements can be modified inside the function).

Some examples follow that illustrate the passing of a complete array to a function.

Example 4.6

Write a program to store the numbers 1 to 10 in an array called *Numbers*. Then call a function named *Average* to calculate the average of these numbers.

Solution 4.6

The required program listing is shown in Figure 4.8. Function *Average* receives the elements of array *Numbers* and calculates the average of the array elements.

```
/*******************************************************************

                    PASSING AN ARRAY TO A FUNCTION
                    =============================

This program stores numbers 1 to 10 in an array called Numbers. Function
Average is then called to calculate the average of these numbers.


Programmer:   Dogan Ibrahim
File:         AVERAGE.C
Date:         May, 2007
*******************************************************************/

/* Function to calculate the average */
float Average(int A[ ])
{
        float Sum = 0.0, k;
        unsigned char j;

        for(j=0; j<10; j++)
        {
                Sum = Sum + A[j];
        }
        k = Sum / 10.0;
        return k;
}


/* Start of the main program */
void main()
{
        unsigned char j;
        float Avrg;
        int Numbers[10];

        for(j=0; j<10; j++)Numbers[j] = j+1;
        Avrg = Average(Numbers);
}
```

Figure 4.8: Program passing an array to a function

Example 4.7

Repeat Example 4.6, but this time define the array size at the beginning of the program and then pass the array size to the function.

Solution 4.7

The required program listing is shown in Figure 4.9.

```
/******************************************************************

                    PASSING AN ARRAY TO A FUNCTION
                    ===================================

This program stores numbers 1 to N in an array called Numbers where N is
defined at the beginning of the program. Function Average is then called to
calculate the average of these numbers.


Programmer:  Dogan Ibrahim
File:        AVERAGE2.C
Date:        May, 2007
******************************************************************/

#define Array_Size 20

/* Function to calculate the average */
float Average(int A[ ], int N)
{
        float Sum = 0.0, k;
        unsigned char j;

        for(j=0; j<N; j++)
        {
                Sum = Sum + A[j];
        }
        k = Sum / N;
        return k;
}


/* Start of the main program */
void main()
{
        unsigned char j;
        float Avrg;
        int Numbers[Array_Size];

        for(j=0; j<Array_Size; j++)Numbers[j] = j+1;
        Avrg = Average(Numbers, Array_Size);
}
```

Figure 4.9: Another program passing an array to a function

It is also possible to pass a complete array to a function using pointers. The address of the first element of the array is passed to the function, and the function can then manipulate the array as required using pointer operations. An example follows.

Example 4.8

Repeat Example 4.6, but this time use a pointer to pass the array elements to the function.

Solution 4.8

The required program listing is given in Figure 4.10. An integer pointer is used to pass the array elements to the function, and the function elements are manipulated using pointer operations. Notice that the address of the first element of the array is passed as an integer with the statement: &Numbers[0].

4.1.3   Passing Variables by Reference to Functions

By default, arguments to functions are passed by value. Although this method has many distinct advantages, there are occasions when it is more appropriate and also more efficient to pass the address of the arguments instead, that is, to pass the argument by reference. When the address of an argument is passed, the original value of that argument can be modified by the function; thus the function does not have to return any variables. An example follows which illustrates how the address of arguments can be passed to a function and how the values of these arguments can be modified inside the function.

Example 4.9

Write a function named *Swap* to accept two integer arguments and then to swap the values of these arguments. Use this function in a main program to swap the values of two variables.

Solution 4.9

The required program listing is shown in Figure 4.11. Function *Swap* is defined as *void* since it does not return any value. It has two arguments, a and b, and in the function header two integer pointers are used to pass the addresses of these variables. Inside the function body, the value of an argument is accessed by inserting the "*" character in

```
/*********************************************************************

                    PASSING AN ARRAY TO A FUNCTION
                    ================================

This program stores numbers 1 to 10 in an array called Numbers. Function
Average is then called to calculate the average of these numbers.


Programmer:  Dogan Ibrahim
File:        AVERAGE3.C
Date:        May, 2007
*********************************************************************/

/* Function to calculate the average */
float Average(int *A)
{
        float Sum = 0.0, k;
        unsigned char j;

        for(j=0; j<10; j++)
        {
                Sum = Sum + *(A + j);
        }
        k = Sum / 10.0;
        return k;
}


/* Start of the main program */
void main()
{
        unsigned char j;
        float Avrg;
        int Numbers[10];

        for(j=0; j<10; j++)Numbers[j] = j+1;
        Avrg = Average(&Numbers[0]);
}
```

Figure 4.10: Program passing an array using pointers

front of the argument. Inside the main program, the addresses of the variables are
passed to the function using the "&" character in front of the variable names. At the end
of the program, variables *p* and *q* are set to 20 and 10 respectively.

4.1.4   Variable Number of Arguments

The ellipsis character ("...") consists of three successive periods with no spaces
between them. An ellipsis can be used in the argument lists of function prototypes to

```
/************************************************************

                PASSING VARIABLES BY REFERENCE
                ================================

This program shows how the address of variables can be passed to functions.
The function in this program swaps the values of two integer variables.


Programmer:   Dogan Ibrahim
File:         SWAP.C
Date:         May, 2007
************************************************************/

/* Function to swap two integers */
void Swap(int *a, int *b)
{
        int temp;

        temp = *a;                      // Store a in temp
        *a = *b;                        // Copy b to a
        *b = temp;                      // Copy temp to b
}



/* Start of the main program */
void main()
{
        int p, q;

        p = 10;                         // Set p = 10
        q = 20;                         // Set q = 20
        swap(&p, &q);                   // Swap p and q (p=20, q=10)
}
```

Figure 4.11: Passing variables by reference to a function

indicate a variable number of arguments or arguments with varying types. An example of a declaration of a function prototype with ellipsis follows. In this declaration, when the function is called we must supply at least two integer type arguments, and we can also supply any number of additional arguments:

unsigned char MyFunc(int a, int b,...);

The header file *stdarg.h* must be included at the beginning of a program that uses a variable number of arguments. This header file defines a new data type called *va_list*, which is essentially a character pointer. In addition, macro *va_start()* initializes an object of type va_list to point to the address of the first additional argument presented to the function. To extract the arguments, successive calls to the macro *va_arg()* must be made, with the character pointer and the type of the parameter as the arguments of va_arg().

An example program is given in Figure 4.12. In this program the function header declares only one parameter of type *int*, and an ellipsis is used to declare a variable

```
/***********************************************************************

                    PASSING VARIABLE NUMBER OF ARGUMENTS
                    =======================================

        This program shows how variable number of arguments can be passed to a
        function. The function header declares one integer variable and an ellipsis is
        used to declare variable number of parameters. The function adds all the
        arguments and returns the sum as an integer. The number of arguments is
        supplied by the calling program as the first argument to the function.

        Programmer:  Dogan Ibrahim
        File:        VARIABLE.C
        Date:        May, 2007
        ***********************************************************************/

        #include <stdarg.h>

        /* Function with variable number of parameters */
        int Sum(int num,...)
        {
                unsigned char j;
                va_list ap;
                int temp = 0;

                va_start(ap, num);

                for(j = 0; j < num; j++)
                {
                        temp = temp + va_arg(ap, int);
                }
                va_end(ap);
                return temp;
        }


        /* Start of the main program */
        void main()
        {
                int p;

                p = Sum(2, 3, 5);               // 2 arguments. p=3+5=8
                p = Sum(3, 2, 5, 6);            // 3 arguments, p=2+5+6=13
        }
```

Figure 4.12: Passing variable number of arguments to a function

number of parameters. Variable *num* is the argument count passed by the calling program. The arguments are read by using the macro *va_arg(ap, int)* and then summed using variable *temp* and returned by the function.

4.1.5   Function Reentrancy

The mikroC compiler supports only a limited function reentrancy. Functions that have no arguments and local variables can be called both from the interrupt service routines and from the main program. Functions that have arguments and/or local variables can only be called from the interrupt service routines or from the main program.

4.1.6   Static Function Variables

Normally, variables declared at the beginning of a program, before the main program, are global, and their values can be accessed and modified by all parts of the program. Declaring a variable used in a function as global ensures that its value is retained from one call of the function to another, but this also undermines the variable's privacy and reduces the portability of the function to other applications. A better approach is to declare such variables as *static*. Static variables are mainly used in function definitions. When a variable is declared as *static*, its value is retained from one call of the function to another. In the example code that follows, variable k is declared as static and initialized to zero. This variable is then incremented before exiting from the function, and the value of k remains in existence and holds its last value on the next call to the function (i.e., on the second call to the function the value of k will be 1):

```
void Cnt(void)
{
      static int k = 0;     // Declare k as static
      .................
      .................
      k++;                  // increment k
}
```

4.2   mikroC Built-in Functions

The mikroC compiler provides a set of built-in functions which can be called from the program. These functions are listed in Table 4.1, along with a brief description of each. Most of these functions can be used in a program without having to include header files.

Table 4.1: mikroC built-in functions

Function	Description
Lo	Returns the lowest byte of a number (bits 0 to 7)
Hi	Returns next to the lowest byte of a number (bits 8 to 15)
Higher	Returns next to the highest byte of a number (bits 16 to 23)
Highest	Returns the highest byte of a number (bits 24 to 31)
Delay_us	Creates software delay in microsecond units
Delay_ms	Creates constant software delay in millisecond units
Vdelay_ms	Creates delay in milliseconds using program variables
Delay_Cyc	Creates delay based on microcontroller clock
Clock_Khz	Returns microcontroller clock in KHz
Clock_Mhz	Returns microcontroller clock in MHz

The exceptions are functions *Lo*, *Hi*, *Higher*, and *Highest*, which require the header file *built_in.h*. Further details about using these functions are available in the mikroC manuals.

Functions *Delay_us* and *Delay_ms* are frequently used in programs where delays are required (e.g., when flashing an LED). The following example illustrates the use of the Delay_ms function:

Example 4.10

An LED is connected to bit 0 of PORTB (pin RB0) of a PIC18FXXX microcontroller through a current-limiting resistor as shown in Figure 4.13. Choose a suitable value for the resistor and write a program that will flash the LED ON and OFF continuously at one-second intervals.

Solution 4.10

LEDs can be connected to a microcontroller in two modes: current sinking and current sourcing. In current sinking mode (see Figure 4.14) one leg of the LED is connected to the +5V and the other leg is connected to the microcontroller output port pin through a current limiting resistor R.

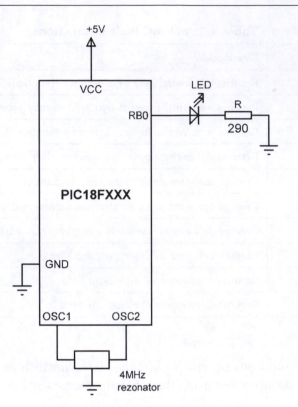

Figure 4.13: LED connected to port RB0 of a PIC microcontroller

Under normal working conditions, the voltage across an LED is about 2V and the current through the LED is about 10mA (some low-power LEDs can operate at as low as 1mA current). The maximum current that can be sourced or sinked at the output port of a PIC microcontroller is 25mA.

The value of the current limiting resistor R can be calculated as follows. In current sinking mode the LED will be turned ON when the output port of the microcontroller is at logic 0 (i.e., at approximately 0V). The required resistor is then:

$$R = \frac{5V - 2V}{10mA} = 0.3K$$

The nearest resistor to choose is 290 Ohm (a slightly higher resistor can be chosen for a lower current and slightly less brightness).

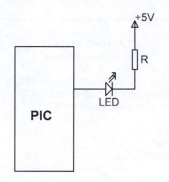

Figure 4.14: Connecting the LED in current sinking mode

In current sourcing mode (see Figure 4.15) one leg of the LED is connected to the output port of the microcontroller and the other leg is connected to the ground through a current limiting resistor. The LED will be turned ON when the output port of the microcontroller is at logic 1 (i.e., at approximately 5V). The same value of resistor can be used in both current sinking and current sourcing modes.

The required program listing is given in Figure 4.16 (program FLASH.C). At the beginning of the program PORTB is configured as output using the TRISB = 0 statement. An endless loop is then formed with the *for* statement, and inside this loop the LED is turned ON and OFF with one-second delays between outputs.

The program given in Figure 4.16 can be made more user-friendly and easier to follow by using *define* statements as shown in Figure 4.17 (program FLASH2.C).

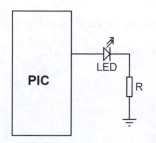

Figure 4.15: Connecting the LED in current sourcing mode

```
/*****************************************************************

                    FLASHING AN LED
                    ================

This program flashes an LED connected to port RB0 of a microcontroller
with one second intervals. mikroC built-in function Delay_ms is used to
create a 1 second delay between the flashes.

Programmer: Dogan Ibrahim
File:         FLASH.C
Date:         May, 2007
*****************************************************************/

void main()
{
        TRISB = 0;                    // Configure PORTB as output
        for(; ;)                      // Endless loop
        {
                PORTB = 1;            // Turn ON LED
                Delay_ms(1000);       // 1 second delay
                PORTB = 0;            // Turn OFF LED
                Delay_ms(1000);       // 1 second delay
        }
}
```

Figure 4.16: Program to flash an LED

4.3   mikroC Library Functions

A large set of library functions is available with the mikroC compiler. These library functions can be called from anywhere in a program, and they do not require that header files are included in the program. The mikroC user manual gives a detailed description of each library function, with examples. In this section, the available library functions are identified, and the important and commonly used library functions are described in detail, with examples.

Table 4.2 gives a list of the mikroC library functions, organized in functional order.

Some of the frequently used library functions are:

- EEPROM library

- LCD library

- Software UART library

- Hardware USART library

```
/*******************************************************************

                        FLASHING AN LED
                        ===============

This program flashes an LED connected to port RB0 of a microcontroller
with one second intervals. mikroC built-in function Delay_ms is used to
create a 1 second delay between the flashes.

Programmer:   Dogan Ibrahim
File:         FLASH2.C
Date:         May, 2007
*******************************************************************/

#define LED PORTB.0
#define ON  1
#define OFF 0
#define One_Second_Delay Delay_ms(1000)

void main()
{
        TRISB = 0;                      // Configure PORTB as output

        for(; ;)                        // Endless loop
        {
                LED = ON;               // Turn ON LED
                One_Second_Delay;       // 1 second delay
                LED = OFF;              // Turn OFF LED
                One_Second_Delay;       // 1 second delay
        }
}
```

Figure 4.17: Another program to flash an LED

- Sound library

- ANSI C library

- Miscellaneous library

4.3.1   EEPROM Library

The EEPROM library includes functions to read data from the on-chip PIC microcontroller nonvolatile EEPROM memory, or to write data to this memory. Two functions are provided:

- Eeprom_Read

- Eeprom_Write

Table 4.2: mikroC library functions

Library	Description
ADC	Analog-to-digital conversion functions
CAN	CAN bus functions
CANSPI	SPI-based CAN bus functions
Compact Flash	Compact flash memory functions
EEPROM	EEPROM memory read/write functions
Ethernet	Ethernet functions
SPI Ethernet	SPI-based Ethernet functions
Flash Memory	Flash memory functions
Graphics LCD	Standard graphics LCD functions
T6963C Graphics LCD	T6963-based graphics LCD functions
I^2C	I^2C bus functions
Keypad	Keypad functions
LCD	Standard LCD functions
Manchester Code	Manchester code functions
Multi Media	Multimedia functions
One Wire	One wire functions
PS/2	PS/2 functions
PWM	PWM functions
RS-485	RS-485 communication functions
Sound	Sound functions
SPI	SPI bus functions
USART	USART serial communication functions
Util	Utilities functions
SPI Graphics LCD	SPI-based graphics LCD functions
Port Expander	Port expander functions

(Continued)

Table 4.2: mikroC library functions (cont'd)

Library	Description
SPI LCD	SPI-based LCD functions
ANSI C Ctype	C Ctype functions
ANSI C Math	C Math functions
ANSI C Stdlib	C Stdlib functions
ANSI C String	C String functions
Conversion	Conversion functions
Trigonometry	Trigonometry functions
Time	Time functions

The *Eeprom_Read* function reads a byte from a specified address of the EEPROM. The address is of type integer, and thus the function supports PIC microcontrollers with more than 256 bytes. A 20ms delay should be used between successive reads from the EEPROM to guarantee the return of correct data. In the following example, the byte at address 0x1F of the EEPROM is read and stored in variable *Temp*:

```
Temp = Eeprom_Read(0x1F);
```

The *Eeprom_Write* function writes a byte to a specified address of the EEPROM. The address is of type integer and thus the function supports PIC microcontrollers with more than 256 bytes. A 20ms delay should be used between successive reads or writes to the EEPROM to guarantee the correct transfer of data to the EEPROM. In the following example, number 0x05 is written to address 0x2F of the EEPROM:

```
Eeprom_Write(0x2F, 0x05);
```

Example 4.11

Write a program to read the contents of EEPROM from address 0 to 0x2F and then send this data to PORTB of a PIC microcontroller.

Solution 4.11

The required program is given in Figure 4.18. A *for* loop is used to read data from the EEPROM and then send it to PORT B of the microcontroller. Notice that a 20ms delay is used between each successive read.

```
/**************************************************************************

                    READING FROM THE EEPROM
                    =========================

This program reads data from addresses 0 to 0x2F of the EEPROM and then
sends this data to PORTB of the microcontroller.

Programmer:   Dogan Ibrahim
File:         EEPROM.C
Date:         May, 2007
**************************************************************************/

void main()
{
        unsigned int j;
        unsigned char Temp;

        TRISB = 0;                                  // Configure PORTB as output

        for(j=0; j <= 0x2F; j++)
        {
                Temp = Eeprom_Read(j);
                PORTB = Temp;
                Delay_ms(20);
        }
}
```

Figure 4.18: Program to read from the EEPROM

4.3.2   LCD Library

One thing all microcontrollers lack is some kind of video display. A video display would make a microcontroller much more user-friendly, enabling text messages, graphics, and numeric values to be output in a more versatile manner than with 7-segment displays, LEDs, or alphanumeric displays. Standard video displays require complex interfaces and their cost is relatively high. LCDs are alphanumeric (or graphic) displays which are frequently used in microcontroller-based applications. These display devices come in different shapes and sizes. Some LCDs have forty or more character lengths with the capability to display several lines. Others can be programmed to display graphic images. Some modules offer color displays, while others incorporate backlighting so they can be viewed in dimly lit conditions.

There are basically two types of LCDs as far as the interfacing technique is concerned: parallel and serial. Parallel LCDs (e.g., the Hitachi HD44780 series) are connected to the microcontroller circuitry such that data is transferred to the LCD using more than one line, usually four or eight data lines. Serial LCDs are connected to a microcontroller using one data line only, and data is transferred using the RS232 asynchronous data communications protocol. Serial LCDs are generally much easier to work with but more costly than parallel ones. In this book only parallel LCDs are discussed, as they are used more often in microcontroller-based projects.

Low-level programming of a parallel LCD is usually a complex task and requires a good understanding of the internal operation of the LCD, including the timing diagrams. Fortunately, mikroC language provides functions for both text-based and graphic LCDs, simplifying the use of LCDs in PIC-microcontroller-based projects.

The HD44780 controller is a common choice in LCD-based microcontroller applications. A brief description of this controller and information on some commercially available LCD modules follows.

The HD44780 LCD Controller

The HD44780 is one of the most popular LCD controllers, being used both in industrial and commercial applications and also by hobbyists. The module is monochrome and comes in different shapes and sizes. Modules with 8, 16, 20, 24, 32, and 40 characters are available. Depending on the model, the display provides a 14-pin or 16-pin connector for interfacing. Table 4.3 shows the pin configuration and pin functions of a typical 14-pin LCD.

V_{SS} is the 0V supply or ground. The V_{DD} pin should be connected to the positive supply. Although the manufacturers specify a 5V DC supply, the modules usually work with as low as 3V or as high as 6V.

Pin 3 is named as V_{EE} and is the contrast control pin. It is used to adjust the contrast of the display and should be connected to a DC supply. A potentiometer is usually connected to the power supply with its wiper arm connected to this pin and the other leg of the potentiometer connected to the ground. This way the voltage at the V_{EE} pin, and hence the contrast of the display, can be adjusted as desired.

Pin 4 is the register select (RS) and when this pin is LOW, data transferred to the LCD is treated as commands. When RS is HIGH, character data can be transferred to and from the module.

Pin 5 is the read/write (R/W) pin. This pin is pulled LOW in order to write commands or character data to the LCD module. When this pin is HIGH, character data or status information can be read from the module.

Table 4.3: Pin configuration of the HD44780 LCD module

Pin no.	Name	Function
1	V_{SS}	Ground
2	V_{DD}	+ve supply
3	V_{EE}	Contrast
4	RS	Register select
5	R/W	Read/write
6	EN	Enable
7	D0	Data bit 0
8	D1	Data bit 1
9	D2	Data bit 2
10	D3	Data bit 3
11	D4	Data bit 4
12	D5	Data bit 5
13	D6	Data bit 6
14	D7	Data bit 7

Pin 6 is the enable (EN) pin, which is used to initiate the transfer of commands or data between the module and the microcontroller. When writing to the display, data is transferred only on the HIGH to LOW transition of this pin. When reading from the display, data becomes available after the LOW to HIGH transition of the enable pin, and this data remains valid as long as the enable pin is at logic HIGH.

Pins 7 to 14 are the eight data bus lines (D0 to D7). Data can be transferred between the microcontroller and the LCD module using either a single 8-bit byte or two 4-bit nibbles. In the latter case, only the upper four data lines (D4 to D7) are used. The 4-bit mode has the advantage of requiring fewer I/O lines to communicate with the LCD.

The mikroC LCD library provides a large number of functions to control text-based LCDs with 4-bit and 8-bit data interfaces, and for graphics LCDs. The most common are the 4-bit-interface text-based LCDs. This section describes the available mikroC functions for these LCDs. Further information on other text- or graphics-based LCD functions are available in the mikroC manual.

The following are the LCD functions available for 4-bit-interface text-based LCDs:

- Lcd_Config
- Lcd_Init
- Lcd_Out
- Lcd_Out_Cp
- Lcd_Chr
- Lcd_Chr_Cp
- Lcd_Cmd

Lcd_Config   The *Lcd_Config* function is used to configure the LCD interface. The default connection between the LCD and the microcontroller is:

```
LCD         Microcontroller port pin

RS              2
EN              3
D4              4
D5              5
D6              6
D7              7
```

The R/W pin of the LCD is not used and should be connected to the ground.

This function should be used to change the default connection. It should be called with the parameters in the following order:

```
port name, RS pin, EN pin, R/W pin, D7 pin, D6 pin, D5 pin, D4 pin
```

The port name should be specified by passing its address. For example, if the RS pin is connected to RB0, EN pin to RB1, D7 pin to RB2, D6 pin to RB3, D5 pin to RB4, and the D4 pin to RB5, then the function should be called as follows:

```
Lcd_Config(&PORTB, 0, 1, 2, 3, 4, 5);
```

Lcd_Init   The *Lcd_Init* function is called to configure the interface between the microcontroller and the LCD when the default connections are made as just illustrated. The port name should be specified by passing its address. For example, assuming that the LCD is connected to PORTB and the preceding default connections are used, the function should be called as:

```
Lcd_Init(&PORTB);
```

Lcd_Out   The *Lcd_Out* function displays text at the specified row and column position of the LCD. The function should be called with the parameters in the following order:

```
row, column, text
```

For example, to display text "Computer" at row 1 and column 2 of the LCD we should call the function as:

```
Lcd_Out(1, 2, "Computer");
```

Lcd_Out_Cp   The *Lcd_Out_Cp* function displays text at the current cursor position. For example, to display text "Computer" at the current cursor position the function should be called as:

```
Lcd_Out_Cp("Computer");
```

Lcd_Chr   The *Lcd_Chr* function displays a character at the specified row and column position of the cursor. The function should be called with the parameters in the following order:

```
row, column, character
```

For example, to display character "K" at row 2 and column 4 of the LCD we should call the function as:

```
LCD_Chr(2, 4, 'K');
```

Lcd_Chr_Cp   The *Lcd_Chr_Cp* function displays a character at the current cursor position. For example, to display character "M" at the current cursor position the function should be called as:

```
Lcd_Chr_Cp('M');
```

Lcd_Cmd   The *Lcd_Cmd* function is used to send a command to the LCD. With the commands we can move the cursor to any required row, clear the LCD, blink the cursor,

Table 4.4: LCD commands

LCD command	Description
LCD_CLEAR	Clear display
LCD_RETURN_HOME	Return cursor to home position
LCD_FIRST_ROW	Move cursor to first row
LCD_SECOND_ROW	Move cursor to second row
LCD_THIRD_ROW	Move cursor to third row
LCD_FOURTH_ROW	Move cursor to fourth row
LCD_BLINK_CURSOR_ON	Blink cursor
LCD_TURN_ON	Turn display on
LCD_TURN_OFF	Turn display off
LCD_MOVE_CURSOR_LEFT	Move cursor left
LCD_MOVE_CURSOR_RIGHT	Move cursor right
LCD_SHIFT_LEFT	Shift display left
LCD_SHIFT_RIGHT	Shift display right

shift display, etc. A list of the most commonly used LCD commands is given in Table 4.4. For example, to clear the LCD we should call the function as:

```
Lcd_Cmd(Lcd_Clear);
```

An example illustrates initialization and use of the LCD.

Example 4.12

A text-based LCD is connected to a PIC18F452 microcontroller in the default mode as shown in Figure 4.19. Write a program to send the text "My Computer" to row 1, column 4 of the LCD.

Solution 4.12

The required program listing is given in Figure 4.20 (program LCD.C). At the beginning of the program PORTB is configured as output with the TRISB = 0

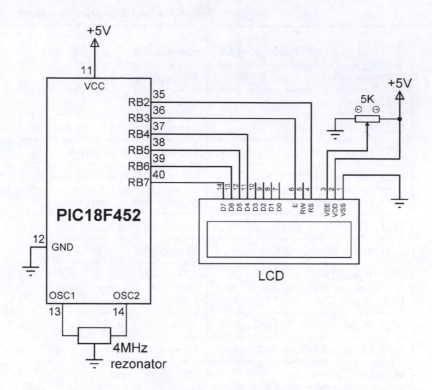

Figure 4.19: Connecting an LCD to a PIC microcontroller

```
/***********************************************************************

                      WRITING TEXT TO AN LCD
                      =========================

A text based LCD is connected to a PIC microcontroller in the default mode.
This program displays the text "My Computer" on the LCD.

Programmer:   Dogan Ibrahim
File:         LCD.C
Date:         May, 2007
***********************************************************************/

void main()
{
        TRISB = 0;                              // Configure PORTB as output

        Lcd_Init(&PORTB);                       // Initialize the LCD
        Lcd_Cmd(LCD_CLEAR);                     // Clear the LCD
        Lcd_Out(1, 4, "My Computer");           // Display text on LCD
}
```

Figure 4.20: LCD program listing

statement. The LCD is then initialized, the display is cleared, and the text message "My Computer" is displayed on the LCD.

4.3.3 Software UART Library

Universal asynchronous receiver transmitter (UART) software library is used for RS232-based serial communication between two electronic devices. In serial communication, only two cables (plus a ground cable) are required to transfer data in either direction. Data is sent in serial format over the cable bit by bit. Normally, the receiving device is in idle mode with its transmit (TX) pin at logic 1, also known as MARK. Data transmission starts when this pin goes to logic 0, also known as SPACE. The first bit sent is the start bit at logic 0. Following this bit, 7 or 8 data bits are sent, followed by an optional parity bit. The last bit sent is the stop bit at logic 1. Serial data is usually sent as a 10-bit frame consisting of a start bit, 8 data bits, and a stop bit, and no parity bits. Figure 4.21 shows how character "A" can be sent using serial communication. Character "A" has the ASCII bit pattern 01000001. As shown in the figure, first the start bit is sent, this is followed by 8 data bits 01000001, and finally the stop bit is sent.

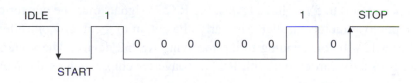

Figure 4.21: Sending character "A" in serial communication

The bit timing is very important in serial communication, and the transmitting (TX) and receiving (RX) devices must have the same bit timings. The bit timing is measured by the baud rate, which specifies the number of bits transmitted or received each second. Typical baud rates are 4800, 9600, 19200, 38400, and so on. For example, when operating at 9600 baud rate with a frame size of 10 bits, 960 characters are transmitted or received each second. The timing between bits is then about 104ms.

In RS232-based serial communication the two devices are connected to each other (see Figure 4.22) using either a 25-way connector or a 9-way connector. Normally only the TX, RX, and GND pins are required for communication. The required pins for both types of connectors are given in Table 4.5.

The voltage levels specified by the RS232 protocol are $\pm12V$. A logic HIGH signal is at $-12V$ and a logic LOW signal is at $+12V$. PIC microcontrollers, on the other hand,

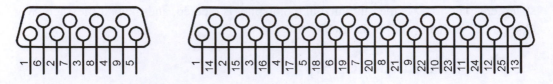

Figure 4.22: 25-way and 9-way RS232 connectors

Table 4.5: Pins required for serial communication

Pin	9-way connector	25-way connector
TX	2	2
RX	3	3
GND	5	7

normally operate at 0 to 5V voltage levels, the RS232 signals must be converted to 0 to 5V when input to a microcontroller. Similarly, the output of the microcontroller must be converted to ±12V before sending to the receiving RS232 device. The voltage conversion is usually carried out with RS232 converter chips, such as the MAX232, manufactured by Maxim Inc.

Serial communication may be implemented in hardware using a specific pin of a microcontroller, or the required signals can be generated in software from any required pin of a microcontroller. Hardware implementation requires either an on-chip UART (or USART) circuit or an external UART chip that is connected to the microcontroller. Software-based UART is more commonly used and does not require any special circuits. Serial data is generated by delay loops in software-based UART applications. In this section only the software-based UART functions are described.

The mikroC compiler supports the following software UART functions:

- Soft_Uart_Init
- Soft_Uart_Read
- Soft_Uart_Write

Soft_Uart_Init

The *Soft_Uart_Init* function specifies the serial communications parameters and does so in the following order:

```
port, rx pin, tx pin, baud rate, mode
```

port is the port used as the software UART (e.g., PORTB), *rx* is the receive pin number, *tx* is the transmit pin number, *baud rate* is the chosen baud rate where the maximum value depends on the clock rate of the microcontroller, and *mode* specifies whether or not the data should be inverted at the output of the port. A 0 indicates that it should not be inverted, and a 1 indicates that it should be inverted. When an RS232 voltage level converter chip is used (e.g., MAX232) then the mode must be set to 0. Soft_Uart_Init must be the first function called before software-based serial communication is established.

The following example configures the software UART to use PORTB as the serial port, with RB0 as the RX pin and RB1 as the TX pin. The baud rate is set to 9600 with the mode noninverted:

```
Soft_Uart_Init(PORTB, 0, 1, 9600, 0);
```

Soft_Uart_Read

The *Soft_Uart_Read* function receives a byte from a specified serial port pin. The function returns an error condition and the data is read from the serial port. The function does not wait for data to be available at the port, and therefore the error parameter must be tested if a byte is expected. The error is normally 1 and becomes 0 when a byte is read from the port.

The following example illustrates reading a byte from the serial port configured by calling the function Soft_Uart_Init. The received byte is stored in variable *Temp*:

```
do
      Temp = Soft_Uart_Read(&Rx_Error);
while (Rx_Error);
```

Soft_Uart_Write

The *Soft_Uart_Write* function transmits a byte to a configured serial port pin. The data to be sent must be specified as a parameter in the call to the function.

For example, to send character "A" to the serial port pin:

```
char MyData = 'A';
Soft_Uart_Write(MyData);
```

The following example illustrates the use of software UART functions.

Example 4.13

The serial port of a PC (e.g., COM1) is connected to a PIC18F452 microcontroller, and terminal emulation software (e.g., HyperTerminal) is operated on the PC to use the serial port. Pins RB0 and RB1 of the microcontroller are the RX and TX pins respectively. The required baud rate is 9600.

Write a program to read data from the terminal, then increase this data by one and send it back to the terminal. For example, if the user enters character "A," then character "B" should be displayed on the terminal. Assume that a MAX232-type voltage level converter chip is converting the microcontroller signals to RS232 levels. Figure 4.23 shows the circuit diagram of this example.

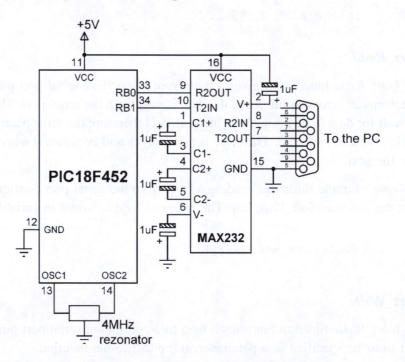

Figure 4.23: Circuit diagram of Example 4.13

Solution 4.13

The MAX232 chip receives the TX signal from pin RB1 of the microcontroller and converts it to RS232 levels. Comparably, the serial data received by the MAX232 chip is converted into microcontroller voltage levels and then sent to pin RB0. Note that correct operation of the MAX232 chip requires four capacitors to be connected to the chip.

The required program listing is shown in Figure 4.24 (program SERIAL.C). At the beginning of the program, function Soft_Uart_Init is called to configure the serial port. Then an endless loop is formed using a *for* statement. The Soft_Uart_Read function is called to read a character from the terminal. After reading a character, the data byte is incremented by one and then sent back to the terminal by calling function Soft_Uart_Write.

```
/***********************************************************************

                    READING AND WRITING TO SERIAL PORT
                    ==================================

    In this program PORTB pins RB0 and RB1 are configured as serial RX and
    TX pins respectively. The baud rate is set to 9600. A character is received from a
    serial terminal, incremented by one and then sent back to the terminal. Thus, if
    character "A" is entered on the keyboard, character "B" will be displayed.

    Programmer:  Dogan Ibrahim
    File:        SERIAL.C
    Date:        May, 2007
***********************************************************************/

void main()
{
    unsigned char MyError, Temp;

    Soft_Uart_Init(PORTB, 0, 1, 9600, 0);          // Configure serial port
    for(; ;)                                       // Endless loop
    {
        do
        {
            Temp = Soft_Uart_Read(&MyError);       // Read a byte
        } while(MyError);
        Temp++;                                    // Increment byte
        Soft_Uart_Write(Temp);                     // Send the byte
    }
}
```

Figure 4.24: Program listing of Example 4.13

4.3.4   Hardware USART Library

The universal synchronous asynchronous receiver transmitter (USART) hardware library contains a number of functions to transmit and receive serial data using the USART circuits built on the PIC microcontroller chips. Some PIC18F-series microcontrollers have only one USART (e.g., PIC18F452), while others have two USART circuits (e.g., PIC18F8520). Hardware USART has an advantage over software-implemented USART, in that higher baud rates are generally available and the microcontroller can perform other operations while data is sent to the USART.

The hardware USART library provides the following functions:

- Usart_Init
- Usart_Data_Ready
- Usart_Read
- Usart_Write

Usart_Init

The *Usart_Init* function initializes the hardware USART with the specified baud rate. This function should be called first, before any other USART functions. The only parameter required by this function is the baud rate. The following example call sets the baud rate to 9600:

```
Usart_Init(9600);
```

Usart_Data_Ready

The *Usart_Data_Ready* function can be called to check whether or not a data byte has been received by the USART. The function returns a 1 if data has been received and a 0 if no data has been received. The function has no parameters. The following code checks if a data byte has been received or not:

```
if(Usart_Data_Ready( ))
```

Usart_Read

The *Usart_Read* function is called to read a data byte from the USART. If data has not been received, a 0 is returned. Note that reading data from the USART is nonblocking (i.e., the function always returns whether or not the USART has received a data byte). The Usart_Read function should be called after calling the function *Usart_Data_Ready* to make sure that data is available at the USART. Usart_Read has no parameters. In the following example, USART is checked and if a data byte has been received it is copied to variable *MyData*:

```
char MyData;
if(Usart_Data_Read( )) MyData = Usart_Read();
```

Usart_Write

The *Usart_Write* function sends a data byte to the USART, and thus a serial data is sent out of the USART. The data byte to be sent must be supplied as a parameter to the function. In the following example, character "A" is sent to the USART:

```
char Temp = 'A';
Usart_Write(Temp);
```

The following example illustrates how the hardware USART functions can be used in a program.

Example 4.14

The serial port of a PC (e.g., COM1) is connected to a PIC18F452 microcontroller, and terminal emulation software (e.g., HyperTerminal) is operated on the PC to use the serial port. The microcontroller's hardware USART pins RC7 (USART receive pin, RX) and RC6 (USART transmit pin, TX) are connected to the PC via a MAX232-type RS232 voltage level converter chip. The required baud rate is 9600. Write a program to read data from the terminal, then increase this data by one and send it back to the terminal. For example, if the user enters character "A," then character "B" should be displayed on the terminal. Figure 4.25 shows the circuit diagram of this example.

Solution 4.14

The required program listing is shown in Figure 4.26 (program SERIAL2.C). At the beginning of the program, function *Usart_Init* is called to set the baud rate to 9600. Then an endless loop is formed using a *for* statement. The *Usart_Data_Ready* function is called to check whether a character is ready, and the character is read by calling function *Usart_Read*. After reading a character, the data byte is incremented by one and then sent back to the terminal by calling function *Usart_Write*.

In PIC microcontrollers that have more than one USART, the second USART is accessed by appending a "2" to the end of the function (e.g., *Usart_Write2*, *Usart_Read2*, etc.).

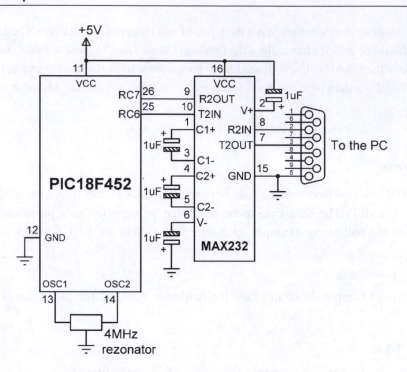

Figure 4.25: Circuit diagram of Example 4.14

4.3.5 Sound Library

Functions in the sound library make it possible to generate sounds in our applications. A speaker (e.g., a piezo speaker) should be connected to the required microcontroller port.

The following functions are offered by the sound library:

- Sound_Init

- Sound_Play

Sound_Init

The *Sound_Init* function initializes the sound library and requires two parameters: the name and the bit number of the port where the speaker is connected. The address of the port name should be passed to the function. For example, if the speaker is connected to bit 3 of PORTB, then the function should be called as:

```
Sount_Init(&PORTB, 3);
```

```
/********************************************************************

                READING AND WRITING TO SERIAL PORT VIA USART
                =============================================

    In this program a PIC18F452 microcontroller is used and USART I/O pins are
    connected to a terminal through a MAX232 voltage converter chip. The baud rate is
    set to 9600. A character is received from a serial terminal, incremented by one and
    then sent back to the terminal. Thus, if character "A" is entered on the keyboard,
    character "B" will be displayed.

    Programmer:   Dogan Ibrahim
    File:         SERIAL2.C
    Date:         May, 2007
    ********************************************************************/

    void main()
    {
          unsigned char MyError, Temp;

          Usart_Init(9600);                         // Set baud rate
          for(; ;)                                  // Endless loop
          {
                while (!User_Data_Ready());         // Wait for data byte
                Temp = Usart_Read();                // Read data byte
                Temp++;                             // Increment data byte
                Usart_Write(Temp);                  // Send the byte byte
          }
    }
```

Figure 4.26: Program listing of Example 4.14

Sound_Play

The *Sound_Play* function plays a sound at a specified port pin. The function receives two arguments: the period divided by 10 (TDIV) and the number of periods (N). The first parameter is the period in microcontroller cycles divided by 10. The second parameter specifies the duration (number of clock periods) of the sound.

The following formula calculates the value used as the first parameter:

$$\text{TDIV} = \frac{f}{40F}$$

where

TDIV   is the number to be used as the first parameter

F   is the required sound frequency (Hz)

f   is the microcontroller clock frequency (Hz)

Example 4.15

Write a program to play a sound at 1KHz, assuming the clock frequency is 4MHz. The required duration of the sound is 250 periods.

Solution 4.15

The first parameter is calculated as follows:

$$\text{TDIV} = \frac{f}{40F} = \frac{4 \times 10^6}{40 \times 10^3} = 100$$

Since the required duration is 250 periods, the function is called with the parameters:

```
Sound_Play(100, 250);
```

4.3.6   ANSI C Library

The ANSI C library consists of the following libraries (further details on these libraries are available in the mikroC user manual):

- Ctype library
- Math library
- Stdlib library
- String library

Ctype Library

The functions in the Ctype library are mainly used for testing or data conversion. Table 4.6 lists the commonly used functions in this library.

Math Library

The functions in the Math library are used for floating point mathematical operations. Table 4.7 lists the commonly used functions in this library.

Table 4.6: Commonly used Ctype library functions

Function	Description
isalnum	Returns 1 if the specified character is alphanumeric (a–z, A–Z, 0–9)
isalpha	Returns 1 if the specified character is alphabetic (a–z, A–Z)
isntrl	Returns 1 if the specified character is a control character (decimal 0–31 and 127)
isdigit	Returns 1 if the specified character is a digit (0–9)
islower	Returns 1 if the specified character is lowercase
isprint	Returns 1 if the specified character is printable (decimal 32–126)
isupper	Returns 1 if the specified character is uppercase
toupper	Convert a character to uppercase
tolower	Convert a character to lowercase

Stdlib Library

The Stdlib library contains standard library functions. Table 4.8 lists the commonly used functions in this library.

Example 4.16

Write a program to calculate the trigonometric sine of the angles from 0° to 90° in steps of 1° and store the result in an array called *Trig_Sine*.

Solution 4.16

The required program listing is shown in Figure 4.27 (program SINE.C). A loop is created using a *for* statement, and inside this loop the sine of the angles are calculated and stored in array Trig_Sine. Note that the angles must be converted into radians before they are used in function *sin*.

String Library

The functions in the String library are used to perform string and memory manipulation operations. Table 4.9 lists the commonly used functions in this library.

Table 4.7: Commonly used Math library functions

Function	Description
acos	Returns in radians the arc cosine of its parameter
asin	Returns in radians the arc sine of its parameter
atan	Returns in radians the arc tangent of its parameter
atan2	Returns in radians the arc tangent of its parameter where the signs of both parameters are used to determine the quadrant of the result
cos	Returns the cosine of its parameter in radians
cosh	Returns the hyperbolic cosine of its parameter
exp	Returns the exponential of its parameter
fabs	Returns the absolute value of its parameter
log	Returns the natural logarithm of its parameter
Log10	Returns the logarithm to base 10 of its parameter
pow	Returns the power of a number
sin	Returns the sine of its parameter in radians
sinh	Returns the hyperbolic sine of its parameter
sqrt	Returns the square root of its parameter
tan	Returns the tangent of its parameter in radians
tanh	Returns the hyperbolic sine of its parameter

Example 4.17

Write a program to illustrate how the two strings "MY POWERFUL" and "COMPUTER" can be joined into a new string using String library functions.

Solution 4.17

The required program listing is shown in Figure 4.28 (program JOIN.C). The mikroC String library function *strcat* is used to join the two strings pointed to by p1 and p2 into a new string stored in a character array called *New_String*.

Table 4.8: Commonly used Stdlib library functions

Function	Description
abs	Returns the absolute value
atof	Converts ASCII character into floating point number
atoi	Converts ASCII character into integer number
atol	Converts ASCII character into long integer
max	Returns the greater of two integers
min	Returns the lesser of two integers
rand	Returns a random number between 0 and 32767; function *srand* must be called to obtain a different sequence of numbers
srand	Generates a seed for function *rand* so a new sequence of numbers is generated
xtoi	Convert input string consisting of hexadecimal digits into integer

```
/******************************************************************

                 TRIGONOMETRIC SINE OF ANGLES 0 to 90 DEGREES
                 =============================================

This program calculates the trigonometric sine of angles from 0 degrees to
90 degrees in steps of 1 degree. The results are stored in an array called
Trig_Sine.

Programmer:  Dogan Ibrahim
File:         SINE.C
Date:         May, 2007
******************************************************************/

void main()
{
        unsigned char j;
        double PI = 3.14159, rads;

        for(j = 0; j <= 90; j++)
        {
                rads = j * PI /180.0;
                angle = sin(rad);
                Trig_Sine[j] = angle;
        }
}
```

Figure 4.27: Calculating the sine of angles 0° to 90°

Table 4.9: Commonly used String library functions

Function	Description
strcat, strncat	Append two strings
strchr, strpbrk	Locate the first occurrence of a character in a string
strcmp, strncmp	Compare two strings
strcpy, strncpy	Copy one string into another one
strlen	Return the length of a string

```
/******************************************************************

                         JOINING TWO STRINGS
                         ===================

This program shows how two strings can be joined to obtain a new string.
mikroC library function strcat is used to join the two strings pointed to by
p1 and p2 into a new string stored in character array New_String.

Programmer:  Dogan Ibrahim
File:        JOIN.C
Date:        May, 2007
******************************************************************/

void main()
{
        const char *p1 = "MY POWERFUL ";      // First string
        const char *p2 = "COMPUTER";          // Second string
        char New_String[80];

        strcat(strcat(New_String, p1), p2);   // join the two strings
}
```

Figure 4.28: Joining two strings using function strcat

4.3.7   Miscellaneous Library

The functions in the Miscellaneous library include routines to convert data from one
type to another type, as well as to perform some trigonometric functions. Table 4.10
lists the commonly used functions in this library.

The following general programs illustrate the use of various library routines available
with the mikroC language.

Table 4.10: Commonly used Miscellaneous library functions

Function	Description
ByteToStr	Convert a byte into string
ShortToStr	Convert a short into string
WordToStr	Convert an unsigned word into string
IntToStr	Convert an integer into string
LongToStr	Convert a long into string
FloatToStr	Convert a float into string
Bcd2Dec	Convert a BCD number into decimal
Dec2Bcd	Convert a decimal number into BCD

Example 4.18

Write a function to convert the string pointed to by p into lowercase or uppercase, depending on the value of a mode parameter passed to the function. If the mode parameter is nonzero, then convert to lowercase, otherwise convert it to uppercase. The function should return a pointer to the converted string.

Solution 4.18

The required program listing is given in Figure 4.29 (program CASE.C). The program checks the value of the mode parameter, and if this parameter is nonzero the string is converted to lowercase by calling function *ToLower*, otherwise the function *ToUpper* is called to convert the string to uppercase. The program returns a pointer to the converted string.

Example 4.19

Write a program to define a complex number structure, then write functions to add and subtract two complex numbers. Show how you can use these functions in a main program.

Solution 4.19

Figure 4.30 shows the required program listing (program COMPLEX.C). At the beginning of the program, a data type called *complex* is created as a structure having a real part and an imaginary part. A function called *Add* is then defined to add two complex numbers and return the sum as a complex number. Similarly, the function

```
/*************************************************************************

                    CONVERT A STRING TO LOWER/UPPERCASE
                    =========================================

This program receives a string pointer and a mode parameter. If the mode is 1
Then the string is converted to lowercase, otherwise the string is converted to
uppercase.

Programmer:  Dogan Ibrahim
File:        CASE.C
Date:        May, 2007
*************************************************************************/

unsigned char *Str_Convert(unsigned char *p, unsigned char mode)
{
        unsigned char *ptr = p;

        if (mode != 0)
        {
                while(*p != '\0')  *p++ = ToLower(*p);
        }
        else
        {
                while(*p != '\0') *p++ = ToUpper(*p);
        }
        return ptr;
}
```

Figure 4.29: Program for Example 4.18

Subtract is defined to subtract two complex numbers and return the result as a complex number. The main program uses two complex numbers, a and b, where,

```
a = 2.0 - 3.0j
b = 2.5 + 2.0j
```

Two other complex numbers, c and d, are also declared, and the following complex number operations are performed:

```
The program calculates, c = a + b and, d = a - b
```

Example 4.20

A projectile is fired at an angle of θ degrees at an initial velocity of v meters per second. The distance traveled by the projectile (d), the flight time (t), and the maximum height reached (h) are given by the following formulas:

$$h = \frac{v^2 \sin(\theta)}{g} \quad t = \frac{2v \sin(\theta)}{g} \quad d = \frac{v^2 \sin(2\theta)}{g}$$

```
/**************************************************************

              COMPLEX NUMBER ADDITION AND SUBTRACTION
              ============================================

This program creates a data structure called complex having a real part and
an imaginary part. Then, functions are defined to add or subtract two complex
numbers and store the result in another complex number.

The first complex number is,      a = 2.0 − 2.0j
The second complex number is, b = 2.5 + 2.0j

The program calculates, c = a + b
and,                    d = a − b

Programmer:   Dogan Ibrahim
File:         COMPLEX.C
Date:         May, 2007
**************************************************************/

/* Define a new data type called complex */
typedef struct
{
        float real;
        float imag;
} complex;

/* Define a function to add two complex numbers and return the result as
   a complex number */
complex Add(complex i, complex j)
{
        complex z;

        z.real = i.real + j.real;
        z.imag = i.imag + j.imag

        return z;
}


/* Define a function to subtract two complex numbers and return the result as
   a complex number */
complex Subtract(complex i, complex j)
{
        complex z;

        z.real = i.real − j.real;
        z.imag = i.imag − j.imag;

        return z;
```

Figure 4.30: Program for Example 4.19

(Continued)

```
        }

        /* Main program */
        void main()
        {
                complex a,b,c, d;

                a.real = 2.0;  a.imag =−3.0;          // First complex number
                b.real = 2.5;  b.imag = 2.0;          // second complex number

                c = Add(a, b);                        // Add numbers
                d = Subtract(a, b);                   // Subtract numbers
        }
```

Figure 4.30: (Cont'd)

Write functions to calculate the height, flight time, and distance traveled. Assuming that $g = 9.81 m/s^2$, $v = 12$ m/s, and $\theta = 45°$, call the functions to calculate the three variables. Figure 4.31 shows the projectile pattern.

Solution 4.20

The required program is given in Figure 4.32 (program PROJECTILE.C). Three functions are defined: *Height* calculates the maximum height of the

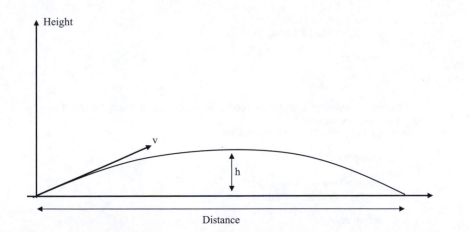

Figure 4.31: Projectile pattern

```
/*************************************************************************

                        PROJECTILE CALCULATION
                        =========================

This program calculates the maximum height, distance traveled, and the flight
time of a projectile. Theta is the firing angle, and v is the initial velocity of the
projectile respectively.

Programmer:  Dogan Ibrahim
File:        PROJECTILE.C
Date:        May, 2007
*************************************************************************/

#define gravity 9.81

/* This function converts degrees to radians */
float Radians(float y)
{
        float rad;

        rad = y * 3.14159 / 180.0;

        return rad;
}


/* Flight time of the projectile */
float Flight_time(float theta, float v)
{
        float t, rad;

        rad = Radians(theta);
        t = (2.0*v*sin(rad)) / gravity;

        return t;
}

float Height(float theta, float v)
{
        float h, rad;

        rad = Radians(theta);
        h = (v*v*sin(rad)) / gravity;

        return h;
}

float Distance(float theta, float v)
{
```

Figure 4.32: Program for Example 4.20

(Continued)

```
            float d, rad;

            rad = radians(theta);
            d = (v*v*sin(2*rad)) / gravity;

            return d;
    }
/* Main program */
void main()
{
            float theta, v, h, d, t;

            theta = 45.0;
            v = 12.0;
            h = Height(theta, v);
            d = Distance(theta, v);
            t = Flight_time(theta, v);
    }
```

Figure 4.32: (Cont'd)

projectile, *Flight_time* calculates the flight time, and *Distance* calculates the distance traveled. In addition, a function called *Radians* converts degrees into radians to use in the trigonometric function *sine*. The height, distance traveled, and flight time are calculated and stored in floating point variables h, d, and t respectively.

4.4 Summary

This chapter has discussed the important topics of functions and libraries. Functions are useful when part of a code must be repeated several times from different points of a program. They also make programs more readable and easier to manage and maintain. A large program can be split into many functions that are tested independently and, once all of them are working, are combined to produce the final program.

The mikroC language library functions have also been described briefly, along with examples of how to use several of these functions in main programs. Library functions simplify programmers' tasks by providing ready and tested routines that can be called and used in their programs.

4.5   Exercises

1. Write a function to calculate the circumference of a rectangle. The function should receive the two sides of the rectangle as floating point numbers and return the circumference as a floating point number.

2. Write a main program to use the function you developed in Exercise 1. Find the circumference of a rectangle whose sides are 2.3cm and 5.6cm. Store the result in a floating point number called MyResult.

3. Write a function to convert inches to centimeters. The function should receive inches as a floating point number and then calculate the equivalent centimeters.

4. Write a main program to use the function you developed in Exercise 3. Convert 12.5 inches into centimeters and store the result as a floating point number.

5. An LED is connected to port pin RB0 of a PIC18F452-type microcontroller through a current limiting resistor in current sinking mode. Write a program to flash the LED in five-second intervals.

6. Eight LEDs are connected to PORTB of a PIC18F452-type microcontroller. Write a program so that the LEDs count up in binary sequence with a one-second delay between outputs.

7. An LED is connected to port pin RB7 of a PIC18F452 microcontroller. Write a program to flash the LED such that the ON time is five seconds, and the OFF time is three seconds.

8. A text-based LCD is connected to a PIC18F452-type microcontroller in 4-bit data mode. Write a program that will display a count from 0 to 255 on the LCD with a one-second interval between counts.

9. A text-based LCD is connected to a PIC microcontroller as in Exercise 8. Write a program to display the text "Exercise 9" on the first row of the LCD.

10. Repeat Exercise 9 but display the message on the first row, starting from column 3 of the LCD.

11. A two-row text-based LCD is connected to a PIC18F452-type microcontroller in 4-bit-data mode. Write a program to display the text "COUNTS:" on row 1 and then to count repeatedly from 1 to 100 on row 2 with two-second intervals.

12. Write a program to calculate the trigonometric cosine of angles from 0° to 45° in steps of 1° and store the results in a floating point array.

13. Write a function to calculate and return the length of the hypotenuse of a right-angle triangle, given its two sides. Show how you can use the function in a main program to calculate the hypotenuse of a right-angle triangle whose two sides are 4.0cm and 5.0cm.

14. Write a program to configure port pin RB2 of a PIC18F452 microcontroller as the RS232 serial output port. Send character "X" to this port at 4800 baud.

15. Port RB0 of a PIC18F452 microcontroller is configured as the RS232 serial output port. Write a program to send out string "SERIAL" at 9600 baud.

16. Repeat Exercise 15 but use the hardware USART available on the microcontroller chip.

17. Explain the differences between software-implemented serial data communication and USART hardware-based serial communication.

18. Write a function to add two arrays that are passed to the function as arguments. Store the sum in one of these arrays.

19. Write a function to perform the following operations on two-dimensional matrices:

    a) Add matrices

    b) Subtract matrices

    c) Multiply matrices

20. Write a function to convert between polar and rectangular coordinates.

21. Write functions to convert temperature expressed in Celsius to Fahrenheit and vice versa. Show how these functions can be called from main programs to convert 20°C to °F and also 100°F to °C.

22. Write a program to store the value of function f(x) in an array as x is varied from 0 to 10 in steps of 0.5. Assume that:

$$f(x) = 1.3x^3 - 2.5x^2 + 3.1x - 4.5$$

PIC18 Development Tools

The development of a microcontroller-based system is a complex process. Development tools are hardware and software tools designed to help programmers develop and test systems in a relatively short time. There are many such tools, and a discussion of all of them is beyond the scope of this book. This chapter offers a brief review of the most common tools.

The tools for developing software and hardware for microcontroller-based systems include editors, assemblers, compilers, debuggers, simulators, emulators, and device programmers. A typical development cycle starts with writing the application program using a text editor. The program is then translated into an executable code with the help of an assembler or compiler. If the program has several modules, a linker is used to combine them into a single application. Any syntax errors are detected by the assembler or compiler and must be corrected before the executable code can be generated. Next, a simulator is used to test the application program without the target hardware. Simulators are helpful in checking the correctness of an algorithm or a program with limited or no input-outputs, and most errors can be removed during simulation. Once the program seems to be working and the programmer is happy with it, the executable code is loaded to the target microcontroller chip using a device programmer, and the system logic is tested. Software and hardware tools such as in-circuit debuggers and in-circuit emulators can analyze the program's operation and display the variables and registers in real time with the help of breakpoints set in the program.

5.1   Software Development Tools

Software development tools are computer programs, usually run on personal computers, that allow the programmer (or system developer) to create, modify, and test applications programs. Some common software development tools are:

- Text editors

- Assemblers/compilers

- Simulators

- High-level language simulators

- Integrated development environments (IDEs)

5.1.1   Text Editors

A text editor is used to create or edit programs and text files. The Windows operating system comes with a text editor program called Notepad. Using Notepad, we can create a new program file, modify an existing file, or display or print the contents of a file. It is important to realize that programs used for word processing, such as Microsoft Word, cannot be used for this purpose, since they embed word formatting characters such as bold, italic, and underline within the text.

Most assemblers and compilers come with built-in text editors, making it possible to create a program and then assemble or compile it without having to exit from the editor. These editors provide additional features as well, such as automatic keyword highlighting, syntax checking, parenthesis matching, and comment line identification. Different parts of a program can be shown in different colors to make the program more readable (e.g., comments in one color and keywords in another). Such features help to eliminate syntax errors during the programming stage, thus speeding up the development process.

5.1.2   Assemblers and Compilers

Assemblers generate executable code from assembly language programs, and that generated code can then be loaded into the flash program memory of a PIC18-based microcontroller. Compilers generate executable code from high-level language programs. The compilers used most often for PIC18 microcontrollers are BASIC, C, and PASCAL.

Assembly language is used in applications where processing speed is critical and the microcontroller must respond to external and internal events in the shortest possible time. However, it is difficult to develop complex programs using assembly language, and assembly language programs are not easy to maintain.

High-level languages, on the other hand, are easier to learn, and complex programs can be developed and tested in a much shorter time. High-level programs are also maintained more easily than assembly language programs.

Discussions of programming in this book are limited to the C language. Many different C language compilers are available for developing PIC18 microcontroller-based programs. Some of the popular ones are:

- CCS C (http://www.ccsinfo.com)

- Hi-Tech C (http://htsoft.com)

- C18 C (http://www.microchip.com)

- mikroC C (http://www.mikroe.com)

- Wiz-C C (http://www.fored.co.uk)

Although most C compilers are essentially the same, each one has its own additions or modifications to the standard language. The C compiler used in this book is mikroC, developed by mikroElektronika.

5.1.3   Simulators

A simulator is a computer program that runs on a PC without the microcontroller hardware. It simulates the behavior of the target microcontroller by interpreting the user program instructions using the microcontroller instruction set. Simulators can display the contents of registers, memory, and the status of input-output ports as the user program is interpreted. Breakpoints can be set to stop the program and check the contents of various registers at desired locations. In addition, the user program can be executed in a single-step mode, so the memory and registers can be examined as the program executes one instruction at a time as a key is pressed.

Some assembler programs contain built-in simulators. Three popular PIC18 microcontroller assemblers with built-in simulators are:

- MPLAB IDE (http://www.microchip.com)

- Oshon Software PIC18 simulator (http://www.oshonsoft.com)

- Forest Electronics PIC18 assembler (http://www.fored.co.uk)

5.1.4   High-Level Language Simulators

High-level language simulators, also known as source-level debuggers, are programs that run on a PC and locate errors in high-level programs. The programmer can set breakpoints in high-level statements, execute the program up to a breakpoint, and then view the values of program variables, the contents of registers, and memory locations at that breakpoint.

A source-level debugger can also invoke hardware-based debugging using a hardware debugger device. For example, the user program on the target microcontroller can be stopped and the values of various variables and registers can be examined.

Some high-level language compilers, including the following three, have built-in source-level debuggers:

- C18 C

- Hi-Tech PIC18 C

- mikroC C

5.1.5   Integrated Development Environments (IDEs)

Integrated development environments (IDEs) are powerful PC-based programs which include everything to edit, assemble, compile, link, simulate, and source-level debug a program, and then download the generated executable code to the physical microcontroller chip using a programmer device. These programs are in graphical user interface (GUI), where the user can select various options from the program without having to exit it. IDEs can be extremely useful when developing microcontroller-based systems. Most PIC18 high-level language compilers are IDEs, thus enabling the programmer to do most tasks within a single software development tool.

5.2   Hardware Development Tools

Numerous hardware development tools are available for the PIC18 microcontrollers. Some of these products are manufactured by Microchip Inc., and some by third-party companies. The most ones are:

- Development boards

- Device programmers

- In-circuit debuggers

- In-circuit emulators

- Breadboards

5.2.1   Development Boards

Development boards are invaluable microcontroller development tools. Simple development boards contain just a microcontroller and the necessary clock circuitry. Some sophisticated development boards contain LEDs, LCD, push buttons, serial ports, USB port, power supply circuit, device programming hardware, and so on.

This section is a survey of various commercially available PIC18 microcontroller development boards and their specifications.

LAB-XUSB Experimenter Board

The LAB-XUSB Experimenter board (see Figure 5.1), manufactured by microEngineering Labs Inc., can be used in 40-pin PIC18-based project development. The board is available either assembled or as a bare board.

The board contains:

- 40-pin ZIF socket for PIC microcontroller

- 5-volt regulator

- 20MHz oscillator

- Reset button

- 16-switch keypad

- Two potentiometers

- Four LEDs

- 2-line by 20-character LCD module

- Speaker

Figure 5.1: LAB-XUSB Experimenter board

- RC servo connector

- RS232 interface

- USB connector

- Socket for digital-to-analog converter (device not included)

- Socket for I2C serial EEPROM (device not included)

- Socket for Dallas DS1307 real-time clock (device not included)

- Pads for Dallas DS18S20 temperature sensors (device not included)

- In-circuit programming connector

- Prototyping area for additional circuits

PICDEM 2 Plus

Th PICDEM 2 Plus kit (see Figure 5.2), manufactured by Microchip Inc., can be used in the development of PIC18 microcontroller-based projects.

Figure 5.2: PICDEM 2 Plus development board

The board contains:

- 2 × 16 LCD display

- Piezo sounder driven by PWM signal

- Active RS 232 port

- On-board temperature sensor

- Four LEDs

- Two push-button switches and master reset

- Sample PIC18F4520 and PIC16F877A flash microcontrollers

- MPLAB REAL ICE/MPLAB ICD 2 connector

- Source code for all programs

- Demonstration program displaying a real-time clock and ambient temperature

- Generous prototyping area

- Works off of a 9V battery or DC power pack

PICDEM 4

The PICDEM 4 kit (see Figure 5.3), manufactured by Microchip Inc., can be used in the development of PIC18 microcontroller-based projects.

The board contains:

- Three different sockets supporting 8-, 14-, and 18-pin DIP devices

- On-board +5V regulator for direct input from 9V, 100 mA AC/DC wall adapter

- Active RS-232 port

- Eight LEDs

- 2 × 16 LCD display

- Three push-button switches and master reset

- Generous prototyping area

- I/O expander

- Supercapacitor circuitry

- Area for an LIN transceiver

Figure 5.3: PICDEM 4 development board

- Area for a motor driver

- MPLAB ICD 2 connector

PICDEM HPC Explorer Board

The PICDEM HPC Explorer development board (see Figure 5.4), manufactured by Microchip Inc., can be used in the development of high pin count PIC18-series microcontroller-based projects.

The main features of this board are:

- PIC18F8722, 128K flash, 80-pin TQFP microcontroller

- Supports PIC18 J-series devices with plug-in modules

- 10MHz crystal oscillator (to be used with internal PLL to provide 40MHz operation)

- Power supply connector and programmable voltage regulator, capable of operation from 2.0 to 5.5V

- Potentiometer (connected to 10-bit A/D, analog input channel)

- Temperature sensor demo included

Figure 5.4: PICDEM HPC Explorer development board

- Eight LEDs (connected to PORTD with jumper disable)

- RS-232 port (9-pin D-type connector, UART1)

- Reset button

- 32KHz crystal for real-time clock demonstration

MK-1 Universal PIC Development Board

The MK-1 Universal PIC development board (see Figure 5.5), manufactured by Baji Labs, can be used for developing PIC microcontroller-based projects with up to 40 pins. The board has a key mechanism which allows any peripheral device to be mapped to any pin of the processor, making the board very flexible. A small breadboard area is also provided, enabling users to design and test their own circuits.

Figure 5.5: MK-1 Universal PIC development board

The board has the following features:

- On-board selectable 3.3V or 5V

- 16×2 LCD character display (8- or 4-bit mode supported)

- 4-digit multiplexed 7-segment display

- Ten LED bar graph (can be used as individual LEDs)

- Eight-position dip switch

- Socketed oscillator for easy change of oscillators

- Stepper motor driver with integrated driver

- I^2C real-time clock with crystal and battery backup support

- I^2C temperature sensor with 0.5 degree C precision

- Three potentiometers for direct A/D development

- 16-button telephone keypad wired as 4×4 matrix

- RS232 driver with standard DB9 connector

- Socketed SPI and I^2C EEPROM

- RF Xmit and receive sockets

- IR Xmit and receive

- External drive buzzer

- Easy access to pull up resistors

- AC adapter included

SSE452 Development Board

The SSE452 development board (see Figure 5.6), manufactured by Shuan Shizu Electronic Laboratory, can be used for developing PIC18-based microcontroller projects, especially the PIC18FXX2 series of microcontrollers, and also for programming the microcontrollers.

The main features of this board are:

- One PCB suitable for any 28- or 40-pin PIC18 devices

- Three external interrupt pins

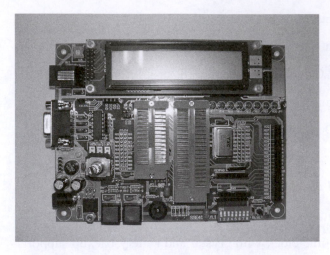

Figure 5.6: SSE452 development board

- Two input-capture/output-compare/pulse-width modulation modules (CCP)

- Support SPI, I^2C functions

- 10-bit analog-to-digital converter

- RS-232 connector

- Two debounced push-button switches

- An 8-bit DIP-switch for digital input

- 4 × 4 keypad connector

- Rotary encoder with push button

- TC77 SPI temperature sensor

- EEPROM (24LC04B)

- 2 × 20 bus expansion port

- ICD2 connector

- On-board multiple digital signals from 1Hz to 8MHz

- Optional devices are 2 × 20 character LCD, 48/28-pin ZIF socket

SSE8720 Development Board

The SSE8720 development board (see Figure 5.7), manufactured by Shuan Shizu Electronic Laboratory, can be used for the development of PIC18-based microcontroller projects. A large amount of memory and I/O interface is provided, and the board can also be used to program microcontrollers.

The main features of this board are:

- 20MHz oscillator with socket

- One DB9 connector provides EIA232 interface

- In-circuit debugger (ICD) connector

- Four debounced switches, and one reset switch

- 4 × 4 keypad connector

- One potentiometer for analog-to-digital conversion

- Eight red LEDs

- 8-bit DIP switch for digital inputs

- 2 × 20 character LCD module

- Twenty-four different digital signals, from 1Hz to 16MHz

Figure 5.7: SSE8720 development board

- On-board 5V regulator

- One I^{2}C EEPROM with socket

- SPI-compatible digital temperature sensor

- SPI-compatible real-time clock

- CCP1 output via an NPN transistor

SSE8680 Development Board

The SSE8680 development board (see Figure 5.8), manufactured by Shuan Shizu Electronic Laboratory, can be used for developing PIC18-based microcontroller projects. The board supports CAN network, and a large amount of memory and I/O interface is provided. The board can also be used to program microcontrollers.

The main features of this board are:

- 20MHz oscillator with socket

- One DB9 connector provides EIA232 interface

- In-circuit debugger (ICD) connector

- Four debounced switches, and one reset switch

- 4 × 4 keypad connector

Figure 5.8: SSE8680 development board

- One potentiometer for analog-to-digital conversion

- 8 red LEDs

- 8-bit DIP switch for digital inputs

- 2 × 20 character LCD module

- Twenty-four different digital signals, from 1Hz to 16MHz

- On-board 5V regulator

- One I^2C EPROM with socket

- SPI-compatible digital temperature sensor

- SPI-compatible real-time clock

- CCP1 output via an NPN transistor

- Rotary encoder

- CAN transceiver

PIC18F4520 Development Kit

The PIC18F4520 development kit (see Figure 5.9), manufactured by Custom Computer Services Inc., includes a C compiler (PCWH), a prototyping board with PIC18F4520 microcontroller, an in-circuit debugger, and a programmer.

Figure 5.9: PIC18F4520 development kit

The main features of this development kit are:

- PCWH compiler

- PIC18F4520 prototyping board

- Breadboard area

- 93LC56 serial EEPROM chip

- DS1631 digital thermometer chip

- NJU6355 real-time clock IC with attached 32.768KHz crystal

- Two-digit 7-segment LED module

- In-circuit debugger/programmer

- DC adapter and cables

Custom Computer Services manufactures a number of other PIC18 microcontroller-based development kits and prototyping boards, such as development kits for CAN, Ethernet, Internet, USB, and serial buses. More information is available on the company's web site.

BIGPIC4 Development Kit

The BIGPIC4 is a sophisticated development kit (Figure 5.10) that supports the latest 80-pin PIC18 microcontrollers. The kit comes already assembled, with a

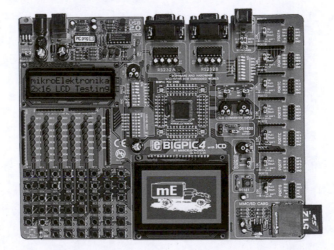

Figure 5.10: BIGPIC4 development kit

PIC18F8520 microcontroller installed and working at 10MHz. It includes an on-board USB port, an on-board programmer, and an in-circuit debugger. The microcontroller on the board can be replaced easily.

The main features of this development kit are:

- Forty-six buttons
- Forty-six LEDs
- USB connector
- External or USB power supply
- Two potentiometers
- Graphics LCD
- 2 × 16 text LCD
- MMC/SD memory card slot
- Two serial RS232 ports
- In-circuit debugger
- Programmer
- PS2 connector
- Digital thermometer chip (DS1820)
- Analog inputs
- Reset button

The BIGPIC4 is used in some of the projects in this book.

FUTURLEC PIC18F458 Training Board

The FUTURLEC PIC18F458 training board is a very powerful development kit (see Figure 5.11) based on the PIC18F458 microcontroller and developed by Futurlec (www.futurlec.com). The kit comes already assembled and tested. One of its biggest advantages is its low cost, at under $45.

Figure 5.11: FUTURLEC PIC18F458 training board

Its main features are:

- PIC18F458 microcontroller with 10MHz crystal

- RS232 communication

- Test LED

- Optional real-time clock chip with battery backup

- LCD connection

- Optional RS485/RS422 with optional chip

- CAN and SPI controller

- I^2C expansion

- In-circuit programming

- Reset button

- Speaker

- Relay socket

- All port pins are available at connectors

5.2.2 Device Programmers

After the program is written and translated into executable code, the resulting HEX file is loaded to the target microcontroller's program memory with the help of a device programmer. The type of device programmer depends on the type of microcontroller to be programmed. For example, some device programmers can only program PIC16 series, some can program both PIC16 and PIC18 series, while some are designed to program other microcontroller models (e.g., the Intel 8051 series).

Some microcontroller development kits include on-board device programmers, so the microcontroller chip does not need to be removed and inserted into a separate programming device. This section describes some of the popular device programmers used to program PIC18 series of microcontrollers.

Forest Electronics USB Programmer

The USB programmer, manufactured by Forest Electronics (see Figure 5.12), can be used to program most PIC microcontrollers with up to 40 pins, including the PIC18 series. The device is connected to the USB port of a PC and takes its power from this port.

Figure 5.12: Forest Electronics USB programmer

Mach X Programmer

The Mach X programmer (Figure 5.13), manufactured by Custom Computer Services Inc., can program microcontrollers of the PIC12, PIC14, PIC16, and PIC18 series ranging from 8 to 40 pins. It can also read the program inside a microcontroller and then generate a HEX file. In-circuit debugging is also supported by this programmer.

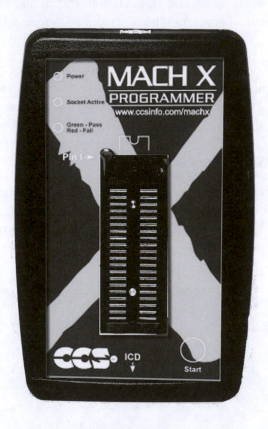

Figure 5.13: Mach X programmer

Melabs U2 Programmer

The Melabs U2 device programmer (see Figure 5.14), manufactured by microEngineering Labs Inc., can be used to program most PIC microcontroller chips having from 8 to 40 pins. The device is USB-based and receives its power from the USB port of a PC.

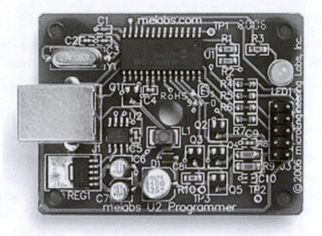

Figure 5.14: Melabs U2 programmer

EasyProg PIC Programmer

The EasyProg PIC is a low-cost programmer (Figure 5.15) used with microcontrollers of the PIC16 and PIC18 series having up to 40 pins. It connects to a PC via a 9-pin serial cable.

Figure 5.15: EasyProg programmer

PIC Prog Plus Programmer

The PIC Prog Plus is another low-cost programmer (Figure 5.16) that can be used to program most PIC microcontrollers. The device is powered from an external 12V DC supply.

5.2.3   In-Circuit Debuggers

An in-circuit debugger is hardware connected between a PC and the target microcontroller test system used to debug real-time applications quickly and easily. With in-circuit debugging, a monitor program runs in the PIC microcontroller in the test circuit. The programmer can set breakpoints on the PIC, run code, single-step the program, and examine variables and registers on the real device and, if required, change their values. An in-circuit debugger uses some memory and I/O pins of the target PIC microcontroller during debugging operations. Some in-circuit debuggers only debug assembly language programs. Other, more powerful debuggers can debug high-level language programs.

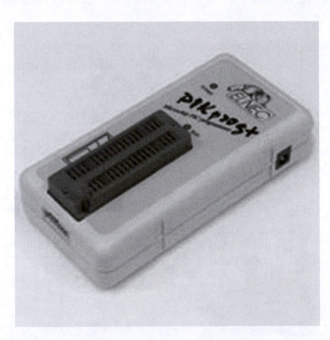

Figure 5.16: PIC Prog Plus programmer

This section discusses some of the popular in-circuit debuggers used in PIC18 microcontroller-based system applications.

ICD2

The ICD2, a low-cost in-circuit debugger (see Figure 5.17) manufactured by Microchip Inc., can debug most PIC microcontroller-based systems. With the ICD2, programs are downloaded to the target microcontroller chip and executed in real time. This debugger supports both assembly language and C language programs.

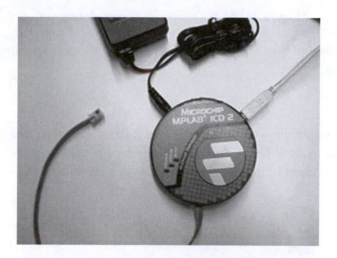

Figure 5.17: ICD2 in-circuit debugger

The ICD2 connects to a PC through either a serial RS232 or a USB interface. The device acts like an intelligent interface between the PC and the test system, allowing the programmer to set breakpoints, look into the test system, view registers and variables at breakpoints, and single-step through the user program. It can also be used to program the target PIC microcontroller.

ICD-U40

The ICD-U40 is an in-circuit debugger (see Figure 5.18) manufactured by Custom Computer Services Inc. to debug programs developed with their CCS C compiler. The device operates with a 40MHz clock frequency, is connected to a PC via the USB interface, and is powered from the USB port. The company also manufactures

Figure 5.18: ICD-U40 in-circuit debugger

a serial-port version of this debugger called ICD-S40, which is powered from the target test system.

PICFlash 2

The PICFlash 2 in-circuit debugger (see Figure 5.19) is manufactured by mikroElektronika and can be used to debug programs developed in mikroBasic, mikroC, or mikroPascal languages. The device is connected to a PC through its USB

Figure 5.19: PICFlash 2 in-circuit debugger

interface. Power is drawn from the USB port so the debugger requires no external power supply. The PICFlash 2 is included in the BIGPIC4 development kit. Details on the use of this in-circuit debugger are discussed later in this chapter.

5.2.4 In-Circuit Emulators

The in-circuit emulator (ICE) is one of the oldest and the most powerful devices for debugging a microcontroller system. It is also the only tool that substitutes its own internal processor for the one in the target system. Like all in-circuit debuggers, the emulator's primary function is target access—the ability to examine and change the contents of registers, memory, and I/O. Since the emulator replaces the CPU, it does not require a working CPU in the target system. This makes the in-circuit emulator by far the best tool for troubleshooting new or defective systems.

In general, each microcontroller family has its own set of in-circuit emulators. For example, an in-circuit emulator designed for the PIC16 microcontrollers cannot be used for PIC18 microcontrollers. Moreover, the cost of in-circuit emulators is usually quite high. To keep costs down, emulator manufacturers provide a base board which can be used with most microcontrollers in a given family, for example, with all PIC microcontrollers, and also make available probe cards for individual microcontrollers. To emulate a new microcontroller in the same family, then, only the specific probe card has to be purchased.

Several models of in-circuit emulators are available on the market. The following four are some of the more popular ones.

MPLAB ICE 4000

The MPLAB ICE 4000 in-circuit emulator (Figure 5.20), manufactured by Microchip Inc., can be used to emulate microcontrollers in the PIC18 series. It consists of an emulator pod connected with a flex cable to device adapters for the specific microcontroller. The pod is connected to the PC via its parallel port or USB port. Users can insert an unlimited number of breakpoints in order to examine register values.

RICE3000

The RICE3000 is a powerful in-circuit emulator (Figure 5.21), manufactured by Smart Communications Ltd, for the PIC16 and PIC18 series of microcontrollers.

Figure 5.20: MPLAB ICE 4000

Figure 5.21: RICE3000 in-circuit emulator

The device consists of a base unit with different probe cards for the various members of the PIC microcontroller family. It provides full-speed real-time emulation up to 40MHz, supports observation of floating point variables and complex variables such as arrays and structures, and provides source level and symbolic debugging in both assembly and high-level languages.

ICEPIC 3

The ICEPIC 3 is a modular in-circuit emulator (see Figure 5.22), manufactured by RF Solutions, for the PIC12/16 and PIC18 series of microcontrollers. It connects to the PC via its USB port and consists of a mother board with additional daughter boards for each microcontroller type. The daughter boards are connected to the target system with device adapters. A trace board can be added to capture and analyze execution addresses, opcodes, and external memory read/writes.

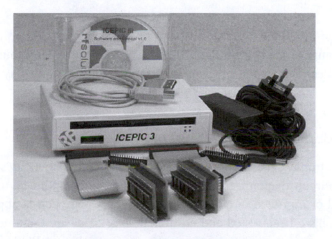

Figure 5.22: ICEPIC 3 in-circuit emulator

PICE-MC

The PICE-MC, a highly sophisticated emulator (see Figure 5.23) manufactured by Phyton Inc., supports most PIC microcontrollers and consists of a main board, pod, and adapters. The main board contains the emulator logic, memory, and an interface to the PC. The pod contains a slave processor that emulates the target microcontroller. The adapters are the mechanical parts that physically connect to the microcontroller sockets of the target system. The PICE-MC provides source-level debugging of

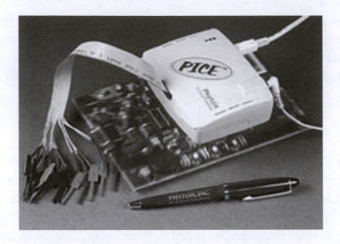

Figure 5.23: PICE-MC in-circuit emulator

programs written in both assembly and high-level languages. A large memory is provided to capture target system data. The user can set up a large number of breakpoints and can access the program and data memories to display or change their contents.

5.2.5   Breadboards

Building an electronic circuit requires connecting the components as shown in the relevant circuit diagram, usually by soldering the components together on a strip board or a printed circuit board (PCB). This approach is appropriate for circuits that have been tested and are functioning as desired, and also when the circuit is being made permanent. However, making a PCB design for just a few applications—for instance, while still developing the circuit—is not economical.

Instead, while the circuit is still under development, the components are usually assembled on a solderless breadboard. A typical breadboard (see Figure 5.24) consists of rows and columns of holes spaced so that integrated circuits and other components can be fitted inside them. The holes have spring actions so the component leads are held tightly in place. There are various types and sizes of breadboards, suitable for circuits of different complexities. Breadboards can also be

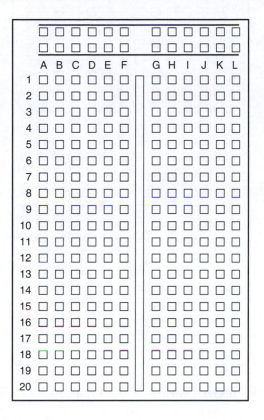

Figure 5.24: A typical breadboard layout

stacked together to make larger boards for very complex circuits. Figure 5.25 shows the internal connection layout of the breadboard in Figure 5.24.

The top and bottom halves of the breadboard are entirely separate. Columns 1 to 20 in rows A to F are connected to each other on a column basis. Rows G to L in columns 1 to 20 are likewise connected to each other on a column basis. Integrated circuits are placed such that the legs on one side are on the top half of the breadboard, and the legs on the other side are on the bottom half. The two columns on the far left of the board are usually reserved for the power and ground connections. Connections between components are usually made with stranded (or solid) wires plugged into the holes to be connected.

Figure 5.26 shows a breadboard holding two integrated circuits and a number of resistors and capacitors.

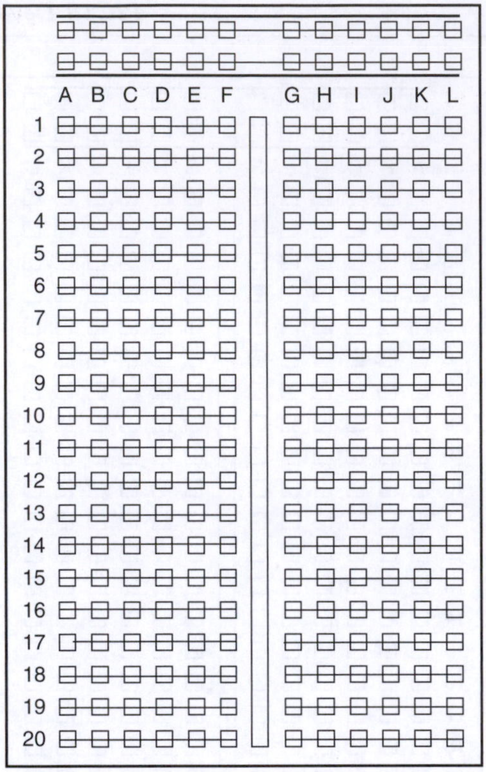

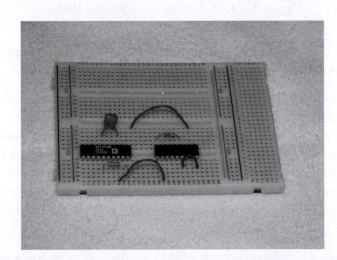

Figure 5.25: Internal wiring of the breadboard in Figure 5.24

Figure 5.26: Picture of a breadboard with some components

The nice thing about breadboard design is that the circuit can be modified easily and quickly, and ideas can be tested without having to solder the components. Once a circuit has been tested and is working satisfactorily, the components are easily removed and the breadboard can be used for other projects.

5.3   mikroC Integrated Development Environment (IDE)

In this book we are using the mikroC compiler developed by mikroElektronika. Before using this compiler, we need to know how the mikroC integrated development environment (IDE) is organized and how to write, compile, and simulate a program in the mikroC language. In this section we will look at the operation of the mikroC IDE in detail.

A free 2K program size limited version of the mikroC IDE, available on the mikroElektronika web site (www.mikroe.com), is adequate for most small or medium-sized applications. Alternatively, you can purchase a license and turn the limited version into a fully working, unlimited IDE to use for projects of any size and complexity.

After installing the mikroC IDE, a new icon should appear by default on your desktop. Double-click this icon to start the IDE.

5.3.1   mikroC IDE Screen

After the mikroC icon is double-clicked to start the IDE, the screen shown in Figure 5.27 is displayed by default.

The screen is divided into four areas: the top-left section, the bottom-left section, the middle section, and the bottom section.

Top-Left Section

The top left, the Code Explorer section, displays every declared item in the source code. In the example in Figure 5.28, *main* is listed under *Functions* and variables *Sum* and *i* are listed under *main*.

Code Explorer

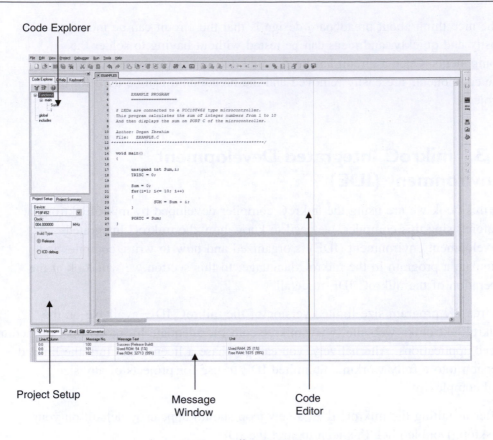

Project Setup   Message   Code
                                 Window   Editor

Figure 5.27: mikroC IDE screen

There are two additional tabs in the Code Explorer. As shown in Figure 5.29, the *QHelp* tab lists all the available built-in functions and library functions for a quick reference.

The *Keyboard* tab lists all the available keyboard shortcuts in mikroC IDE (see Figure 5.30).

Bottom-Left Section

In the bottom-left section, called Project Setup (see Figure 5.31), the microcontroller device type, clock rate, and build type are specified. The build type can be either

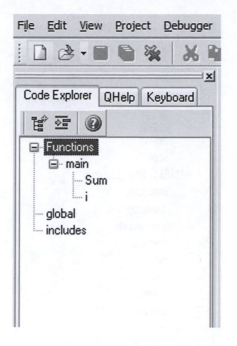

Figure 5.28: Code Explorer form

Release, which is the normal compiler operating mode, or *ICD debug*, if the program is to be debugged using the in-circuit debugger.

The Project Setup section has a tab called *Project Summary* which lists all the types of files used in the project, as shown in Figure 5.32.

Middle Section

The middle section is the Code Editor, an advanced text editor. Programs are written in this section of the screen. The Code Editor supports:

- Code Assistant
- Parameter Assistant
- Code Template
- Auto Correct
- Bookmarks

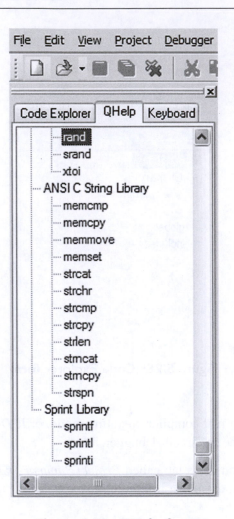

Figure 5.29: QHelp form

The Code Assistant is useful when writing a program. Type the first few letters of an identifier and then press the CTRL+SPACE keys to list all valid identifiers beginning with those letters. In Figure 5.33, for example, to locate identifier *strlen*, the letters *str* are typed and CTRL+SPACE is pressed. *strlen* can be selected from the displayed list of matching valid words by using keyboard arrows and pressing ENTER.

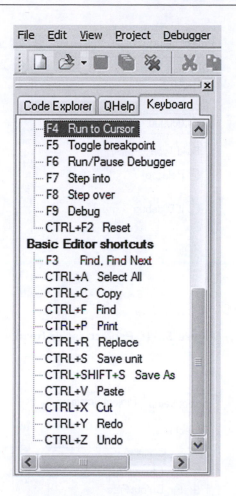

Figure 5.30: Keyboard form

The Parameter Assistant is invoked when a parenthesis is opened after a function or a procedure name. The expected parameters are listed in a small window just above the parenthesis. In Figure 5.34, function *strlen* has been entered, and *unsigned char *s* appears in a small window when a parenthesis is opened.

Code Template is used to generate code in the program. For example, as shown in Figure 5.35, typing *switch* and pressing CTRL+J automatically generates code for the

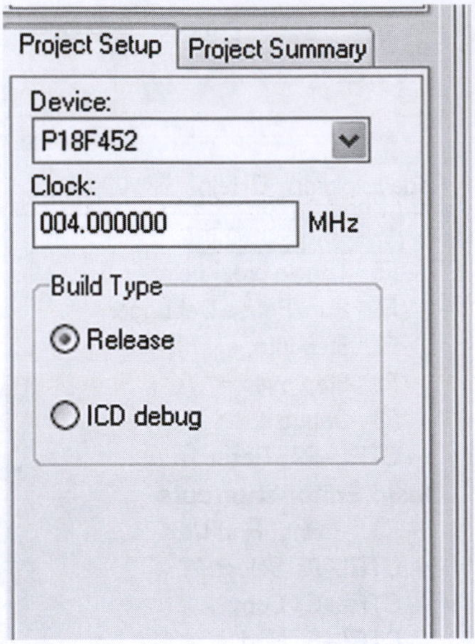

Figure 5.31: Project setup form

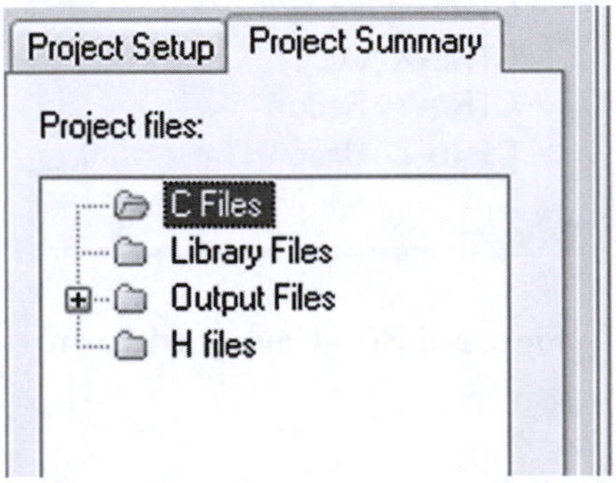

Figure 5.32: Project summary form

switch statement. We can add our own templates by selecting *Tools -> Options -> Auto Complete*. Some of the available templates are *array, switch, for,* and *if*.

Auto Correct corrects typing mistakes automatically. A new list of recognized words can be added by selecting *Tools -> Options -> Auto Correct Tab*.

```
Sum = 0;
for(i=1; i<= 10; i++)
{
        SUM = Sum + i;
}
str
POR
```

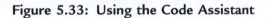

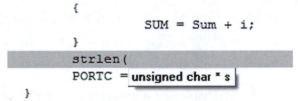

function	void **Strobe**()
function	unsigned char * **strcat**(unsigned char *, unsigned char *)
function	unsigned char * **strchr**(unsigned char *, unsigned char)
function	signed int **strcmp**(unsigned char *, unsigned char *)
function	unsigned char * **strcpy**(unsigned char *, unsigned char *)
function	signed int **strlen**(unsigned char *)
function	unsigned char * **strncat**(unsigned char *, unsigned char *, signed int)
function	unsigned char * **strncpy**(unsigned char *, unsigned char *, signed int)

Figure 5.33: Using the Code Assistant

```
{
        SUM = Sum + i;
}
strlen(
PORTC = unsigned char * s
}
```

Figure 5.34: Using the Parameter Assistant

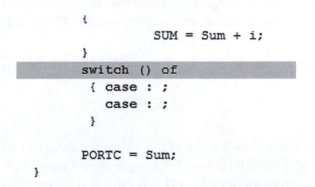

```
{
        SUM = Sum + i;
}
switch () of
 { case : ;
   case : ;
 }

PORTC = Sum;
}
```

Figure 5.35: Using the Code Template

Bookmarks make the navigation easier in large code. We can set bookmarks by entering CTRL+SHIFT+number, and can then jump to the bookmark by pressing CTRL+number, where number is the bookmark number.

Bottom Section

The bottom section of the screen, also called the Message Window, consists of three tabs: *Messages*, *Find*, and *QConverter*. Compilation errors and warnings are reported under the *Messages* tab. Double-clicking on a message line highlights the line where the error occurred. A HEX file can be generated only if the source file contains no errors. Figure 5.36 shows the results of a successful compilation listed in the Message Window. The *QConverter* tab can be used to convert decimal numbers into binary or hexadecimal, and vice versa.

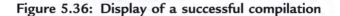

Line/Column	Message No.	Message Text	Unit
0:0	100	Success (Release Build)	
0:0	101	Used ROM: 54 (1%)	Used RAM: 25 (1%)
0:0	102	Free ROM: 32713 (99%)	Free RAM: 1615 (99%)

Figure 5.36: Display of a successful compilation

5.3.2   Creating and Compiling a New File

mikroC files are organized into projects, and all files for a single project are stored in the same folder. By default, a project file has the extension ".ppc". A project file contains the project name, the target microcontroller device, device configuration flags, the device clock, and list of source files with their paths. C source files have the extension ".c".

The following example illustrates step-by-step how to create and compile a program source file.

Example 5.1

Write a C program to calculate the sum of the integer numbers 1 to 10 and then send the result to PORTC of a PIC18F452-type microcontroller. Assume that eight LEDs are connected to the microcontroller's PORTC via current limiting resistors. Draw the circuit diagram and show the steps involved in creating and compiling the program.

Solution 5.1

Figure 5.37 shows the circuit diagram of the project. The LEDs are connected to PORTC using 390 ohm current limiting resistors. The microcontroller is operated from a 4MHz resonator.

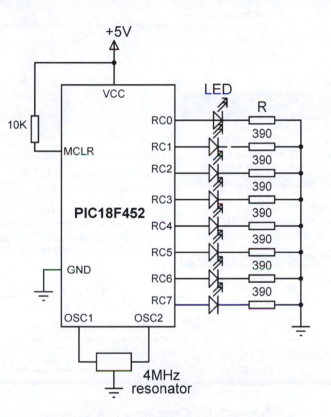

Figure 5.37: Circuit diagram of the project

The program is created and compiled as follows:

Step 1   Double-click the mikroC icon to start the IDE.

Step 2   Create a new project called EXAMPLE. Click *Project -> New Project* and fill in the form, as shown in Figure 5.38, by selecting the device type, the clock, and the configuration fuse.

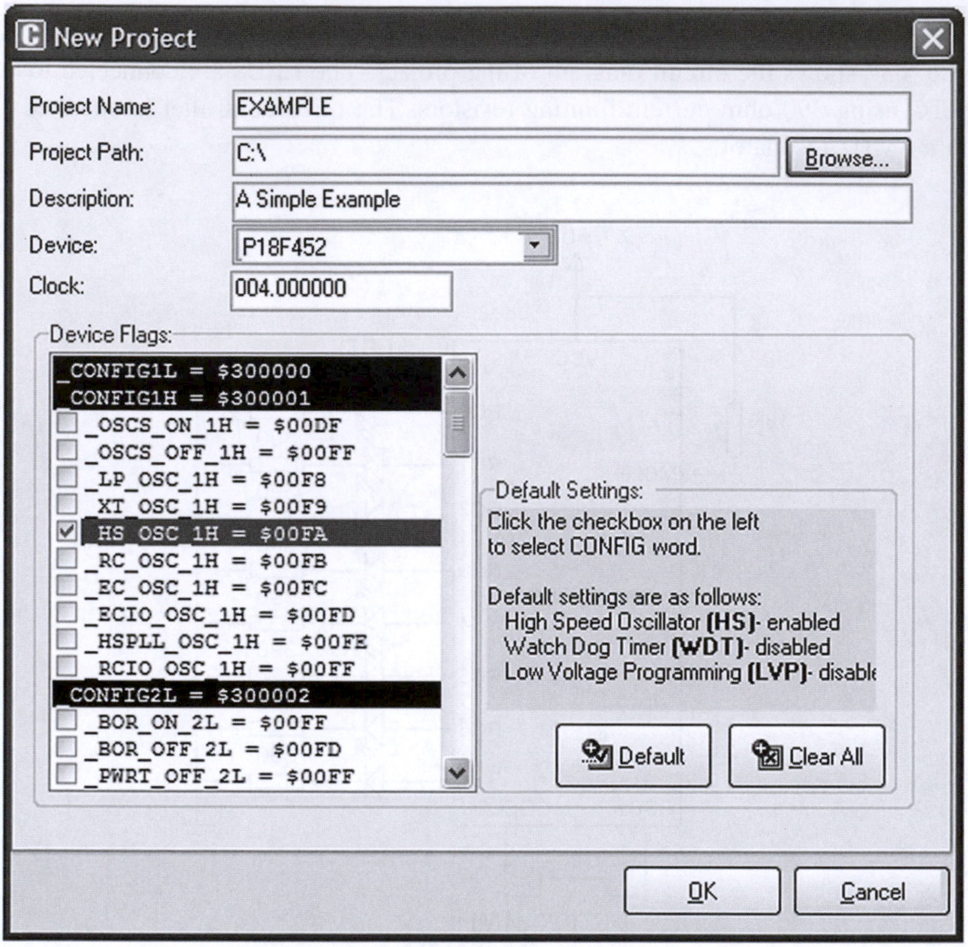

Figure 5.38: Creating a new project

Step 3 Enter the following program into the Code Editor section of the IDE:

```
/***********************************************************************
                         EXAMPLE PROGRAM

8 LEDs are connected to a PIC18F452 type microcontroller.
This program calculates the sum of integer numbers from 1 to 10
And then displays the sum on PORTC of the microcontroller.

Author: Dogan Ibrahim
File:   EXAMPLE.C

***********************************************************************/
```

```
void main ()
{

        unsigned int Sum, i;
        TRISC = 0;

        Sum = 0;
        for (i=1; i<= 10; i++)
        {
                Sum = Sum + i;
        }

        PORTC = Sum;
}
```

Step 4   Save the program with the name EXAMPLE by clicking *File -> Save As*. The program will be saved with the name EXAMPLE.C.

Step 5   Compile the project by pressing CTRL+F9 or by clicking the Build Project button (see Figure 5.39).

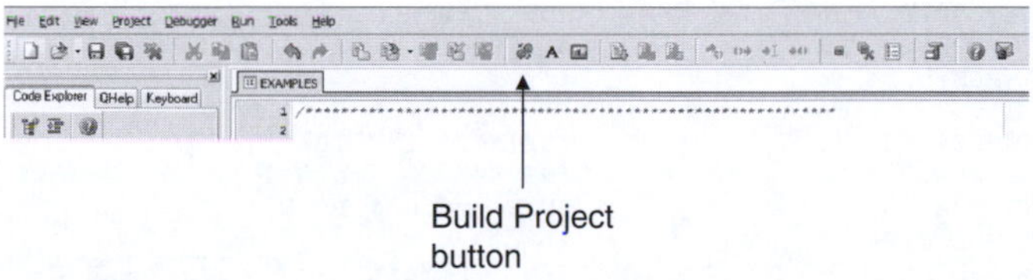

Build Project
button

Figure 5.39: Build Project button

Step 6   If the compilation is successful, a Success message will appear in the Message Window as shown in Figure 5.36. Any program errors will appear in the Message Window and should be corrected before the project proceeds further.

The compiler generates a number of output files, which can be selected by clicking *Tools -> Options -> Output*. The various output files include:

.ASM file   This is the assembly file of the program. Figure 5.40 shows the EXAMPLE. ASM file.

```
;  ASM code generated by mikroVirtualMachine for PIC - V. 6.2.1.0
;  Date/Time: 07/07/2007 16:46:12
;  Info: http://www.mikroelektronika.co.yu


; ADDRESS     OPCODE      ASM
; ---------------------------------------------
$0000 $EF04 F000             GOTO  _main
$0008 $     _main:
;EXAMPLE.c,14 ::             void main()
;EXAMPLE.c,18 ::             TRISC = 0;
$0008 $6A94                  CLRF  TRISC, 0
;EXAMPLES.c,20 ::               Sum = 0;
$000A $6A15                  CLRF  main_Sum_L0, 0
$000C $6A16                  CLRF  main_Sum_L0+1, 0
;EXAMPLE.c,21 ::             for(i=1; i<= 10; i++)
$000E $0E01                  MOVLW 1
$0010 $6E17                  MOVWF main_i_L0, 0
$0012 $0E00                  MOVLW 0
$0014 $6E18                  MOVWF main_i_L0+1, 0
$0016 $     L_main_0:
$0016 $0E00                  MOVLW 0
$0018 $6E00                  MOVWF STACK_0, 0
$001A $5018                  MOVF  main_i_L0+1, 0, 0
$001C $5C00                  SUBWF STACK_0, 0, 0
$001E $E102                  BNZ   L_main_3
$0020 $5017                  MOVF  main_i_L0, 0, 0
$0022 $080A                  SUBLW 10
$0024 $     L_main_3:
$0024 $E307                  BNC   L_main_1
;EXAMPLE.c,23 ::             SUM = Sum + i;
$0026 $5017                  MOVF  main_i_L0, 0, 0
$0028 $2615                  ADDWF main_Sum_L0, 1, 0
$002A $5018                  MOVF  main_i_L0+1, 0, 0
$002C $2216                  ADDWFC      main_Sum_L0+1, 1, 0
;EXAMPLE.c,24 ::             }
$002E $     L_main_2:
;EXAMPLE.c,21 ::             for(i=1; i<= 10; i++)
$002E $4A17                  INFSNZ      main_i_L0, 1, 0
$0030 $2A18                  INCF  main_i_L0+1, 1, 0
;EXAMPLE.c,24 ::             }
$0032 $D7F1                  BRA   L_main_0
$0034 $     L_main_1:
;EXAMPLE.c,26 ::             PORTC = Sum;
$0034 $C015 FF82             MOVFF main_Sum_L0, PORTC
;EXAMPLE.c,27 ::             }
$0038 $D7FF                  BRA   $
```

Figure 5.40: EXAMPLE.ASM

.LST file   This is the listing file of the program. Figure 5.41 shows the EXAMPLE.LST file.

```
;   ASM code generated by mikroVirtualMachine for PIC - V. 6.2.1.0
;   Date/Time: 07/07/2007 17:07:12
;   Info: http://www.mikroelektronika.co.yu


; ADDRESS    OPCODE      ASM
; ----------------------------------------------
$0000 $EF04 F000              GOTO  _main
$0008 $      _main:
;EXAMPLE.c,14 ::              void main()
;EXAMPLE.c,18 ::              TRISC = 0;
$0008 $6A94                   CLRF  TRISC, 0
;EXAMPLE.c,20 ::              Sum = 0;
$000A $6A15                   CLRF  main_Sum_L0, 0
$000C $6A16                   CLRF  main_Sum_L0+1, 0
;EXAMPLE.c,21 ::              for(i=1; i<= 10; i++)
$000E $0E01                   MOVLW 1
$0010 $6E17                   MOVWF main_i_L0, 0
$0012 $0E00                   MOVLW 0
$0014 $6E18                   MOVWF main_i_L0+1, 0
$0016 $      L_main_0:
$0016 $0E00                   MOVLW 0
$0018 $6E00                   MOVWF STACK_0, 0
$001A $5018                   MOVF  main_i_L0+1, 0, 0
$001C $5C00                   SUBWF STACK_0, 0, 0
$001E $E102                   BNZ   L_main_3
$0020 $5017                   MOVF  main_i_L0, 0, 0
$0022 $080A                   SUBLW 10
$0024 $      L_main_3:
$0024 $E307                   BNC   L_main_1
;EXAMPLE.c,23 ::              SUM = Sum + i;
$0026 $5017                   MOVF  main_i_L0, 0, 0
$0028 $2615                   ADDWF main_Sum_L0, 1, 0
$002A $5018                   MOVF  main_i_L0+1, 0, 0
$002C $2216                   ADDWFC     main_Sum_L0+1, 1, 0
;EXAMPLE.c,24 ::              }
$002E $      L_main_2:
;EXAMPLE.c,21 ::              for(i=1; i<= 10; i++)
$002E $4A17                   INFSNZ     main_i_L0, 1, 0
$0030 $2A18                   INCF  main_i_L0+1, 1, 0
;EXAMPLE.c,24 ::              }
$0032 $D7F1                   BRA   L_main_0
$0034 $      L_main_1:
;EXAMPLE.c,26 ::              PORTC = Sum;
$0034 $C015 FF82              MOVFF main_Sum_L0, PORTC
;EXAMPLE.c,27 ::              }
$0038 $D7FF                   BRA   $


//** Procedures locations **
//ADDRESS        PROCEDURE
//----------------------------------------------
$0008        main


//** Labels locations **
//ADDRESS        LABEL
//----------------------------------------------

$0008        _main:
$0016        L_main_0:
$0024        L_main_3:
$002E        L_main_2:
$0034        L_main_1:
```

Figure 5.41: EXAMPLE.LST

(Continued)

```
//** Variables locations **
//ADDRESS        VARIABLE
//---------------------------------------------
$0000          STACK_0
$0001          STACK_1
$0002          STACK_2
$0003          STACK_3
$0004          STACK_4
$0005          STACK_5
$0006          STACK_6
$0007          STACK_7
$0008          STACK_8
$0009          STACK_9
$000A          STACK_10
$000B          STACK_11
$000C          STACK_12
$000D          STACK_13
$000E          STACK_14
$000F          STACK_15
$0010          STACK_16
$0011          STACK_17
$0012          STACK_18
$0013          STACK_19
$0014          STACK_20
$0015          main_Sum_L0
$0017          main_i_L0
$0F82          PORTC
$0F94          TRISC
$0FD8          STATUS
$0FD9          FSR2L
$0FDA          FSR2H
$0FDB          PLUSW2
$0FDC          PREINC2
$0FDD          POSTDEC2
$0FDE          POSTINC2
$0FDF          INDF2
$0FE0          BSR
$0FE1          FSR1L
$0FE2          FSR1H
$0FE3          PLUSW1
$0FE4          PREINC1
$0FE5          POSTDEC1
$0FE6          POSTINC1
$0FE7          INDF1
$0FE8          WREG
$0FE9          FSR0L
$0FEA          FSR0H
$0FEB          PLUSW0
$0FEC          PREINC0
$0FED          POSTDEC0
$0FEE          POSTINC0
$0FEF          INDF0
$0FF3          PRODL
$0FF4          PRODH
$0FF5          TABLAT
$0FF6          TBLPTRL
$0FF7          TBLPTRH
$0FF8          TBLPTRU
$0FF9          PCL
$0FFA          PCLATH
$0FFB          PCLATU
$0FFD          TOSL
$0FFE          TOSH
$0FFF          TOSU
```

Figure 5.41: (Cont'd)

.HEX file This is the most important output file as it is the one sent to the programming device to program the microcontroller. Figure 5.42 shows the EXAMPLE.HEX file.

```
:1000000004EF00F0FFFFFFFF946A156A166A010E05
:10001000176E000E186E000E006E1850005C02E1A4
:1000200017500A0807E317501526185016272174ACA
:10003000182AF1D715C082FFFFD7FFFFFFFFFFFF90
:020000040030CA
:0E000000FFF9FFFEFFFFFBFFFFFFFFFFFFFFFF0B
:00000001FF
```

Figure 5.42: EXAMPLE.HEX

5.3.3   Using the Simulator

The program developed in Section 5.3.2 is simulated following the steps given here, using the simulator in software (release mode). That is, no hardware is used in this simulation.

Example 5.2

Describe the steps for simulating the program developed in Example 5.1. Display the values of various variables and PORTC during the simulation while single-stepping the program. What is the final value displayed on PORTC?

Solution 5.2

The steps are as follows:

Step 1   Start the mikroC IDE, making sure the program developed in Example 5.1 is displayed in the Code Editor window.

Step 2   From the drop-down menu select *Debugger -> Select Debugger -> Software PIC Simulator*, as shown in Figure 5.43.

Step 3   From the drop-down menu select *Run -> Start Debugger*. The debugger form shown in Figure 5.44 will appear.

Step 4   Select the variables we want to see during the simulation. Assuming we want to display the values of variables *Sum*, *i*, and *PORTC*:

- Click on *Select from variable list* and then find and click on the variable name *Sum*

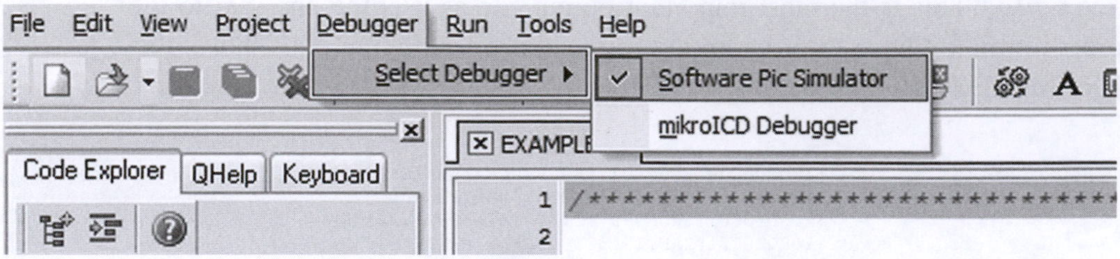

Figure 5.43: Selecting the debugger

- Click *Add* to add this variable to the Watch list

- Repeat these steps for variable *i* and *PORTC*

The debugger window should now look like Figure 5.45.

Step 5   We can now single-step the program and see the variables changing.

Press the F8 key on the keyboard. You should see a blue line to move down. This shows the line where the program is currently executing. Keep pressing F8 until you are inside the loop and you will see that variables *Sum* and *i* have become 1, as shown in Figure 5.46. Recently changed items appear in red. Double-clicking an item in the Watch window opens the Edit Value window, where you can change the value of a variable or register, or display the value in other bases such as decimal, hexadecimal, binary, or as a floating point or character.

Step 6   Keep pressing F8 until the program comes out of the *for* loop and executes the line that sends data to PORTC. A this point, as shown in Figure 5.47, $i = 11$ and $Sum = 55$.

Step 7   Press F8 again to send the value of variable *Sum* to PORTC. As shown in Figure 5.48, in this case PORTC will have the decimal value 55, which is the sum of numbers from 1 to 10.

This is the end of the simulation. Select from drop-down menu *Run -> Stop Debugger*.

In the above simulation example, we single-stepped through the program to the end and then we could see the final value of PORTC. The next example shows how to set breakpoints in the program and then execute up to a breakpoint.

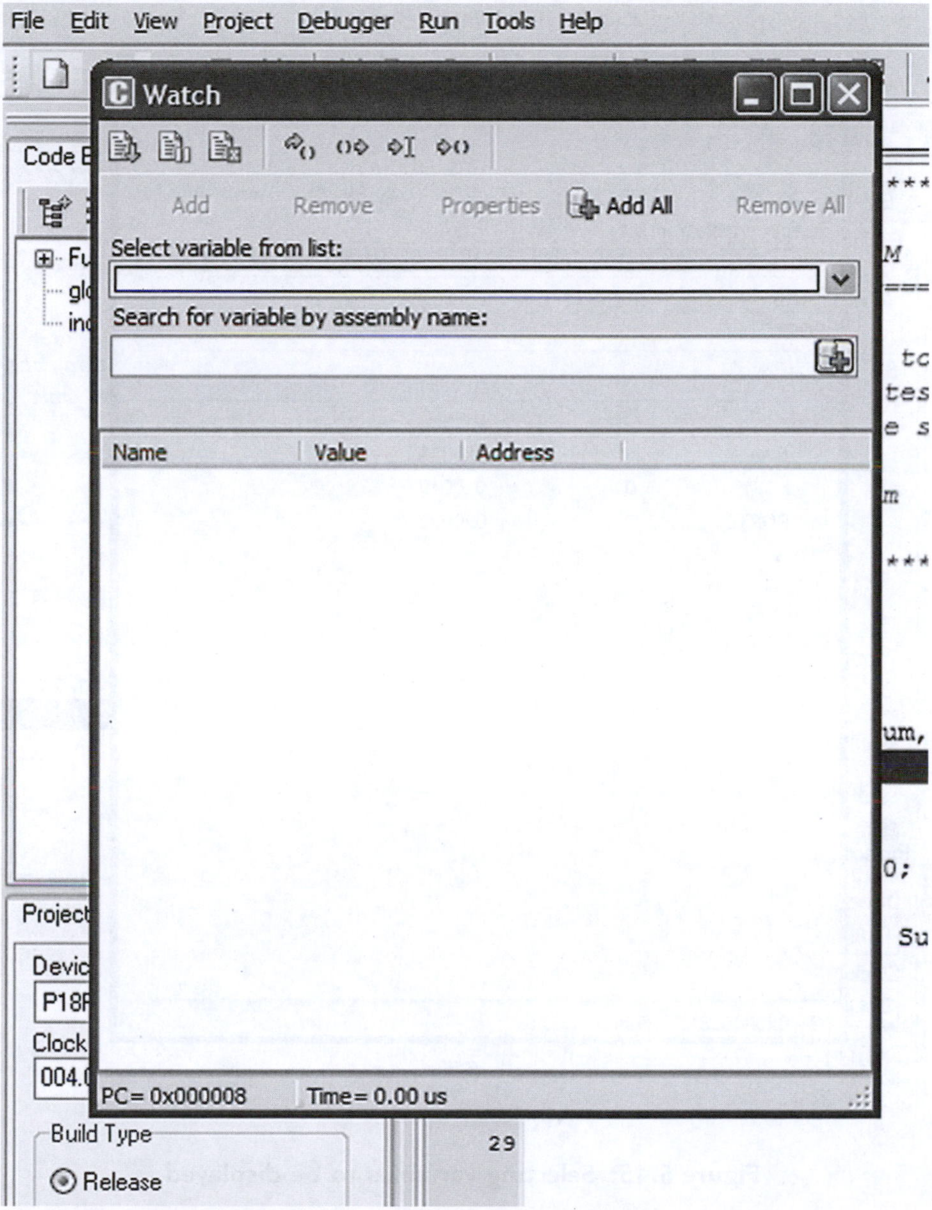

Figure 5.44: Starting the debugger

File   Edit   View   Project   Debugger   Run   Tools   Help

```
C Watch                                                    _ □ ✕

  📑  📑  📑     ⚙()  0⟡  ⟡I  ⟡0

  🔲 Add      Remove      Properties   🔲 Add All   🔲 Remove All

  Select variable from list:
  PORTC                                                    ▼

  Search for variable by assembly name:
  PORTC                                                    🔲


  Name          Value         Address
    Sum           0             0x0015
    i             0             0x0017
    PORTC         0             0x0F82














PC = 0x000008        Time = 0.00 us
```

Figure 5.45: Selecting variables to be displayed

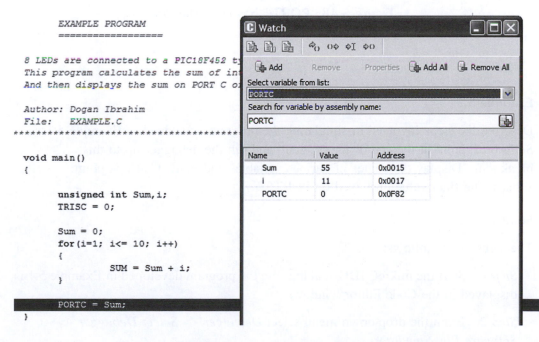

```
 2
 3              EXAMPLE  PROGRAM
 4              ===================
 5
 6     8 LEDs are connected to a PIC18F452 ty
 7     This program calculates the sum of in
 8     And then displays the sum on PORT C o
 9
10     Author: Dogan Ibrahim
11     File:    EXAMPLE.C
12   ************************************
13
14     void main()
15     {
16
17          unsigned int Sum,i;
18          TRISC = 0;
19
20          Sum = 0;
21          for(i=1; i<= 10; i++)
22          {
23                   SUM = Sum + i;
24          }
25
26          PORTC = Sum;
27     }
```

Figure 5.46: Single-stepping through the program

```
              EXAMPLE  PROGRAM
              ===================

     8 LEDs are connected to a PIC18F452 ty
     This program calculates the sum of in
     And then displays the sum on PORT C o

     Author: Dogan Ibrahim
     File:    EXAMPLE.C
   ************************************

     void main()
     {

          unsigned int Sum,i;
          TRISC = 0;

          Sum = 0;
          for(i=1; i<= 10; i++)
          {
                   SUM = Sum + i;
          }

          PORTC = Sum;
     }
```

Figure 5.47: Single-stepping through the program

```
        EXAMPLE PROGRAM
        ==================

8 LEDs are connected to a PIC18F452 t
This program calculates the sum of in
And then displays the sum on PORT C o

Author: Dogan Ibrahim
File:   EXAMPLE.C
****************************************

void main()
{

        unsigned int Sum,i;
        TRISC = 0;

        Sum = 0;
        for(i=1; i<= 10; i++)
        {
                SUM = Sum + i;
        }

        PORTC = Sum;
}
```

Watch

Add   Remove   Properties   Add All   Remove All

Select variable from list:
PORTC

Search for variable by assembly name:
PORTC

Name	Value	Address
Sum	55	0x0015
i	11	0x0017
PORTC	55	0x0F82

Figure 5.48: PORTC has the value 55

Example 5.3

Describe the steps involved in simulating the program developed in Example 5.1.
Set a breakpoint at the end of the program and run the debugger up to this
breakpoint. Display the values of various variables and PORTC at this point.
What is the final value displayed on PORTC?

Solution 5.3

The steps are as follows:

Step 1   Start the mikroC IDE, making sure the program developed in Example 5.1 is
displayed in the Code Editor window.

Step 2   From the drop-down menu select *Debugger -> Select Debugger ->
Software PIC Simulator*.

Step 3 From the drop-down menu select *Run -> Start Debugger*.

Step 4 Select variables *Sum, i,* and *PORTC* from the Watch window as described in Example 5.2.

Step 5 To set a breakpoint at the end of the program, click the mouse at the last closing bracket of the program, which is at line 27, and press F5. As shown in Figure 5.49, you should see a red line at the breakpoint and a little marker in the left column of the Code Editor window.

Step 6 Now, start the debugger, and press F6 key to run the program. The program will stop at the breakpoint, displaying variables as shown in Figure 5.48.

This is the end of the simulation. Select from drop-down menu *Run -> Stop Debugger*.

To clear a breakpoint, move the cursor over the line where the breakpoint is and then press F5. To clear all breakpoints in a program, press the SHIFT+CTRL+F5 keys. To display the breakpoints in a program, press the SHIFT+F4 keys.

The following are some other useful debugger commands:

Step Into [F7] Executes the current instruction and then halts. If the instruction is a call to a routine, the program enters the routine and halts at the first instruction.

```
13
14    void main()
15    {
16
17        unsigned int Sum,i;
18        TRISC = 0;
19
20        Sum = 0;
21        for(i=1; i<= 10; i++)
22        {
23            SUM = Sum + i;
24        }
25
26        PORTC = Sum;
27    }
28
29
```

Figure 5.49: Setting a breakpoint at line 27

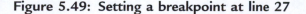

Step Over [F8]   Executes the current instruction and then halts. If the instruction is a call to a routine, it skips it and halts at the first instruction following the call.

Step Out [CTRL+F8]   Executes the current instruction and then halts. If the instruction is within a routine, it executes the instruction and halts at the first instruction following the call.

Run to Cursor [F4]   Executes all instructions between the current instruction and the cursor position.

Jump to Interrupt [F2]   Jumps to the interrupt service routine address (address 0x08 for PIC18 microcontrollers) and executes the procedure located at that address.

5.3.4   Using the mikroICD In-Circuit Debugger

This section discusses how to use the mikroICD in-circuit debugger (also called the PICFlash 2 programmer) to debug the program developed in Example 5.1. First of all, we have to build the hardware and then connect the in-circuit debugger device. In this example, the hardware is built on a breadboard, and a PICFlash 2 mikroICD in-circuit debugger is used to debug the system. Note that pins RB6 and RB7 are used by the mikroICD and are not available for I/O while mikroICD is active.

The Circuit Diagram

The project's circuit diagram is shown in Figure 5.50. The mikroICD in-circuit debugger is connected to the development circuit using the following pins of the microcontroller:

- MCLR
- RB6
- RB7
- +5V
- GND

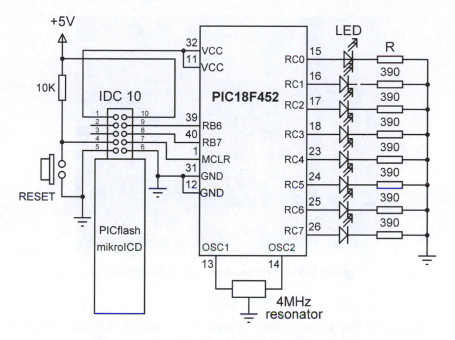

Figure 5.50: Circuit diagram of the project

The mikroICD has two modes of operation. In *inactive mode* all lines from the microcontroller used by the debugger device are connected to the development system. In *active mode* the MCLR, RB6, and RB7 pins are disconnected from the development system and used to program the microcontroller. After the programming, these lines are restored.

The mikroICD debugger device has a 10-way IDC connector and can be connected to the target system with a 10-way IDC header. Once the development is finished and the mikroICD debugger is removed, opposite pairs of the IDC header can be connected with jumpers. Figure 5.51 shows the system built on a breadboard.

Debugging

After building the hardware we are ready to program the microcontroller and test the system's operation with the in-circuit debugger. The steps are as follows:

Step 1   Start the mikroC IDE, making sure the program developed in Example 5.1 is displayed in the Code Editor window.

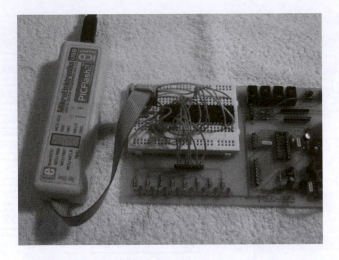

Figure 5.51: System built on a breadboard

Step 2   Click the Edit Project button (Figure 5.52) and set DEBUG_ON as shown in Figure 5.53.

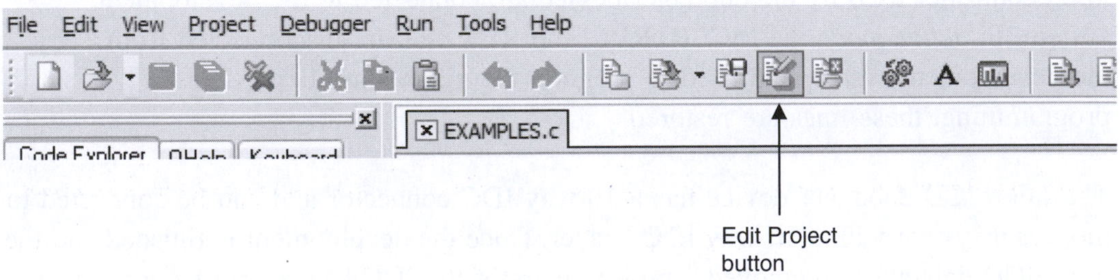

Figure 5.52: Edit Project button

Step 3   Select ICD Debug in the Project Setup window as shown in Figure 5.54.

Step 4   Click the Build Project icon to compile the program with the debugger. After a successful compilation you should see the message *Success (ICD Build)* in the Message Window.

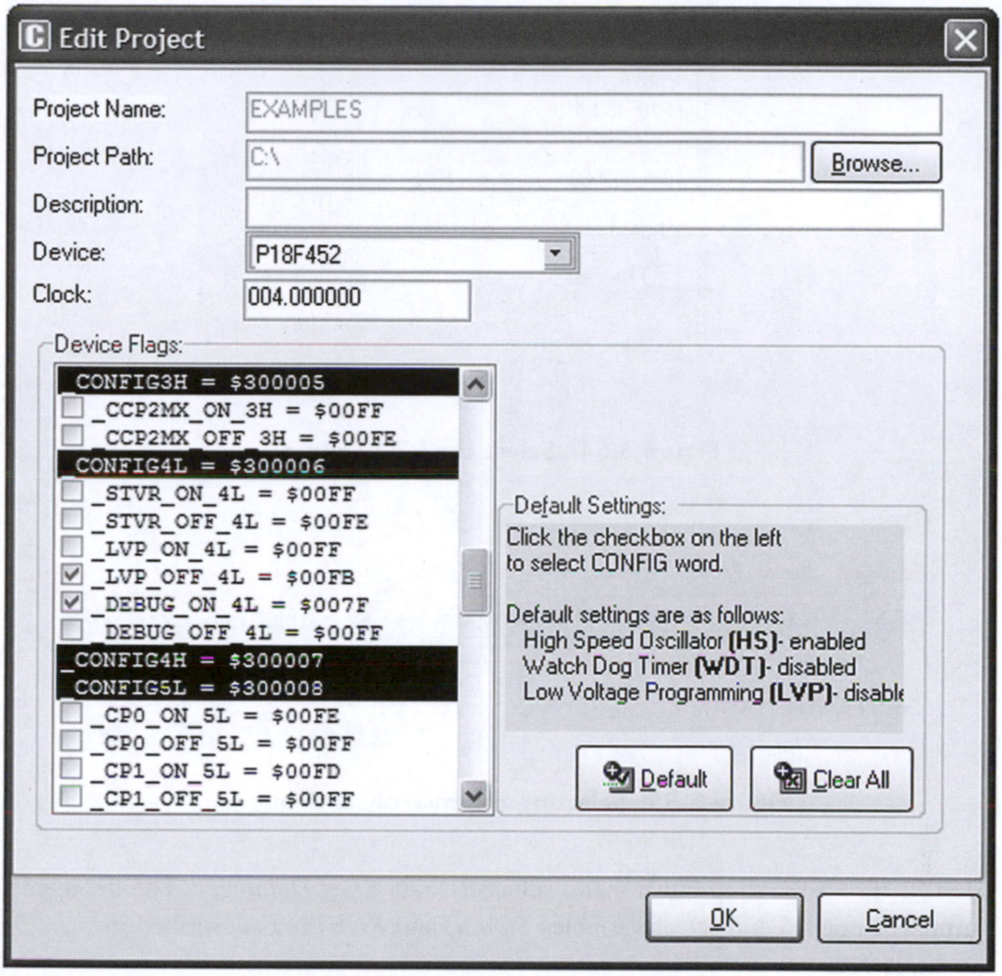

Figure 5.53: Set the DEBUG_ON

Step 5   Make sure the mikroICD debugger device is connected as in Figure 5.50, and select *Tools -> PicFlash Programmer* from the drop-down menu to program the microcontroller.

Step 6   From the drop-down menu select *Debugger -> Select Debugger -> mikroICD Debugger* as shown in Figure 5.55.

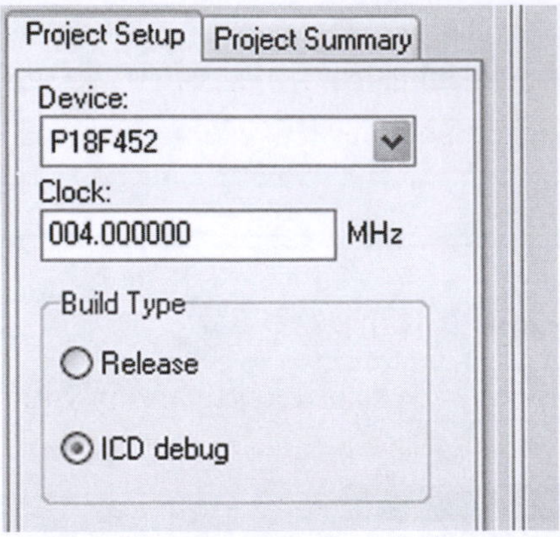

Figure 5.54: Select the ICD Debug

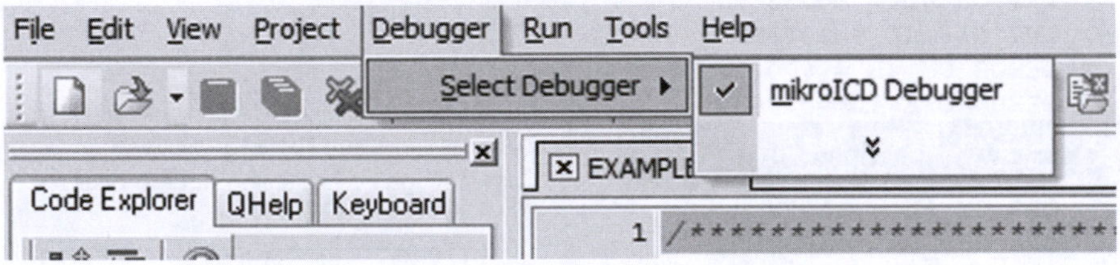

Figure 5.55: Selecting the mikroICD debugger

Step 7 From the drop-down menu select *Run -> Start Debugger*. The debugger form will pop up and select variables *Sum*, *i*, and *PORTC* as described in Example 5.2.

Step 8 Single-step through the program by pressing the F8 key. You should see the values of variables changing. At the end of the program, decimal value 55 will be sent to PORTC, and LEDs 0,1,2,4, and 5 should be turned ON, as shown in Figure 5.56, corresponding to this number.

Figure 5.56: Decimal number 55 shown in LEDs

Step 9   Stop the debugger.

In routines that contain delays, the *Step Into* [F7] and *Step Over* [F8] commands can take a long time. *Run to Cursor* [F4] and breakpoints should be used instead.

5.3.5   Using a Development Board

It is easy to develop microcontroller-based applications with the help of a development board. This section explains how to use the development board BIGPIC4, described earlier in this chapter. The program written in Example 5.1 is compiled and then loaded to the microcontroller using the on-board mikroICD in-circuit emulator. Then the program runs and displays the sum of the numbers 1 to 10 on the LEDs connected to PORTC.

However, before using the development board we need to know how the BIGPIC4 is organized and how to use the various devices on the board.

BIGPIC4 Development Board

Figure 5.57 shows the BIGPIC4 development board with the functions of various devices identified with arrows. The board can be powered either from an external power supply (8- to 16-C AC/DC) or from the USB port of a computer, using a jumper. In this application, the board is powered from a USB port.

A 2-row by 16-column LCD can be connected in the board's upper left corner. The contrast of the LCD can be adjusted with a small potentiometer.

The forty-six LEDs on the board can be connected to the output ports of the microcontroller, selected by switch S2. Figure 5.58 shows how to select the LEDs, using PORTC as an example. 1K resistors are used in series with the LEDs to limit the current. For example, to connect eight LEDs to PORTC we have to set the switch arm marked PORTC of switch S2 to the ON position.

The forty-six push-button switches on the board can be used to program digital inputs to the microcontroller. There is also a push-button switch that acts as the RESET. Jumper J12 determines whether a button press will bring logical 0 or logical 1 to the microcontroller. When the button is not pressed, the pin state is determined by jumper J5.

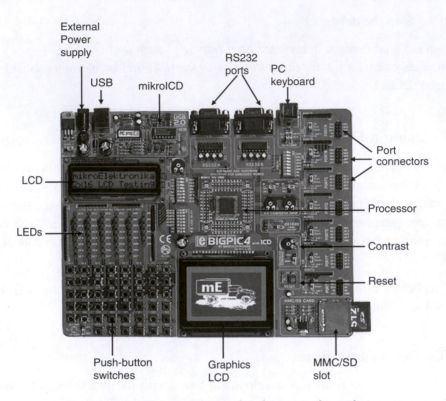

Figure 5.57: BIGPIC4 development board

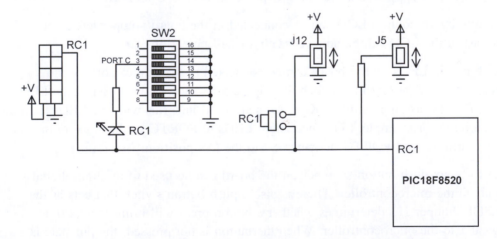

Figure 5.58: LED and push-button switch connections

At the bottom central position, a 128 × 64 pixel graphics LCD can be connected to the board. The contrast of the LCD can be adjusted by a small potentiometer.

The MMC/SD card slot at the bottom right-hand corner of the board supports cards up to 2GB storage capacity.

The RESET button is located just above the MMC/SD card slot.

Above the RESET button are two potentiometers for analog-to-digital converter applications.

All of the microcontroller port pins are available at the connectors situated along the right-hand side of the board. In the top middle portion of the board are two RS232 ports and a connection to a PC keyboard.

The board supports both 64-pin and 80-pin microcontrollers. The board comes with a PIC18F8520 microcontroller connected to the board, operating with a 10MHz crystal.

Further details about the operation of the board can be found in the BIGPIC4 user's manual.

The steps in developing an application using the BIGPIC4 board are as follows:

Step 1   Double-click the mikroC icon to start the IDE.

Step 2   Create a new project called EXAMPLE2 (see Figure 5.59) and select the microcontroller type as PIC18F8520, the clock as 10MHz, and device flags as:

- _OSC_HS_1H
- _WDT_OFF_2H
- _LVP_OFF_4L
- _DEBUG_ON_4L

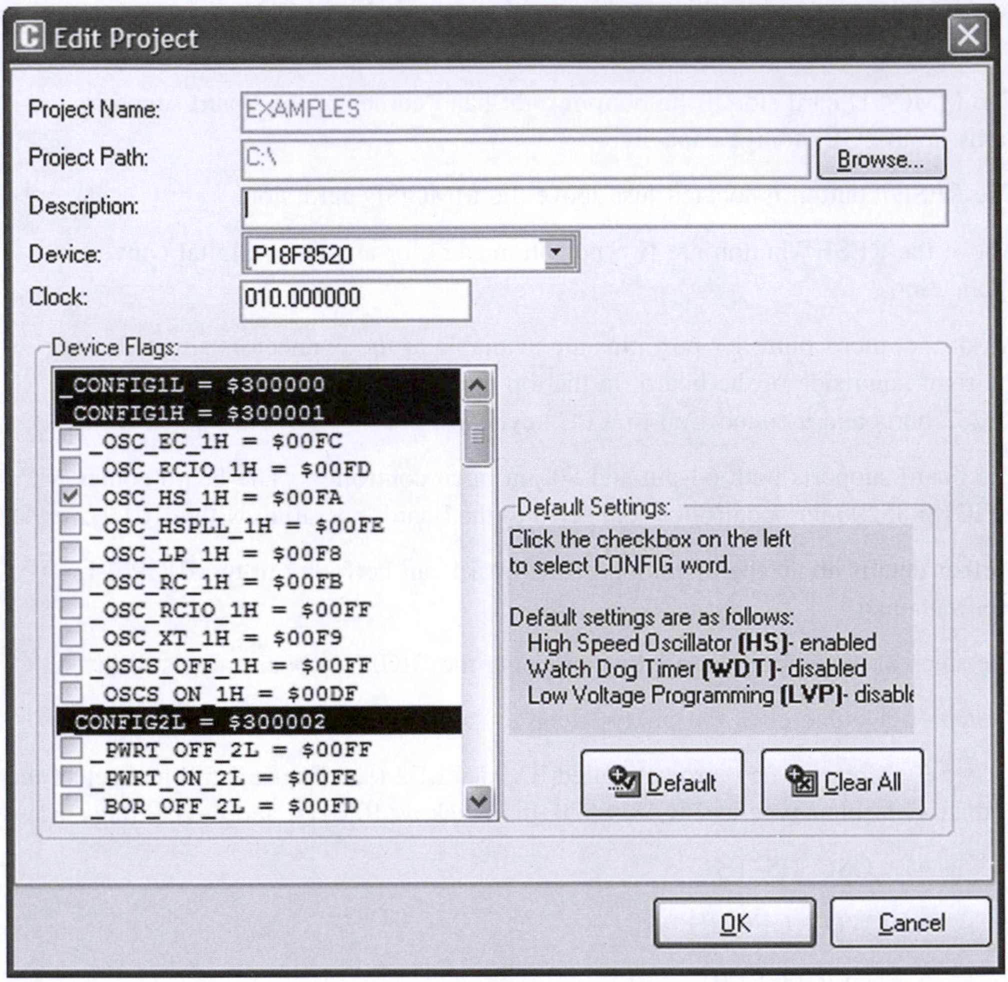

Figure 5.59: Creating a new project

Step 3   Enter the following program into the Code Editor section of the IDE:

```
/****************************************************************
                        EXAMPLE PROGRAM
```

This program uses the PICBIG4 Development Board. 8 LEDs are connected
To PORTC of the microcontroller which is a PIC18F8520 operating at 10MHz.
This program calculates the sum of integer numbers from 1 to 10
And then displays the sum on PORTC of the microcontroller.

```
Author:   Dogan Ibrahim
File:     EXAMPLE2.C
**************************************************************/
void main()
{
        unsigned int Sum,i;
        TRISC = 0;

        Sum = 0;
        for(i=1; i<= 10; i++)
        {
                Sum = Sum + i;
        }

        PORTC = Sum;
}
```

Step 4 Save the program with the name EXAMPLE2 by clicking *File -> Save As*.

Step 5 Tick option *ICD Debug* in the Project Setup window. Compile the project by pressing CTRL+F9 or by clicking the Build Project button.

Step 6 Connect the BIGPIC4 development board to the USB port on the computer. Configure the development board by routing eight LEDs to PORTC: Set the arm marked PORTC on switch S2 to the ON position.

Step 7 Select *Tools -> PicFlash Programmer* from the drop-down menu to program the microcontroller.

Step 8 Select *Debugger -> Select Debugger -> mikroICD Debugger*.

Step 9 Start the debugger by clicking *Run -> Start Debugger* and select variables *Sum*, *i*, and *PORTC* from the Watch window.

Step 10 Single-step through the program until the end by repeatedly pressing F8. At the end of the program, the PORTC LEDs will turn ON to display decimal 55 (i.e., LEDs 0,1,2,4, and 5 will turn ON).

Step 11 Stop the debugger.

View the EEPROM Window The mikroICD EEPROM window is invoked from the mikroC IDE drop-down menu when the mikroICD debug mode is selected and started, and it displays contents of the PIC internal EEPROM memory. To view the

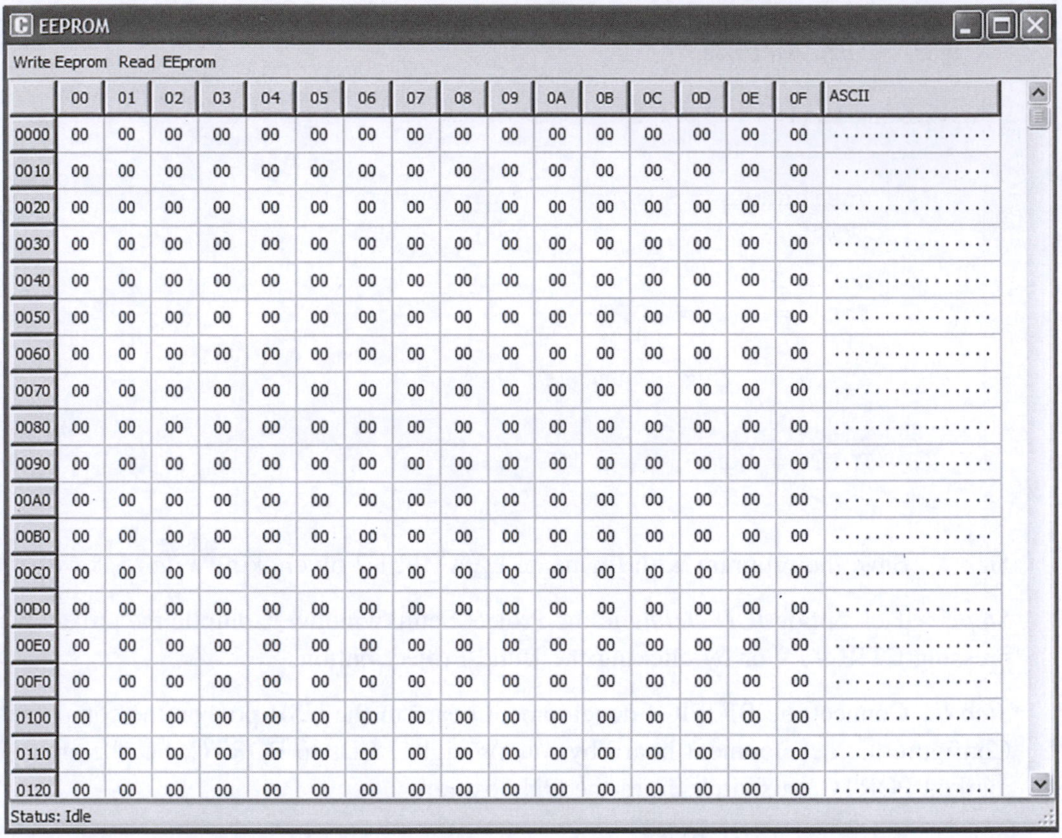

Figure 5.60: Display of EEPROM memory

memory, click *View -> Debug Windows -> View EEPROM*. Figure 5.60 shows an example EEPROM window display.

View the RAM Window   The mikroICD RAM window is invoked from the mikroC IDE drop-down menu when the mikroICD debug mode is selected and started, and it displays contents of the PIC internal RAM memory. To view the memory, click *View -> Debug Windows -> View RAM*. Figure 5.61 shows an example RAM window display.

View the Code Window   The mikroICD Code window is invoked from the mikroC IDE drop-down menu when the mikroICD debug mode is selected and started, and it displays the contents of the PIC internal code memory. To view

C RAM

RAM | History

	00	01	02	03	04	05	06	07	08	09	0A	0B	0C	0D	0E	0F	ASCII
0000	A0	16	24	10	10	4C	1C	11	80	00	B7	91	00	00	30	10	...
0010	81	1B	04	C0	00	51	20	82	50	C0	68	06	51	28	48	0A	...
0020	18	05	01	32	CC	C3	04	01	12	00	62	02	02	44	04	82	...
0030	02	18	88	A2	60	88	40	A2	00	5A	80	C0	41	C9	00	20	...
0040	32	3A	84	FE	C2	40	11	8F	15	0C	01	C2	08	06	08	06	...
0050	1A	01	40	04	00	82	00	CC	17	B4	10	C0	52	03	03	61	...
0060	46	A4	42	60	1B	66	28	05	4A	60	4C	50	02	02	13	37	...
0070	00	76	74	46	50	81	00	07	35	99	20	6A	20	03	99	E0	...
0080	C9	02	C8	02	2A	CD	02	12	0C	A1	A4	12	10	19	55	14	...
0090	88	D6	02	10	D0	02	40	53	9D	00	40	29	A0	08	40	08	...
00A0	36	29	01	09	02	0E	01	40	01	20	12	13	2B	0C	38	E2	...
00B0	60	51	74	04	62	30	34	32	04	C1	90	BD	82	C0	C8	A4	...
00C0	48	32	0C	29	0D	92	00	88	00	FA	CB	86	09	A1	63	90	...
00D0	20	09	81	81	D7	08	0A	10	00	43	44	84	55	02	01	0C	...
00E0	21	04	21	08	04	85	00	4D	99	04	88	11	04	10	00	0C	...
00F0	44	A3	46	45	00	21	28	92	0E	00	C7	30	18	0E	20	2B	...

Figure 5.61: Display of RAM memory

the memory, click *View -> Debug Windows -> View Code*. Figure 5.62 shows an example Code window display.

View the Statistics The Statistics window is invoked from the mikroC IDE drop-down menu and it displays various statistical data about our program. To view the statistics window, click *View -> View Statistics*. Figure 5.63 shows an example Statistics window, which consists of several tabs. The *Memory Usage* tab displays the amount of RAM (data memory) and ROM (code memory) used. The *Procedures* tabs display information about the size and locations of the procedures. The *RAM* and *ROM* tabs display memory usage in detail.

Figure 5.62: Display of Code memory

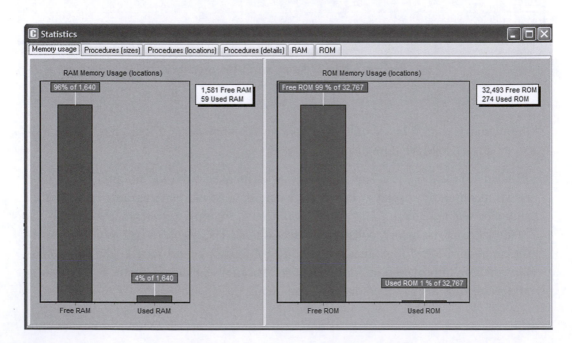

Figure 5.63: Display of Statistics window

5.4   Summary

This chapter has described the PIC microcontroller software development tools (such as text editors, assemblers, compilers, and simulators) and hardware development tools (including development boards and kits, programming devices, in-circuit debuggers, and in-circuit emulators). The mikroC compiler was used in the examples and projects. The steps in developing and testing a mikroC-based C program were presented both with and without a hardware in-circuit debugger, followed by an example of how to use the BIGPIC4 development board, with the on-board in-circuit debugger enabled.

5.5   Exercises

1. Describe the phases of the microcontroller-based system development cycle.

2. Describe briefly the microcontroller development tools.

3. Explain the advantages and disadvantages of assemblers and compilers.

4. Explain why a simulator can be a useful tool while developing a microcontroller-based product.

5. Explain in detail what a device programmer is. Give some examples of device programmers for the PIC18 series of microcontrollers.

6. Describe briefly the differences between in-circuit debuggers and in-circuit emulators. List the advantages and disadvantages of both debugging tools.

7. Enter the following program into the mikroC IDE and compile the program, correcting any syntax errors. Then, using the software ICD, simulate the operation of the program by single-stepping through the code, and observe the values of the variables during the simulation.

    ```
    /*=====================================
                    A SIMPLE LED PROJECT
    ```

    This program flashes the 8 LEDs connected to PORTC of a PIC18F452 microcontroller.

    ```
    =====================================*/

    void main()
    {

            TRISC = 0;      // PORTC is output
    ```

```
        do
        {
                PORTC = 0xFF;        // Turn ON LEDs on PORTC
                PORTC = 0;           // Turn OFF LEDs on PORTC
        } while(1);          // Endless loop
}
```

8. Describe the steps in using the mikroICD in-circuit debugger.

9. The following C program contains some deliberately introduced errors. Compile the program to find and correct the errors.

```
void main()
{
        unsigned char i,j,k
        i = 10;
        j = i + 1;

        for(i = 0; i < 10; i++)
        {
                Sum = Sum + i;
                j++
        }
        }
}
```

10. The following C program contains some deliberately introduced errors. Compile the program to find and correct the errors.

```
int add(int a, int b)
{
        result = a + b
}

void main()
{
        int p,q;
        p = 12;
        q = 10;
        z = add(p, q)
        z++;
        for(i = 0; i < z; i++)p++
}
}
```

Simple PIC18 Projects

In this chapter we will look at the design of simple PIC18 microcontroller-based projects, with the idea of becoming familiar with basic interfacing techniques and learning how to use the various microcontroller peripheral registers. We will look at the design of projects using LEDs, push-button switches, keyboards, LED arrays, sound devices, and so on, and we will develop programs in C language using the mikroC compiler. The hardware is designed on a low-cost breadboard, but development kits such as BIGPIC4 can be used for these projects. We will start with very simple projects and proceed to more complex ones. It is recommended that the reader moves through the projects in their given order. The following are provided for each project:

- Description of the program

- Description of the hardware

- Circuit diagram

- Algorithm description (in PDL)

- Program listing

- Suggestions for further development

The program's algorithm can be described in a variety of graphic and text-based methods, some of the common ones being a flow diagram, a structure chart, and program description language. In this book we are using program description language (PDL).

6.1 Program Description Language (PDL)

Program description language (PDL) is free-format English-like text which describes the flow of control in a program. PDL is not a programming language but rather is a tool which helps the programmer to think about the logic of the program *before* the program has been developed. Commonly used PDL keywords are described as follows.

6.1.1 START-END

Every PDL program description (or subprogram) should begin with a START keyword and terminate with an END keyword. The keywords in a PDL code should be highlighted in bold to make the code more clear. It is also a good practice to indent program statements between PDL keywords in order to enhance the readability of the code.

Example:

```
START
    ......
    ......
END
```

6.1.2 Sequencing

For normal sequencing in a program, write the statements as short English text as if you are describing the program.

Example:

```
Turn on the LED

Wait 1 second

Turn off the LED
```

6.1.3 IF-THEN-ELSE-ENDIF

Use IF, THEN, ELSE, and ENDIF keywords to describe the flow of control in a program.

Example:

```
IF switch = 1 THEN
     Turn on LED 1
ELSE
     Turn on LED 2
     Start the motor
ENDIF
```

6.1.4 DO-ENDDO

Use Do and ENDDO keywords to show iteration in the PDL code.

Example:

To create an unconditional loop in a program we can write:

```
Turn on LED
DO 10 times
     Set clock to 1
     Wait for 10ms
     Set clock to 0
ENDDO
```

A variation of the DO-ENDDO construct is to use other keywords like DO-FOREVER, DO-UNTIL, etc. as shown in the following examples.

Example:

To create a conditional loop in a program we can write:

```
Turn off buzzer
IF switch = 1 THEN
     DO UNTIL Port 1 = 1
          Turn on LED
          Wait for 10ms
          Read Port 1
     ENDDO
ENDIF
```

The following construct can be used when an endless loop is required:

```
DO FOREVER
     Read data from Port 1
     Send data to PORT 2
     Wait for 1 second
ENDDO
```

6.1.5 REPEAT-UNTIL

REPEAT-UNTIL is another control construct used in PDL codes. In the following example the program waits until a switch value is equal to 1.

Example:

```
REPEAT
     Turn on buzzer
     Read switch value
UNTIL switch = 1
```

Notice that the REPEAT-UNTIL loop is always executed at least once, and more than once if the condition at the end of the loop is not met.

PROJECT 6.1—Chasing LEDs

Project Description

In this project eight LEDs are connected to PORTC of a PIC18F452-type microcontroller, and the microcontroller is operated from a 4MHz resonator. When power is applied to the microcontroller (or when the microcontroller is reset), the LEDs turn ON alternately in an anticlockwise manner where only one LED is ON at any time. There is a one-second delay between outputs so the LEDs can be seen turning ON and OFF.

An LED can be connected to a microcontroller output port in two different modes: current sinking and current sourcing.

Current Sinking Mode

As shown in Figure 6.1, in current sinking mode the anode leg of the LED is connected to the +5V supply, and the cathode leg is connected to the microcontroller output port through a current limiting resistor.

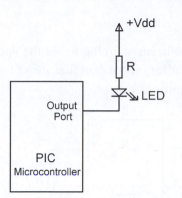

Figure 6.1: LED connected in current sinking mode

The voltage drop across an LED varies between 1.4V and 2.5V, with a typical value of 2V. The brightness of the LED depends on the current through the LED, and this current can vary between 8 and 16mA, with a typical value of 10mA.

The LED is turned ON when the output of the microcontroller is at logic 0 so the current flows through the LED. Assuming the microcontroller output voltage is about 0.4V when the output is low, we can calculate the value of the required resistor as follows:

$$R = \frac{V_S - V_{LED} - V_L}{I_{LED}} \qquad (6.1)$$

where

V_S is the supply voltage (5V)

V_{LED} is the voltage drop across the LED (2V)

V_L is the maximum output voltage when the output port is low (0.4V)

I_{LED} is the current through the LED (10mA)

Substituting the values into Equation (6.1) we get,

$$R = \frac{5 - 2 - 0.4}{10} = 260 \, \text{ohm}$$

The nearest physical resistor is 270 ohms.

Current Sourcing Mode

As shown in Figure 6.2, in current sourcing mode the anode leg of the LED is connected to the microcontroller output port and the cathode leg is connected to the ground through a current limiting resistor.

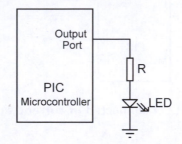

Figure 6.2: LED connected in current sourcing mode

In this mode the LED is turned ON when the microcontroller output port is at logic 1 (i.e., +5V). In practice, the output voltage is about 4.85V and the value of the resistor can be determined as:

$$R = \frac{V_O - V_{LED}}{I_{LED}} \qquad (6.2)$$

where

V_O is the output voltage of the microcontroller port when at logic 1 (+4.85V).

Thus, the value of the required resistor is:

$$R = \frac{4.85 - 2}{10} = 285\,ohm$$

The nearest physical resistor is 290 ohm.

Project Hardware

The circuit diagram of the project is shown in Figure 6.3. LEDs are connected to PORTC in current sourcing mode with eight 290-ohm resistors. A 4MHz resonator is connected between the OSC1 and OSC2 pins. Also, an external reset push button is connected to the MCLR input to reset the microcontroller when required.

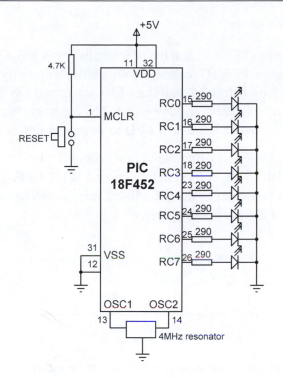

Figure 6.3: Circuit diagram of the project

Project PDL

The PDL of this project is very simple and is given in Figure 6.4.

```
START
        Configure PORTC pins as output
        Initialise J = 1
        DO FOREVER
                Set PORTC = J
                Shift left J by 1 digit
                IF J = 0 THEN
                        J = 1
                ENDIF
                Wait 1 second
        ENDDO
END
```

Figure 6.4: PDL of the project

Project Program

The program is named as LED1.C, and the program listing is given in Figure 6.5. At the beginning of the program PORTC pins are configured as outputs by setting TRISC = 0. Then an endless *for* loop is formed, and the LEDs are turned ON alternately in an anticlockwise manner to create a chasing effect. The program checks continuously so that when LED 7 is turned ON, the next LED to be turned ON is LED 0.

This program can be compiled using the mikroC compiler. Project settings should be configured to 4MHz clock, XT crystal mode, and WDT OFF. The HEX file (LED1.HEX) should be loaded to the PIC18F452 microcontroller using either an in-circuit debugger or a programming device.

```
/************************************************************************************************
                                  CHASING LEDS
                                  ============

In this project 8 LEDs are connected to PORTC of a PIC18F452 microcontroller
and the microcontroller is operated from a 4MHz resonator. The program turns on
the LEDs in an anti-clockwise manner with one second delay between each output.
The net result is that the LEDs seem to be chasing each other.


Author:    Dogan Ibrahim
Date:      July 2007
File:      LED1.C
*************************************************************************************************/

void main()
{
    unsigned char J = 1;

    TRISC = 0;
    for(;;)                         // Endless loop
    {
        PORTC = J;                  // Send J to PORTC
        Delay_ms(1000);             // Delay 1 second
        J = J << 1;                 // Shift left J
        if(J == 0) J = 1;           // If last LED, move to first LED
    }
}
```

Figure 6.5: Program listing

Further Development

The project can be modified such that the LEDs chase each other in both directions. For example, if the LEDs are moving in an anticlockwise direction, the direction can be changed so that when LED RB7 is ON the next LED to turn ON is RB6, when RB6 is ON the next is RB5, and so on.

PROJECT 6.2—LED Dice

Project Description

This is a simple dice project based on LEDs, a push-button switch, and a PIC18F452 microcontroller operating with a 4MHz resonator. The block diagram of the project is shown in Figure 6.6.

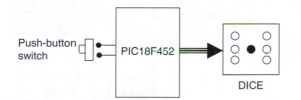

Figure 6.6: Block diagram of the project

As shown in Figure 6.7, the LEDs are organized such that when they turn ON, they indicate numbers as on a real dice. Operation of the project is as follows: The LEDs are all OFF to indicate that the system is ready to generate a new number. Pressing the switch generates a random number between 1 and 6 which is displayed on the LEDs for 3 seconds. After 3 seconds the LEDs turn OFF again.

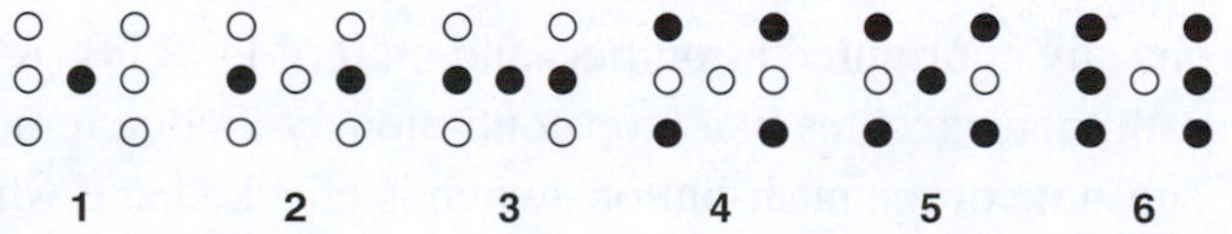

Figure 6.7: LED dice

Project Hardware

The circuit diagram of the project is shown in Figure 6.8. Seven LEDs representing the faces of a dice are connected to PORTC of a PIC18F452 microcontroller in current sourcing mode using 290-ohm current limiting resistors. A push-button switch is connected to bit 0 of PORTB (RB0) using a pull-up resistor. The microcontroller is operated from a 4MHz resonator connected between pins OSC1 and OSC2. The microcontroller is powered from a +9V battery, and a 78L05-type voltage regulator IC is used to obtain the +5V supply required for the microcontroller.

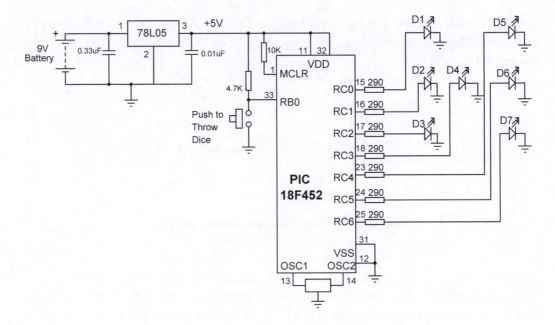

Figure 6.8: Circuit diagram of the project

Project PDL

The operation of the project is described in PDL in Figure 6.9. At the beginning of the program PORTC pins are configured as outputs and bit 0 of PORTB (RB0) is configured as input. The program then executes in a loop continuously and increments a variable between 1 and 6. The state of the push-button switch is checked and when the switch is pressed (switch output at logic 0), the current number is sent to the LEDs. A simple array is used to find out the LEDs to be turned ON corresponding to the dice number.

```
START
        Create DICE table
        Configure PORTC as outputs
        Configure RB0 as input
        Set J = 1
        DO FOREVER
                IF button pressed THEN
                        Get LED pattern from DICE table
                        Turn ON required LEDs
                        Wait 3 seconds
                        Set J = 0
                        Turn OFF all LEDs
                ENDIF
                Increment J
                IF J = 7 THEN
                        Set J = 1
                ENDIF
        ENDDO
END
```

Figure 6.9: PDL of the project

Table 6.1 gives the relationship between a dice number and the corresponding LEDs to be turned ON to imitate the faces of a real dice. For example, to display number 1 (i.e., only the middle LED is ON), we have to turn on D4. Similarly, to display number 4, the LEDs to turn ON are D1, D3, D5, and D7.

Table 6.1: Dice number and LEDs to be turned ON

Required number	LEDs to be turned on
1	D4
2	D2, D6
3	D2, D4, D6
4	D1, D3, D5, D7
5	D1, D3, D4, D5, D7
6	D1, D2, D3, D5, D6, D7

The relationship between the required number and the data to be sent to PORTC to turn on the correct LEDs is given in Table 6.2. For example, to display dice number 2, we have to send hexadecimal 0×22 to PORTC. Similarly, to display number 5, we have to send hexadecimal $0 \times 5D$ to PORTC, and so on.

Table 6.2: Required number and PORTC data

Required number	PORTB data (Hex)
1	0×08
2	0×22
3	0×2A
4	0×55
5	0×5D
6	0×77

Project Program

The program is called LED2.C, and the program listing is given in Figure 6.10. At the beginning of the program *Switch* is defined as bit 0 of PORTB, and *Pressed* is defined as 0. The relationships between the dice numbers and the LEDs to be turned on are stored in an array called *DICE*. Variable J is used as the dice number. Variable *Pattern* is the data sent to the LEDs. Program then enters an endless *for* loop where the value of variable J is incremented very fast between 1 and 6. When the push-button switch is pressed, the LED pattern corresponding to the current value of J is read from the array and sent to the LEDs. The LEDs remain in this state for 3 seconds (using function *Delay_ms* with the argument set to 3000ms), after which they all turn OFF. The system is then ready to generate a new dice number.

Using a Pseudorandom Number Generator

In the preceding project the value of variable J changes very fast among the numbers between 1 and 6, so we can say that the numbers generated are random (i.e., new numbers do not depend on the previous numbers).

A pseudorandom number generator function can also be used to generate the dice numbers. The modified program listing is shown in Figure 6.11. In this program a function called *Number* generates the dice numbers. The function receives the upper limit of the numbers to be generated (6 in this example) and also a seed value which

```
/*********************************************************************************************
                              SIMPLE DICE
                              ===========

In this project 7 LEDs are connected to PORTC of a PIC18F452 microcontroller
and the microcontroller is operated from a 4MHz resonator. The LEDs are organized
as the faces of a real dice. When a push-button switch connected to RB0 is pressed a
dice pattern is displayed on the LEDs. The display remains in this state for 3 seconds
and after this period the LEDs all turn OFF to indicate that the system is ready for the
button to be pressed again.


Author:    Dogan Ibrahim
Date:      July 2007
File:      LED2.C
*********************************************************************************************/

#define Switch PORTB.F0
#define Pressed 0

void main()
{
    unsigned char J = 1;
    unsigned char Pattern;
    unsigned char DICE[] = {0,0x08,0x22,0x2A,0x55,0x5D,0x77};

    TRISC = 0;                      // PORTC outputs
    TRISB = 1;                      // RB0 input
    PORTC = 0;                      // Turn OFF all LEDs

    for(;;)                         // Endless loop
    {
        if(Switch == Pressed)       // Is switch pressed ?
        {
          Pattern = DICE[J];        // Get LED pattern
          PORTC = Pattern;          // Turn on LEDs
          Delay_ms(3000);           // Delay 3 second
          PORTC = 0;                // Turn OFF all LEDs
          J = 0;                    // Initialise J
        }
        J++;                        // Increment J
        if(J == 7) J = 1;           // Back to 1 if > 6
    }
}
```

Figure 6.10: Program listing

```
/***********************************************************************************
                            SIMPLE DICE
                            ===========

In this project 7 LEDs are connected to PORTC of a PIC18F452 microcontroller
and the microcontroller is operated from a 4MHz resonator. The LEDs are organized
as the faces of a real dice. When a push-button switch connected to RB0 is pressed a
dice pattern is displayed on the LEDs. The display remains in this state for 3 seconds
and after this period the LEDs all turn OFF to indicate that the system is ready for the
button to be pressed again.

In this program a pseudorandom number generator function is
used to generate the dice numbers between 1 and 6.


Author:    Dogan Ibrahim
Date:      July 2007
File:      LED3.C
***********************************************************************************/


#define Switch PORTB.F0
#define Pressed 0

//
// This function generates a pseudo random integer number
// between 1 and Lim
//
unsigned char Number(int Lim, int Y)
{
    unsigned char Result;
    static unsigned int Y;

    Y = (Y * 32719 + 3) % 32749;
    Result = ((Y % Lim) + 1);
    return Result;
}


//
// Start of MAIN program
//
void main()
{
    unsigned char J,Pattern,Seed = 1;
    unsigned char DICE[] = {0,0x08,0x22,0x2A,0x55,0x5D,0x77};

    TRISC = 0;                          // PORTC outputs
    TRISB = 1;                          // RB0 input
    PORTC = 0;                          // Turn OFF all LEDs

    for(;;)                             // Endless loop
```

Figure 6.11: Dice program using a pseudorandom number generator

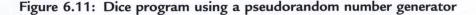

```
    {
        if(Switch == Pressed)                // Is switch pressed ?
        {
          J = Number(6,seed);                // Generate a number between 1 and 6
          Pattern = DICE[J];                 // Get LED pattern
          PORTC = Pattern;                   // Turn on LEDs
          Delay_ms(3000);                    // Delay 3 second
          PORTC = 0;                         // Turn OFF all LEDs
        }
    }
}
```

Figure 6.11: (Cont'd)

defines the number set to be generated. In this example, the seed is set to 1. Every time the function is called, a number between 1 and 6 is generated.

The operation of the program is basically same as in Figure 6.10. When the push-button switch is pressed, function *Number* is called to generate a new dice number between 1 and 6, and this number is used as an index in array *DICE* in order to find the bit pattern to be sent to the LEDs.

PROJECT 6.3—Two-Dice Project

Project Description

This project is similar to Project 2, but here a pair of dice are used—as in many dice games such as backgammon—instead of a single dice.

The circuit shown in Figure 6.8 can be modified by adding another set of seven LEDs for the second dice. For example, the first set of LEDs can be driven from PORTC, the second set from PORTD, and the push-button switch can be connected to RB0 as before. Such a design requires fourteen output ports just for the LEDs. Later on we will see how the LEDs can be combined in order to reduce the input/output requirements. Figure 6.12 shows the block diagram of the project.

Project Hardware

The circuit diagram of the project is shown in Figure 6.13. The circuit is basically same as in Figure 6.8, with the addition of another set of LEDs connected to PORTD.

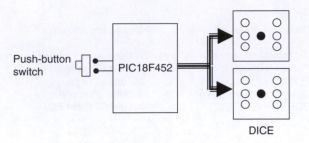

Figure 6.12: Block diagram of the project

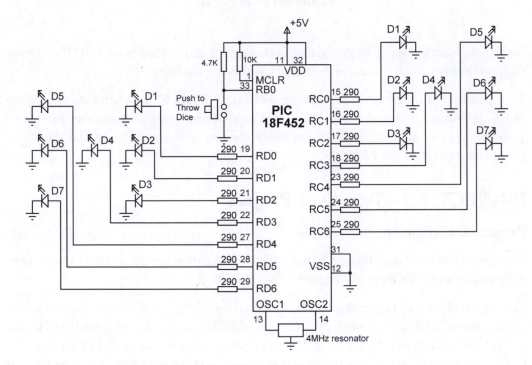

Figure 6.13: Circuit diagram of the project

Project PDL

The operation of the project is very similar to that for Project 2. Figure 6.14 shows the PDL for this project. At the beginning of the program the PORTC and PORTD pins are configured as outputs, and bit 0 of PORTB (RB0) is configured as input. The program then executes in a loop continuously and checks the state of the push-button switch. When the switch is pressed, two pseudorandom numbers between 1 and 6 are

```
START
        Create DICE table
        Configure PORTC as outputs
        Configure PORTD as outputs
        Configure RB0 as input
        DO FOREVER
                IF button pressed THEN
                        Get a random number between 1 and 6
                        Find bit pattern
                        Turn ON LEDs on PORTC
                        Get second random number between 1 and 6
                        Find bit pattern
                        Turn on LEDs on PORTD
                        Wait 3 seconds
                        Turn OFF all LEDs
                ENDIF
        ENDDO
END
```

Figure 6.14: PDL of the project

generated, and these numbers are sent to PORTC and PORTD. The LEDs remain at this state for 3 seconds, after which all the LEDs are turned OFF to indicate that the push-button switch can be pressed again for the next pair of numbers.

Project Program

The program is called LED4.C, and the program listing is given in Figure 6.15. At the beginning of the program *Switch* is defined as bit 0 of PORTB, and *Pressed* is defined as 0. The relationships between the dice numbers and the LEDs to be turned on are stored in an array called *DICE*, as in Project 2. Variable *Pattern* is the data sent to the LEDs. Program enters an endless *for* loop where the state of the push-button switch is checked continuously. When the switch is pressed, two random numbers are generated by calling function *Number*. The bit patterns to be sent to the LEDs are then determined and sent to PORTC and PORTD. The program then repeats inside the endless loop, checking the state of the push-button switch.

PROJECT 6.4—Two-Dice Project Using Fewer I/O Pins

Project Description

This project is similar to Project 3, but here LEDs are shared, which uses fewer input/output pins.

```
/*************************************************************************
                              TWO DICE
                              ========

In this project 7 LEDs are connected to PORTC of a PIC18F452 microcontroller and
7 LEDs to PORTD. The microcontroller is operated from a 4MHz resonator.
The LEDs are organized as the faces of a real dice. When a push-button switch
connected to RB0 is pressed a dice pattern is displayed on the LEDs. The display
remains in this state for 3 seconds and after this period the LEDs all turn OFF to
indicate that the system is ready for the button to be pressed again.

In this program a pseudorandom number generator function is
used to generate the dice numbers between 1 and 6.


Author:    Dogan Ibrahim
Date:      July 2007
File:      LED4.C

*************************************************************************/

#define Switch PORTB.F0
#define Pressed 0


//
// This function generates a pseudo random integer number
// between 1 and Lim
//
unsigned char Number(int Lim, int Y)
{
    unsigned char Result;
    static unsigned int Y;

    Y = (Y * 32719 + 3) % 32749;
    Result = ((Y % Lim) + 1);
    return Result;
}


//
// Start of MAIN program
//
void main()
{
    unsigned char J,Pattern,Seed = 1;
    unsigned char DICE[] = {0,0x08,0x22,0x2A,0x55,0x5D,0x77};

    TRISC = 0;                       // PORTC are outputs
    TRISD = 0;                       // PORTD are outputs
    TRISB = 1;                       // RB0 input
    PORTC = 0;                       // Turn OFF all LEDs
    PORTD = 0;                       // Turn OFF all LEDs
```

Figure 6.15: Program listing

```
for(;;)                          // Endless loop
{
    if(Switch == Pressed)        // Is switch pressed ?
    {
        J = Number(6,seed);      // Generate first dice number
        Pattern = DICE[J];       // Get LED pattern
        PORTC = Pattern;         // Turn on LEDs for first dice
        J = Number(6,seed);      // Generate second dice number
        Pattern = DICE[J];       // Get LED pattern
        PORTD = Pattern;         // Turn on LEDs for second dice
        Delay_ms(3000);          // Delay 3 seconds
        PORTC = 0;               // Turn OFF all LEDs
        PORTD = 0;               // Turn OFF all LEDS
    }
}
}
```

Figure 6.15: (Cont'd)

The LEDs in Table 6.1 can be grouped as shown in Table 6.3. Looking at this table we can say that:

- D4 can appear on its own

- D2 and D6 are always together

- D1 and D3 are always together

- D5 and D7 are always together

Thus, we can drive D4 on its own and then drive the D2,D6 pair together in series, the D1,D3 pair together in series, and also the D5,D7 pair together in series. (Actually, we

Table 6.3: Grouping the LEDs

Required number	LEDs to be turned on
1	D4
2	D2 D6
3	D2 D6 D4
4	D1 D3 D5 D7
5	D1 D3 D5 D7 D4
6	D2 D6 D1 D3 D5 D7

could share D1,D3,D5,D7 but this would require 8 volts to drive if the LEDs are connected in series. Connecting them in parallel would call for even more current, and a driver IC would be required.) Altogether, four lines are needed to drive the seven LEDs of each dice. Thus, a pair of dice can easily be driven from an 8-bit output port.

Project Hardware

The circuit diagram of the project is shown in Figure 6.16. PORTC of a PIC18F452 microcontroller is used to drive the LEDs as follows:

- RC0 drives D2,D6 of the first dice

- RC1 drives D1,D3 of the first dice

- RC2 drives D5,D7 of the first dice

- RC3 drives D4 of the first dice

- RC4 drives D2,D6 of the second dice

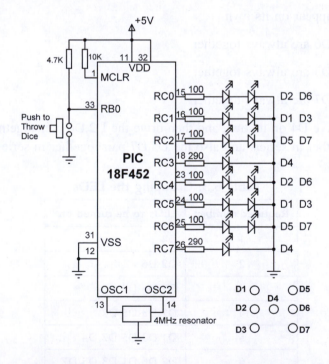

Figure 6.16: Circuit diagram of the project

- RC5 drives D1,D3 of the second dice

- RC6 drives D5,D7 of the second dice

- RC7 drives D4 of the second dice

Since two LEDs are being driven on some outputs, we can calculate the required value of the current limiting resistors. Assuming that the voltage drop across each LED is 2V, the current through the LED is 10mA, and the output high voltage of the microcontroller is 4.85V, the required resistors are:

$$R = \frac{4.85 - 2 - 2}{10} = 85\,ohms$$

We will choose 100-ohm resistors.

We now need to find the relationship between the dice numbers and the bit pattern to be sent to the LEDs for each dice. Table 6.4 shows the relationship between the first dice numbers and the bit pattern to be sent to port pins RC0-RC3. Similarly, Table 6.5 shows the relationship between the second dice numbers and the bit pattern to be sent to port pins RC4-RC7.

Table 6.4: First dice bit patterns

Dice number	RC3 RC2 RC1 RC0	Hex value
1	1 0 0 0	8
2	0 0 0 1	1
3	1 0 0 1	9
4	0 1 1 0	6
5	1 1 1 0	E
6	0 1 1 1	7

We can now find the 8-bit number to be sent to PORTC to display both dice numbers as follows:

- Get the first number from the number generator, call this P

- Index the DICE table to find the bit pattern for low nibble (i.e., L = DICE[P])

- Get the second number from the number generator, call this P

Table 6.5: Second dice bit patterns

Dice number	RC7 RC6 RC5 RC4	Hex value
1	1 0 0 0	8
2	0 0 0 1	1
3	1 0 0 1	9
4	0 1 1 0	6
5	1 1 1 0	E
6	0 1 1 1	7

- Index the DICE table to find the bit pattern for high nibble (i.e., U = DICE[P])

- Multiply high nibble by 16 and add low nibble to find the number to be sent to PORTC (i.e., R = 16*U + L), where R is the 8-bit number to be sent to PORTC to display both dice values.

Project PDL

The operation of this project is very similar to that of Project 2. Figure 6.17 shows the PDL of the project. At the beginning of the program the PORTC pins are

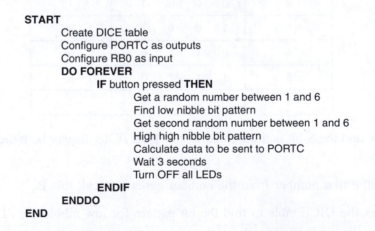

```
START
        Create DICE table
        Configure PORTC as outputs
        Configure RB0 as input
        DO FOREVER
            IF button pressed THEN
                    Get a random number between 1 and 6
                    Find low nibble bit pattern
                    Get second random number between 1 and 6
                    High high nibble bit pattern
                    Calculate data to be sent to PORTC
                    Wait 3 seconds
                    Turn OFF all LEDs
            ENDIF
        ENDDO
END
```

Figure 6.17: PDL of the project

configured as outputs, and bit 0 of PORTB (RB0) is configured as input. The program then executes in a loop continuously and checks the state of the push-button switch. When the switch is pressed, two pseudorandom numbers between 1 and 6 are generated, and the bit pattern to be sent to PORTC is found by the method just described. This bit pattern is then sent to PORTC to display both dice numbers at the same time. The display shows the dice numbers for 3 seconds, and then all the LEDs turn OFF to indicate that the system is waiting for the push-button to be pressed again to display the next set of numbers.

Project Program

The program is called LED5.C, and the program listing is given in Figure 6.18. At the beginning of the program *Switch* is defined as bit 0 of PORTB, and *Pressed* is defined as 0. The relationships between the dice numbers and the LEDs to be turned on are stored in an array called *DICE* as in Project 2. Variable *Pattern* is the data sent to the LEDs. The program enters an endless *for* loop where the state of the push-button switch is checked continuously. When the switch is pressed, two random numbers are generated by calling function *Number*. Variables L and U store the lower and higher nibbles of the bit pattern to be sent to PORTC. The bit pattern to be sent to PORTC is then determined using the method described in the Project Hardware section and stored in variable R. This bit pattern is then sent to PORTC to display both dice numbers at the same time. The dice numbers are displayed for 3 seconds, after which the LEDs are turned OFF to indicate that the system is ready.

Modifying the Program

The program given in Figure 6.18 can made more efficient by combining the two dice nibbles into a single table value as described here.

There are thirty-six possible combinations of the two dice values. Referring to Table 6.4, Table 6.5, and Figure 6.16, we can create Table 6.6 to show all the possible two-dice values and the corresponding numbers to be sent to PORTC.

The modified program (program name LED6.C) is given in Figure 6.19. In this program array *DICE* contains the thirty-six possible dice values. The program enters an endless

```
/*************************************************************************
                    TWO DICE - USING FEWER I/O PINS
                    ===============================
In this project LEDs are connected to PORTC of a PIC18F452 microcontroller
and the microcontroller is operated from a 4MHz resonator. The LEDs are
organized as the faces of a real dice. When a push-button switch connected to
RB0 is pressed a dice pattern is displayed on the LEDs. The display remains
in this state for 3 seconds and after this period the LEDs all turn OFF to indicate
that the system is ready for the button to be pressed again.

In this program a pseudorandom number generator function is
used to generate the dice numbers between 1 and 6.

Author:    Dogan Ibrahim
Date:      July 2007
File:      LED5.C
*************************************************************************/

#define Switch PORTB.F0
#define Pressed 0

//
// This function generates a pseudo random integer number
// between 1 and Lim
//
unsigned char Number(int Lim, int Y)
{
          unsigned char Result;
          static unsigned int Y;

          Y = (Y * 32719 + 3) % 32749;
          Result = ((Y % Lim) + 1);
          return Result;
}



//
// Start of MAIN program
//
void main()
{
    unsigned char J,L,U,R,Seed = 1;
    unsigned char DICE[] = {0,0x08,0x01,0x09,0x06,0x0E,0x07};

    TRISC = 0;                         // PORTC are outputs
    TRISB = 1;                         // RB0 input
    PORTC = 0;                         // Turn OFF all LEDs

    for(;;)                            // Endless loop
```

Figure 6.18: Program listing

```
{
    if(Switch == Pressed)              // Is switch pressed ?
    {
        J = Number(6,seed);            // Generate first dice number
        L = DICE[J];                   // Get LED pattern
        J = Number(6,seed);            // Generate second dice number
        U = DICE[J];                   // Get LED pattern
        R = 16*U + L;                  // Bit pattern to send to PORTC
        PORTC = R;                     // Turn on LEDs for both dice
        Delay_ms(3000);                // Delay 3 seconds
        PORTC = 0;                     // Turn OFF all LEDs
    }
  }
}
```

Figure 6.18: (Cont'd)

Table 6.6: Two-dice combinations and the number to be sent to PORTC

Dice numbers	PORTC value	Dice numbers	PORTC value
1,1	0×88	4,1	0×86
1,2	0×18	4,2	0×16
1,3	0×98	4,3	0×96
1,4	0×68	4,4	0×66
1,5	0×E8	4,5	0×E6
1,6	0×78	4,6	0×76
2,1	0×81	5,1	0×8E
2,2	0×11	5,2	0×1E
2,3	0×91	5,3	0×9E
2,4	0×61	5,4	0×6E
2,5	0×E1	5,5	0×EE
2,6	0×71	5,6	0×7E
3,1	0×89	6,1	0×87
3,2	0×19	6,2	0×17
3,3	0×99	6,3	0×97
3,4	0×69	6,4	0×67
3,5	0×E9	6,5	0×E7
3,6	0×79	6,6	0×77

```
/*********************************************************************************
                    TWO DICE - USING FEWER I/O PINS
                    ================================

In this project LEDs are connected to PORTC of a PIC18F452 microcontroller
and the microcontroller is operated from a 4MHz resonator. The LEDs are
organized as the faces of a real dice. When a push-button switch connected to
RB0 is pressed a dice pattern is displayed on the LEDs. The display remains in
this state for 3 seconds and after this period the LEDs all turn OFF to indicate
that the system is ready for the button to be pressed again.

In this program a pseudorandom number generator function is
used to generate the dice numbers between 1 and 6.

Author:    Dogan Ibrahim
Date:      July 2007
File:      LED6.C
*********************************************************************************/

#define Switch PORTB.F0
#define Pressed 0

//
// Start of MAIN program
//
void main()
{
    unsigned char Pattern, J = 1;
    unsigned char DICE[] = {0,0x88,0x18,0x98,0x68,0xE8,0x78,
                0x81,0x11,0x91,0x61,0xE1,0x71,
                0x89,0x19,0x99,0x69,0xE9,0x79,
                0x86,0x16,0x96,0x66,0xE6,0x76,
                0x8E,0x1E,0x9E,0x6E,0xEE,0x7E,
                0x87,0x17,0x97,0x67,0xE7,0x77};

    TRISC = 0;                              // PORTC are outputs
    TRISB = 1;                              // RB0 input
    PORTC = 0;                              // Turn OFF all LEDs

    for(;;)                                 // Endless loop
    {
        if(Switch == Pressed)               // Is switch pressed ?
        {
            Pattern = DICE[J];              // Number to send to PORTC
            PORTC = Pattern;                // send to PORTC
            Delay_ms(3000);                 // 3 seconds delay
            PORTC = 0;                      // Clear PORTC
        }
        J++;                                // Increment J
        if(J == 37) J = 1;                  // If J = 37, reset to 1
    }
}
```

Figure 6.19: Modified program

for loop, and inside this loop the state of the push-button switch is checked. Also, a variable is incremented from 1 to 36. When the button is pressed, the value of this variable is used as an index to array *DICE* to determine the bit pattern to be sent to PORTC. As before, the program displays the dice numbers for 3 seconds and then turns OFF all LEDs to indicate that it is ready.

PROJECT 6.5—7-Segment LED Counter

Project Description

This project describes the design of a 7-segment LED-based counter which counts from 0 to 9 continuously with a one-second delay between counts. The project shows how a 7-segment LED can be interfaced and used in a PIC microcontroller project.

7-segment displays are used frequently in electronic circuits to show numeric or alphanumeric values. As shown in Figure 6.20, a 7-segment display consists basically of 7 LEDs connected such that the numbers from 0 to 9 and some letters can be displayed. Segments are identified by the letters from a to g, and Figure 6.21 shows the segment names of a typical 7-segment display.

Figure 6.20: Some 7-segment displays

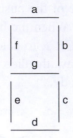

Figure 6.21: Segment names of a 7-segment display

Figure 6.22 shows how the numbers from 0 to 9 are obtained by turning ON different segments of the display.

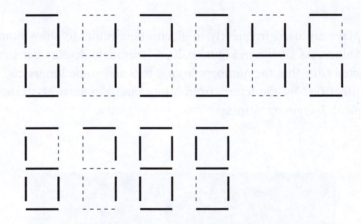

Figure 6.22: Displaying numbers 0 to 9

7-segment displays are available in two different configurations: common cathode and common anode. As shown in Figure 6.23, in common cathode configuration, all the cathodes of all segment LEDs are connected together to the ground. The segments are turned ON by applying a logic 1 to the required segment LED via current limiting resistors. In common cathode configuration the 7-segment LED is connected to the microcontroller in current sourcing mode.

In common anode configuration, the anode terminals of all the LEDs are connected together as shown in Figure 6.24. This common point is then normally connected to the

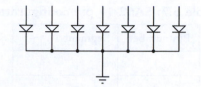

Figure 6.23: Common cathode configuration

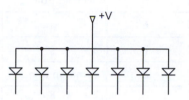

Figure 6.24: Common anode configuration

supply voltage. A segment is turned ON by connecting its cathode terminal to logic 0 via a current limiting resistor. In common anode configuration the 7-segment LED is connected to the microcontroller in current sinking mode.

In this project, a Kingbright SA52-11 red common anode 7-segment display is used. This is a 13mm (0.52 inch) display with ten pins that includes a segment LED for the decimal point. Table 6.7 shows the pin configuration of this display.

Project Hardware

The circuit diagram of the project is shown in Figure 6.25. A PIC18F452 type microcontroller is used with a 4MHz resonator. Segments a to g of the display are connected to PORTC of the microcontroller through 290-ohm current limiting resistors. Before driving the display, we have to know the relationship between the numbers to be displayed and the corresponding segments to be turned ON, and this is shown in Table 6.8. For example, to display number 3 we have to send the hexadecimal number 0×4F to PORTC, which turns ON segments a,b,c,d, and g. Similarly, to display number 9 we have to send the hexadecimal number 0×6F to PORTC which turns ON segments a,b,c,d,f, and g.

Table 6.7: SA52-11 pin configuration

Pin number	Segment
1	e
2	d
3	common anode
4	c
5	decimal point
6	b
7	a
8	common anode
9	f
10	g

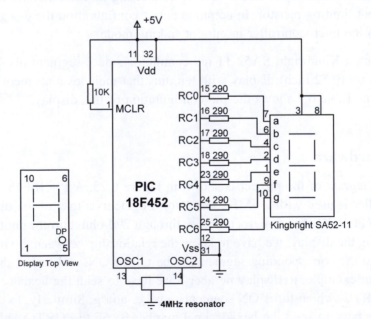

Figure 6.25: Circuit diagram of the project

Table 6.8: Displayed number and data sent to PORTC

Number	x g f e d c b a	PORTC Data
0	0 0 1 1 1 1 1 1	0×3F
1	0 0 0 0 0 1 1 0	0×06
2	0 1 0 1 1 0 1 1	0×5B
3	0 1 0 0 1 1 1 1	0×4F
4	0 1 1 0 0 1 1 0	0×66
5	0 1 1 0 1 1 0 1	0×6D
6	0 1 1 1 1 1 0 1	0×7D
7	0 0 0 0 0 1 1 1	0×07
8	0 1 1 1 1 1 1 1	0×7F
9	0 1 1 0 1 1 1 1	0×6F

x is not used, taken as 0.

Project PDL

The operation of the project is shown in Figure 6.26 with a PDL. At the beginning of the program an array called SEGMENT is declared and filled with the relationships between the numbers 0 and 9 and the data to be sent to PORTC. The PORTC pins are then configured as outputs, and a variable is initialized to 0. The program then enters an

```
START
        Create SEGMENT table
        Configure PORTC as outputs
        Initialize CNT to 0
        DO FOREVER
                Get bit pattern from SEGMENT corresponding to CNT
                Send this bit pattern to PORTC
                Increment CNT between 0 and 9
                Wait 1 second
        ENDDO
END
```

Figure 6.26: PDL of the project

endless loop where the variable is incremented between 0 and 9 and the corresponding bit pattern to turn ON the appropriate segments is sent to PORTC continuously with a one-second delay between outputs.

Project Program

The program is called SEVEN1.C and the listing is given in Figure 6.27. At the beginning of the program character variables *Pattern* and *Cnt* are declared, and *Cnt* is cleared to 0. Then Table 6.8 is implemented using array SEGMENT. After configuring the PORTC pins as outputs, the program enters an endless loop using a *for* statement. Inside the loop the bit pattern corresponding to the contents of *Cnt* is

```
/**********************************************************************
                          7-SEGMENT DISPLAY
                          ==================

In this project a common anode 7-segment LED display is connected to PORTC
of a PIC18F452 microcontroller and the microcontroller is operated from a 4MHz
resonator. The program displays numbers 0 to 9 on the display with a one second
delay between each output.

Author:    Dogan Ibrahim
Date:      July 2007
File:      SEVEN1.C
**********************************************************************/

void main()
{
    unsigned char Pattern, Cnt = 0;
    unsigned char SEGMENT[] = {0x3F,0x06,0x5B,0x4F,0x66,0x6D,
                               0x7D,0x07,0x7F,0x6F};

    TRISC = 0;                          // PORTC are outputs

    for(;;)                             // Endless loop
    {
        Pattern = SEGMENT[Cnt];         // Number to send to PORTC
        Pattern = ~Pattern;             // Invert bit pattern
        PORTC = Pattern;                // Send to PORTC
        Cnt++;
        if(Cnt == 10) Cnt = 0;          // Cnt is between 0 and 9
        Delay_ms(1000);                 // 1 second delay
    }
}
```

Figure 6.27: Program listing

found and stored in variable *Pattern*. Because we are using a common anode display, a segment is turned ON when it is at logic 0 and thus the bit pattern is inverted before it is sent to PORTC. The value of *Cnt* is then incremented between 0 and 9, after which the program waits for a second before repeating the above sequence.

Modified Program

Note that the program can be made more readable if we create a function to display the required number and then call this function from the main program. Figure 6.28 shows the modified program (called SEVEN2.C). A function called *Display* is created with an argument called *no*. The function gets the bit pattern from local array SEGMENT indexed by *no*, inverts it, and then returns the resulting bit pattern to the calling program.

PROJECT 6.6—Two-Digit Multiplexed 7-Segment LED

Project Description

This project is similar to Project 6.5, but here multiplexed two digits are used instead of just one digit and a fixed number. In this project the number 25 is displayed. In multiplexed LED applications (see Figure 6.29) the LED segments of all the digits are tied together and the common pins of each digit are turned ON separately by the microcontroller. When each digit is displayed only for several milliseconds, the eye cannot tell that the digits are not ON all the time. This way we can multiplex any number of 7-segment displays together. For example, to display the number 53, we have to send 5 to the first digit and enable its common pin. After a few milliseconds, number 3 is sent to the second digit and the common point of the second digit is enabled. When this process is repeated continuously, it appears to the user that both displays are ON continuously.

Some manufacturers provide multiplexed multidigit displays, such as 2-, 4-, or 8-digit multiplexed displays, in single packages. The display used in this project is the DC56-11EWA, which is a red 0.56-inch common-cathode two-digit display having 18 pins and the pin configuration as shown in Table 6.9. This display can be controlled from the microcontroller as follows:

- Send the segment bit pattern for digit 1 to segments a to g

- Enable digit 1

```
/*************************************************************************
                      7-SEGMENT DISPLAY
                      ===================

In this project a common anode 7-segment LED display is connected to
PORTC of a PIC18F452 microcontroller and the microcontroller is
operated from a 4MHz resonator. The program displays numbers 0 to 9 on
the display with a one second delay between each output.

In this version of the program a function called "Display" is used to display the
number.

Author:    Dogan Ibrahim
Date:      July 2007
File:      SEVEN2.C
*************************************************************************/

//
// This function displays a number on the 7-segment LED.
// The number is passed in the argument list of the function.
//
unsigned char Display(unsigned char no)
{
    unsigned char Pattern;
    unsigned char SEGMENT[] = {0x3F,0x06,0x5B,0x4F,0x66,0x6D,
                   0x7D,0x07,0x7F,0x6F};

    Pattern = SEGMENT[no];
    Pattern = ~ Pattern;                    // Pattern to return
    return (Pattern);
}

//
// Start of MAIN Program
//
void main()
{
    unsigned char Cnt = 0;

    TRISC = 0;                              // PORTC are outputs

    for(;;)                                 // Endless loop
    {
        PORTC = Display(Cnt);               // Send to PORTC
        Cnt++;
        if(Cnt == 10) Cnt = 0;             // Cnt is between 0 and 9
        Delay_ms(1000);                     // 1 second delay
    }
}
```

Figure 6.28: Modified program listing

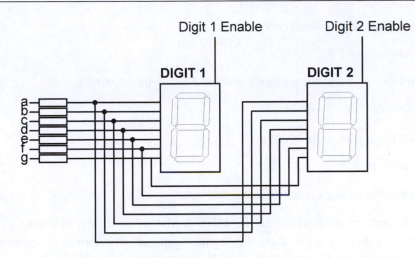

Figure 6.29: Two multiplexed 7-segment displays

Table 6.9: Pin configuration of DC56-11EWA dual display

Pin no.	Segment
1,5	E
2,6	D
3,8	C
14	Digit 1 enable
17,7	G
15,10	B
16,11	A
18,12	F
13	Digit 2 enable
4	Decimal point 1
9	Decimal point 2

- Wait for a few milliseconds

- Disable digit 1

- Send the segment bit pattern for digit 2 to segments a to g

- Enable digit 2

- Wait for a few milliseconds

- Disable digit 2

- Repeat these steps continuously

The segment configuration of the DC56-11EWA display is shown in Figure 6.30. In a multiplexed display application the segment pins of corresponding segments are connected together. For example, pins 11 and 16 are connected as the common *a* segment, pins 15 and 10 are connected as the common *b* segment, and so on.

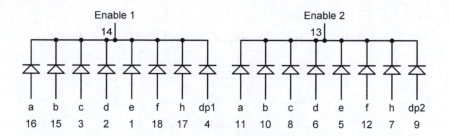

Figure 6.30: DC56-11EWA display segment configuration

Project Hardware

The block diagram of this project is shown in Figure 6.31. The circuit diagram is given in Figure 6.32. The segments of the display are connected to PORTC of a PIC18F452-type microcontroller, operated with a 4MHz resonator. Current limiting resistors are used on each segment of the display. Each digit is enabled using a BC108-type transistor switch connected to port pins RB0 and RB1 of the microcontroller. A segment is turned on when a logic 1 is applied to the base of the corresponding segment transistor.

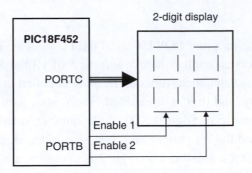

Figure 6.31: Block diagram of the project

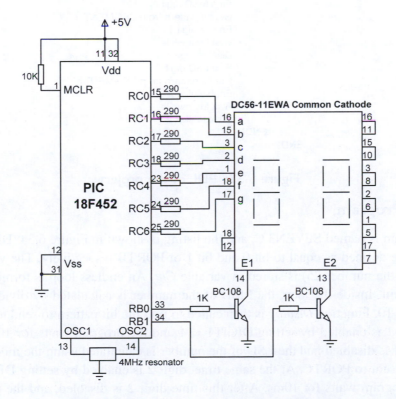

Figure 6.32: Circuit diagram of the project

Project PDL

At the beginning of the program PORTB and PORTC pins are configured as outputs. The program then enters an endless loop where first of all the Most Significant Digit (MSD) of the number is calculated, function *Display* is called to find the bit pattern and then sent to the display, and digit 1 is enabled. Then, after a small delay, digit 1 is disabled, the Least Significant Digit (LSD) of the number is calculated, function *Display* is called to find the bit pattern and then sent to the display, and digit 2 is enabled. Then again after a small delay, digit 2 is disabled, and this process repeats indefinitely. Figure 6.33 shows the PDL of the project.

```
START
        Create SEGMENT table
        Configure PORTB as outputs
        Configure PORTC as outputs
        Initialize CNT to 25
        DO FOREVER
            Find MSD digit
            Get bit pattern from SEGMENT
            Enable digit 1
            Wait for a while
            Disable digit 1
            Find LSD digit
            Get bit pattern from SEGMENT
            Enable digit 2
            Wait for a while
            Disable digit 2
        ENDDO
END
```

Figure 6.33: PDL of the project

Project Program

The program is named SEVEN3.C, and the listing is shown in Figure 6.34. DIGIT1 and DIGIT2 are defined as equal to bit 0 and bit 1 of PORTB respectively. The value to be displayed (the number 25) is stored in variable *Cnt*. An endless loop is formed using a *for* statement. Inside the loop, the MSD of the number is calculated by dividing the number by 10. Function *Display* is then called to find the bit pattern to send to PORTC. Then digit 1 is enabled by setting DIGIT1 = 1 and the program waits for 10ms. After this, digit 1 is disabled and the LSD of the number is calculated using the mod operator ("%") and sent to PORTC. At the same time, digit 2 is enabled by setting DIGIT2 = 1 and the program waits for 10ms. After this time digit 2 is disabled, and the program repeats forever.

```
/************************************************************************
                        Dual 7-SEGMENT DISPLAY
                        ======================

In this project two common cathode 7-segment LED displays are connected to
PORTC of a PIC18F452 microcontroller and the microcontroller is operated
from a 4MHz resonator. Digit 1 (left digit) enable pin is connected to port pin
RB0 and digit 2 (right digit) enable pin is connected to port pin RB1 of the
microcontroller. The program displays number 25 on the displays.

Author:    Dogan Ibrahim
Date:      July 2007
File:      SEVEN3.C
************************************************************************/
#define DIGIT1 PORTB.F0
#define DIGIT2 PORTB.F1

//
// This function finds the bit pattern to be sent to the port to display a number
// on the 7-segment LED. The number is passed in the argument list of the function.
//
unsigned char Display(unsigned char no)
{
    unsigned char Pattern;
    unsigned char SEGMENT[] = {0x3F,0x06,0x5B,0x4F,0x66,0x6D,
                               0x7D,0x07,0x7F,0x6F};

    Pattern = SEGMENT[no];                       // Pattern to return
    return (Pattern);
}


//
// Start of MAIN Program
//
void main()
{
    unsigned char Msd, Lsd, Cnt = 25;

    TRISC = 0;                                   // PORTC are outputs
    TRISB = 0;                                   // RB0, RB1 are outputs

    DIGIT1 = 0;                                  // Disable digit 1
    DIGIT2 = 0;                                  // Disable digit 2

    for(;;)                                      // Endless loop
    {
```

Figure 6.34: Program listing

(Continued)

```
        Msd = Cnt / 10;                      // MSD digit
            PORTC = Display(Msd);            // Send to PORTC
            DIGIT1 = 1;                      // Enable digit 1
            Delay_Ms(10);                    // Wait a while

            DIGIT1 = 0;                      // Disable digit 1
            Lsd = Cnt % 10;                  // LSD digit
            PORTC = Display(Lsd);            // Send to PORTC
            DIGIT2 = 1;                      // Enable digit 2
            Delay_Ms(10);                    // Wait a while
            DIGIT2 = 0;                      // Disable digit 2
    }
}
```

Figure 6.34: (Cont'd)

PROJECT 6.7—Two-Digit Multiplexed 7-Segment LED Counter with Timer Interrupt

Project Description

This project is similar to Project 6 but here the microcontroller's timer interrupt is used to refresh the displays. In Project 6 the microcontroller was busy updating the displays every 10ms and could not perform any other tasks. For example, the program given in Project 6 cannot be used to make a counter with a one-second delay between counts, as the displays cannot be updated while the program waits for one second.

In this project a counter is designed to count from 0 to 99, and the display is refreshed every 5ms inside the timer interrupt service routine. The main program can then perform other tasks, in this example incrementing the count and waiting for one second between counts.

In this project Timer 0 is operated in 8-bit mode. The time for an interrupt is given by:

$$\text{Time} = (4 \times \text{clock period}) \times \text{Prescaler} \times (256 - \text{TMR0L})$$

where Prescaler is the selected prescaler value, and TMR0L is the value loaded into timer register TMR0L to generate timer interrupts every Time period.

In our application the clock frequency is 4MHz, that is, clock period = 0.25µs, and Time = 5ms. Selecting a prescaler value of 32, the number to be loaded into TMR0L can be calculated as follows:

$$TMR0L = 256 - \frac{Time}{4 * clockperiod * prescaler}$$

or

$$TMR0L = 256 - \frac{5000}{4 * 0.25 * 32} = 100$$

Thus, TMR0L should be loaded with 100. The value to be loaded into TMR0 control register T0CON can then be found as:

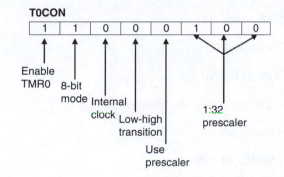

Thus, T0CON register should be loaded with hexadecimal 0×C4. The next register to be configured is the interrupt control register INTCON, where we will disable priority based interrupts and enable the global interrupts and TMR0 interrupts:

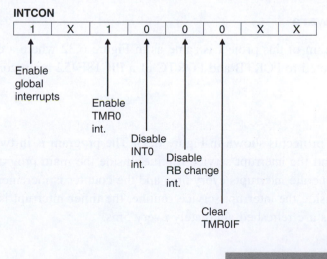

Taking the don't-care entries (X) as 0, the hexadecimal value to be loaded into register INTCON is 0×A0.

When an interrupt occurs, the program automatically jumps to the interrupt service routine. Inside this routine we have to reload register TMR0L, reenable the TMR0 interrupts, and clear the TMR0 interrupt flag bit. Setting INTCON register to 0×20 reenables the TMR0 interrupts and at the same time clears the TMR0 interrupt flag.

The operations to be performed can thus be summarized as follows:

In the main program:

- Load TMR0L with 100
- Set T0CON to 0×C4
- Set INTCON to 0×A0
- Increment the counter with 1-second delays

In the interrupt service routine:

- Re-load TMR0L to 100
- Refresh displays
- Set INTCON to 0×20 (reenable TMR0 interrupts and clear timer interrupt flag)

Project Hardware

The circuit diagram of this project is same as in Figure 6.32 where a dual 7-segment display is connected to PORTB and PORTC of a PIC18F452 microcontroller.

Project PDL

The PDL of the project is shown in Figure 6.35. The program is in two sections: the main program and the interrupt service routine. Inside the main program, TMR0 is configured to generate interrupts every 5ms and the counter is incremented with a one-second delay. Inside the interrupt service routine, the timer interrupt is reenabled and the display digits are refreshed alternately every 5ms.

```
MAIN PROGRAM:

    START
            Configure PORTB as outputs
            Configure PORTC as outputs
            Clear variable Cnt to 0
            Configure TMR0 to generate interrupts every 5ms
            DO FOREVER
                    Increment Cnt between 0 and 99
                    Delay 1 second
            ENDO
    END


INTERRUPT SERVICE ROUTINE:
    START
            Re-configure TMR0
            IF Digit 1 updated THEN
                    Update digit 2
            ELSE
                    Update digit 1
            END
    END
```

Figure 6.35: PDL of the project

Project Program

The program is called SEVEN4.C, and the program listing is given in Figure 6.36. At the beginning of the main program PORTB and PORTC are configured as outputs. Then register T0CON is loaded with $0 \times C4$ to enable the TMR0 and set the prescaler to 32. TMR0L register is loaded with 100 so that an interrupt is generated after 5ms. The program then enters an endless loop where the value of *Cnt* is incremented every second.

Inside the interrupt service routine, register TMR0L is reloaded, TMR0 interrupts are reenabled, and the timer interrupt flag is cleared so that further timer interrupts can be generated. The display digits are then updated alternately. A variable called *Flag* is used to determine which digit to update. Function *Display* is called, as in Project 6, to find the bit pattern to be sent to PORTC.

Modifying the Program

In Figure 6.36 the display counts as 00 01…09 10 11…99 00 01… (i.e., the first digit is shown as 0 for numbers less than 10). The program could be modified so the first

```
/**********************************************************************
                     Dual 7-SEGMENT DISPLAY COUNTER
                     ================================
```

In this project two common cathode 7-segment LED displays are connected to
PORTC of a PIC18F452 microcontroller and the microcontroller is operated
from a 4MHz resonator. Digit 1 (left digit) enable pin is connected to port pin RB0
and digit 2 (right digit) enable pin is connected to port pin RB1 of the microcontroller.
The program counts up from 0 to 99 with one second delay between each count.

The display is updated in a timer interrupt service routine at
every 5ms.

```
Author:    Dogan Ibrahim
Date:      July 2007
File:      SEVEN4.C
**********************************************************************/
#define DIGIT1 PORTB.F0
#define DIGIT2 PORTB.F1

unsigned char Cnt = 0;
unsigned char Flag = 0;

//
// This function finds the bit pattern to be sent to the port to display a number
// on the 7-segment LED. The number is passed in the argument list of the function.
//
unsigned char Display(unsigned char no)
{
    unsigned char Pattern;
    unsigned char SEGMENT[] = {0x3F,0x06,0x5B,0x4F,0x66,0x6D,
                  0x7D,0x07,0x7F,0x6F};

    Pattern = SEGMENT[no];                              // Pattern to return
    return (Pattern);
}


//
// TMR0 timer interrupt service routine. The program jumps to the ISR at
// every 5ms.
//
void interrupt ()
{
    unsigned char Msd, Lsd;
    TMR0L = 100;                                        // Re-load TMR0
    INTCON = 0x20;                                      // Set T0IE and clear T0IF
    Flag = ~ Flag;                                      // Toggle Flag
    if(Flag == 0)                                       // Do digit 1
    {
```

Figure 6.36: Program of the project

```
        DIGIT2 = 0;
        Msd = Cnt / 10;                                 // MSD digit
        PORTC = Display(Msd);                           // Send to PORTC
        DIGIT1 = 1;                                     // Enable digit 1
      }
    else
      {
        DIGIT1 = 0;                                     // Do digit 2
        Lsd = Cnt % 10;                                 // Disable digit 1
        PORTC = Display(Lsd);                           // LSD digit
        DIGIT2 = 1;                                     // Send to PORTC
      }                                                 // Enable digit 2
}


//
// Start of MAIN Program. configure PORTB and PORTC as outputs.
// In addition, configure TMR0 to interrupt at every 10ms
//
void main()
{
    TRISC = 0;                                          // PORTC are outputs
    TRISB = 0;                                          // RB0, RB1 are outputs

    DIGIT1 = 0;                                         // Disable digit 1
    DIGIT2 = 0;                                         // Disable digit 2
//
// Configure TMR0 timer interrupt
//
    T0CON = 0xC4;                                       // Prescaler = 32
    TMR0L = 100;                                        // Load TMR0L with 100
    INTCON = 0xA0;                                      // Enable TMR0 interrupt
    Delay_ms(1000);

    for(;;)                                             // Endless loop
    {
      Cnt++;                                            // Increment Cnt
      if(Cnt == 100) Cnt = 0;                           // Count between 0 and 99
      Delay_ms(1000);                                   // Wait 1 second
    }

}
```

Figure 6.36: (Cont'd)

digit is blanked if the number to be displayed is less than 10. The modified program (called SEVEN5.C) is shown in Figure 6.37. Here, the first digit (MSD) is not enabled if the number to be displayed is 0.

```
/***********************************************
          Dual 7-SEGMENT DISPLAY COUNTER
          ===============================

In this project two common cathode 7-segment LED displays are
connected to PORTC of a PIC18F452 microcontroller and the
microcontroller is operated from a 4MHz resonator. Digit 1 (left
digit) enable pin is connected to port pin RB0 and digit 2
(right digit) enable pin is connected to port pin RB1 of the
microcontroller. The program counts up from 0 to 99 with one
second delay between each count.

The display is updated in a timer interrupt service routine at
every 5ms.

In this version of the program the first digit is blanked if the
number is 0.

Author:    Dogan Ibrahim
Date:      July 2007
File:      SEVEN5.C
 ***********************************************/
#define DIGIT1 PORTB.F0
#define DIGIT2 PORTB.F1

unsigned char Cnt = 0;
unsigned char Flag = 0;

//
// This function finds the bit pattern to be sent to the port to display a number
// on the 7-segment LED. The number is passed in the argument list of the function.
//
unsigned char Display(unsigned char no)
{
    unsigned char Pattern;
    unsigned char SEGMENT[] = {0x3F,0x06,0x5B,0x4F,0x66,0x6D,
                  0x7D,0x07,0x7F,0x6F};

    Pattern = SEGMENT[no];                              // Pattern to return
    return (Pattern);
}


//
// TMR0 timer interrupt service routine. The program jumps to the
// ISR at every 5ms.
//
void interrupt ()
{
        unsigned char Msd, Lsd;
```

Figure 6.37: Modified program

```
    TMR0L = 100;                        // Re-load TMR0
    INTCON = 0x20;                      // Set T0IE and clear T0IF
    Flag = ~ Flag;                      // Toggle Flag
    if(Flag == 0)                       // Do digit 1
    {
      DIGIT2 = 0;
      Msd = Cnt / 10;                   // MSD digit
      if(Msd != 0)
      {
        PORTC = Display(Msd);           // Send to PORTC
        DIGIT1 = 1;                     // Enable digit 1
      }
    }
    else                                // Do digit 2
    {
      DIGIT1 = 0;                       // Disable digit 1
      Lsd = Cnt % 10;                   // LSD digit
      PORTC = Display(Lsd);             // Send to PORTC
      DIGIT2 = 1;                       // Enable digit 2
    }
}


//
// Start of MAIN Program. configure PORTB and PORTC as outputs.
// In addition, configure TMR0 to interrupt at every 10ms
//
void main()
{
    TRISC = 0;                          // PORTC are outputs
    TRISB = 0;                          // RB0, RB1 are outputs

    DIGIT1 = 0;                         // Disable digit 1
    DIGIT2 = 0;                         // Disable digit 2
//
// Configure TMR0 timer interrupt
//
    T0CON = 0xC4;                       // Prescaler = 32
    TMR0L = 100;                        // Load TMR0 with 100
    INTCON = 0xA0;                      // Enable TMR0 interrupt
    Delay_ms(1000);

    for(;;)                             // Endless loop
    {
      Cnt++;                            // Increment Cnt
      if(Cnt == 100) Cnt = 0;          // Count between 0 and 99
      Delay_ms(1000);                  // Wait 1 second
    }

}
```

Figure 6.37: (Cont'd)

PROJECT 6.8—Voltmeter with LCD Display

Project Description

In this project a voltmeter with LCD display is designed. The voltmeter can be used to measure voltages 0–5V. The voltage to be measured is applied to one of the analog inputs of a PIC18F452-type microcontroller. The microcontroller reads the analog voltage, converts it into digital, and then displays it on an LCD.

In microcontroller systems the output of a measured variable is usually displayed using LEDs, 7-segment displays, or LCD displays. LCDs make it possible to display alphanumeric or graphical data. Some LCDs have forty or more character lengths with the capability to display several lines. Other LCD displays can be used to display graphics images. Some modules offer color displays, while others incorporate backlighting so they can be viewed in dimly lit conditions.

There are basically two types of LCDs as far as the interface technique is concerned: parallel and serial. Parallel LCDs (e.g., Hitachi HD44780) are connected to a microcontroller by more than one data line and the data is transferred in parallel form. Both four and eight data lines are commonly used. A four-wire connection saves I/O pins but is slower since the data is transferred in two stages. Serial LCDs are connected to the microcontroller by only one data line, and data is usually sent to the LCD using the standard RS-232 asynchronous data communication protocol. Serial LCDs are much easier to use, but they cost more than the parallel ones.

The programming of a parallel LCD is a complex task and requires a good understanding of the internal operation of the LCD controllers, including the timing diagrams. Fortunately, the mikroC language provides special library commands for displaying data on alphanumeric as well as graphic LCDs. All the user has to do is connect the LCD to the microcontroller, define the LCD connection in the software, and then send special commands to display data on the LCD.

HD44780 LCD Module

The HD44780 is one of the most popular alphanumeric LCD modules and is used both in industry and by hobbyists. This module is monochrome and comes in different sizes.

Modules with 8, 16, 20, 24, 32, and 40 columns are available. Depending on the model chosen, the number of rows may be 1, 2, or 4. The display provides a 14-pin (or 16-pin) connector to a microcontroller. Table 6.10 gives the pin configuration and pin functions of a 14-pin LCD module. The following is a summary of the pin functions:

V_{SS} is the 0V supply or ground. The V_{DD} pin should be connected to the positive supply. Although the manufacturers specify a 5V DC supply, the modules will usually work with as low as 3V or as high as 6V.

Pin 3, named V_{EE}, is the contrast control pin. This pin is used to adjust the contrast of the display and should be connected to a variable voltage supply. A potentiometer is normally connected between the power supply lines with its wiper arm connected to this pin so that the contrast can be adjusted.

Table 6.10: Pin configuration of HD44780 LCD module

Pin no.	Name	Function
1	V_{SS}	Ground
2	V_{DD}	+ ve supply
3	V_{EE}	Contrast
4	RS	Register select
5	R/W	Read/write
6	E	Enable
7	D0	Data bit 0
8	D1	Data bit 1
9	D2	Data bit 2
10	D3	Data bit 3
11	D4	Data bit 4
12	D5	Data bit 5
13	D6	Data bit 6
14	D7	Data bit 7

Pin 4 is the register select (RS), and when this pin is LOW, data transferred to the display is treated as commands. When RS is HIGH, character data can be transferred to and from the module.

Pin 5 is the read/write (R/W) line. This pin is pulled LOW in order to write commands or character data to the LCD module. When this pin is HIGH, character data or status information can be read from the module.

Pin 6 is the enable (E) pin, which is used to initiate the transfer of commands or data between the module and the microcontroller. When writing to the display, data is transferred only on the HIGH-to-LOW transition of this line. When reading from the display, data becomes available after the LOW-to-HIGH transition of the enable pin, and this data remains valid as long as the enable pin is at logic HIGH.

Pins 7 to 14 are the eight data bus lines (D0 to D7). Data can be transferred between the microcontroller and the LCD module using either a single 8-bit byte or as two 4-bit nibbles. In the latter case, only the upper four data lines (D4 to D7) are used. The 4-bit mode means that four fewer I/O lines are used to communicate with the LCD. In this book we are using only an alphanumeric-based LCD and only the 4-bit interface.

Connecting the LCD

The mikroC compiler assumes by default that the LCD is connected to the microcontroller as follows:

```
LCD              Microcontroller port

D7               Bit 7 of the port
D6               Bit 6 of the port
D5               Bit 5 of the port
D4               Bit 4 of the port
E                Bit 3 of the port
RS               Bit 2 of the port
```

where *port* is the port name specified using the *Lcd_Init* statement.

For example, we can use the statement *Lcd_Init(&PORTB)* if the LCD is connected to PORTB with the default connection.

It is also possible to connect the LCD differently, using the command *Lcd_Config* to define the connection.

Project Hardware

Figure 6.38 shows the block diagram of the project. The microcontroller reads the analog voltage, converts it to digital, formats it, and then displays it on the LCD.

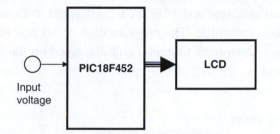

Figure 6.38: Block diagram of the project

The circuit diagram of the project is shown in Figure 6.39. The voltage to be measured (between 0 and 5V) is applied to port AN0 where this port is configured as an analog

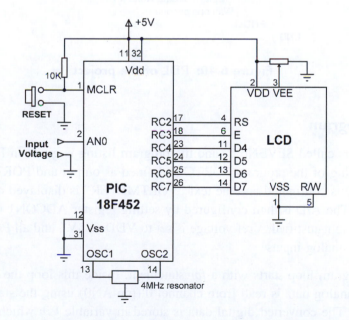

Figure 6.39: Circuit diagram of the project

input in software. The LCD is connected to PORTC of the microcontroller as in the default four-wire connection. A potentiometer is used to adjust the contrast of the LCD display.

Project PDL

The PDL of the project is shown in Figure 6.40. At the beginning of the program PORTC is configured as output and PORTA is configured as input. Then the LCD and the A/D converter are configured. The program then enters an endless loop where analog input voltage is converted to digital and displayed on the LCD. The process is repeated every second.

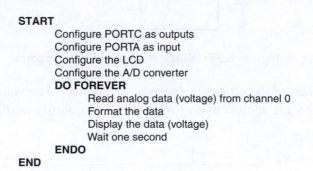

```
START
        Configure PORTC as outputs
        Configure PORTA as input
        Configure the LCD
        Configure the A/D converter
        DO FOREVER
                Read analog data (voltage) from channel 0
                Format the data
                Display the data (voltage)
                Wait one second
        ENDO
END
```

Figure 6.40: PDL of the project

Project Program

The program is called SEVEN6.C, and the program listing is given in Figure 6.41. At the beginning of the program PORTC is defined as output and PORTA as input. Then the LCD is configured and the text "VOLTMETER" is displayed on the LCD for two seconds. The A/D is then configured by setting register ADCON1 to 0×80 so the A/D result is right-justified, Vref voltage is set to VDD (+5V), and all PORTA pins are configured as analog inputs.

The main program loop starts with a *for* statement. Inside this loop the LCD is cleared, and analog data is read from channel 0 (pin AN0) using the statement *Adc_Read(0)*. The converted digital data is stored in variable *Vin* which is declared as an *unsigned long*. The A/D converter is 10-bits wide and thus there are 1024 steps

```
/****************************************************************
                  VOLTMETER WITH LCD DISPLAY
                  ==============================
```

In this project an LCD is connected to PORTC. Also, input port AN0 is used as analog input. Voltage to be measured is applied to AN0. The microcontroller reads the analog voltage, converts into digital, and then displays on the LCD.

Analog input range is 0 to 5V. A PIC18F452 type microcontroller is used in this project, operated with a 4MHz resonator.

Analog data is read using the Adc_Read built-in function. This function uses the internal RC clock for A/D timing.

The LCD is connected to the microcontroller as follows:

Microcontroller   LCD

```
   RC7          D7
   RC6          D6
   RC5          D5
   RC4          D4
   RC3          Enable
   RC2          RS
```

Author: Dogan Ibrahim
Date: July 2007
File: SEVEN6.C

```
****************************************************************/

//
// Start of MAIN Program. Configure LCD and A/D converter
//
void main()
{
    unsigned long Vin, mV;
    unsigned char op[12];
    unsigned char i,j,lcd[5];

    TRISC = 0;                            // PORTC are outputs (LCD)
    TRISA = 0xFF;                         // PORTA is input

//
// Configure LCD
//
    Lcd_Init(&PORTC);                     // LCD is connected to PORTC
    Lcd_Cmd(LCD_CLEAR);
    Lcd_Out(1,1,"VOLTMETER");
    Delay_ms(2000);
//
```

Figure 6.41: Program listing

(Continued)

```
        // Configure A/D converter. AN0 is used in this project
        //
            ADCON1 = 0x80;                              // Use AN0 and Vref=+5V
        //
        // Program loop
        //
            for(;;)                                     // Endless loop
            {
              Lcd_Cmd(LCD_CLEAR);
              Vin = Adc_Read(0);                        // Read from channel 0 (AN0)
              Lcd_Out(1,1,"mV = ");                     // Display "mV = "
              mV = (Vin * 5000) >> 10;                  // mv = Vin x 5000 / 1024
              LongToStr(mV,op);                         // Convert to string in "op"
        //
        // Remove leading blanks
        //
              j=0;
              for(i=0;i<=11;i++)
              {
                if(op[i] != ' ')                        // If a blank
                {
                  lcd[j]=op[i];
                  j++;
                }
              }
        //
        // Display result on LCD
        //
              Lcd_Out(1,6,lcd);                         // Output to LCD
              Delay_ms(1000);                           // Wait 1 second
            }

        }
```

Figure 6.41: (Cont'd)

(0 to 1023) corresponding to the reference voltage of 5000mV. Each step corresponds to 5000mV/1024 = 4.88mV. Inside the loop, variable *Vin* is converted into millivolts by multiplying by 5000 and dividing into 1024. The division is done by shifting right by 10 digits. At this point variable *mV* contains the converted data in millivolts.

Function *LongToStr* is called to convert *mV* into a string in character array *op*. *LongToStr* converts a long variable into a string having a fixed width of eleven characters. If the resulting string is fewer than eleven characters, the left column of the data is filled with space characters.

The leading blanks are then removed and the data is stored in a variable called *lcd*. Function *Lcd_Out* is called to display the data on the LCD starting from column 5 of row 1. For example, if the measured voltage is 1267mV, it is displayed on the LCD as:

$$mV = 1267$$

A More Accurate Display

The voltage displayed in Figure 6.41 is not very accurate, since integer arithmetic has been performed in the calculation and the voltage is calculated by multiplying the A/D output by 5000 and then dividing the result by 1024 using integer division. Although the multiplication is accurate, the accuracy of the measurement is lost when the number is divided by 1024. A more accurate result can be obtained by scaling the number before it is displayed, as follows.

First, multiply the number *Vin* by a factor to remove the integer division. For example, since 5000/1024 = 4.88, we can multiply *Vin* by 488. For the display, we can calculate the integer part of the result by dividing the number into 100, and then the fractional part can be calculated as the remainder. The integer part and the fractional part can be displayed with a decimal point in between. This technique has been implemented in program SEVEN7.C as shown in Figure 6.42. In this program variables *Vdec* and *Vfrac* store the integer and the fractional parts of the number respectively. The decimal part is then converted into a string using function *LongToStr* and leading blanks are removed. The parts of the fractional number are called *ch1* and *ch2*. These are converted into characters by adding 48 (i.e., character "0") and then displayed at the next cursor positions using the LCD command *Lcd_Chr_Cp*.

We could also calculate and display more accurate results by using floating point arithmetic, but since this uses huge amounts of memory it should be avoided if possible.

PROJECT 6.9—Calculator with Keypad and LCD

Project Description

Keypads are small keyboards used to enter numeric or alphanumeric data into microcontroller systems. Keypads are available in a variety of sizes and styles, from 2 × 2 to 4 × 4 or even bigger.

This project uses a 4 × 4 keypad (shown in Figure 6.43) and an LCD to design a simple calculator.

```
/****************************************************************
                   VOLTMETER WITH LCD DISPLAY
                   ==============================
```

In this project an LCD is connected to PORTC. Also, input port
AN0 is used as analog input. Voltage to be measured is applied
to AN0. The microcontroller reads the analog voltage, converts
into digital, and then displays on the LCD.

Analog input range is 0 to 5V. A PIC18F452 type microcontroller
is used in this project, operated with a 4MHz resonator.

Analog data is read using the Adc_Read built-in function. This
function uses the internal RC clock for A/D timing.

The LCD is connected to the microcontroller as follows:

Microcontroller   LCD

    RC7   D7
    RC6   D6
    RC5   D5
    RC4   D4
    RC3   Enable
    RC2   RS

This program displays more accurate results than program SEVEN6.C.
The voltage is displayed as follows:

    mV = nnnn.mm

Author: Dogan Ibrahim
Date: July 2007
File: SEVEN7.C
**/

//
// Start of MAIN Program. Configure LCD and A/D converter
//
void main()
{
    unsigned long Vin, mV,Vdec,Vfrac;
    unsigned char op[12];
    unsigned char i,j,lcd[5],ch1,ch2;

    TRISC = 0;   // PORTC are outputs (LCD)
    TRISA = 0xFF;   // PORTA is input

//
// Configure LCD
```

### Figure 6.42: A more accurate program

```

```
//
    Lcd_Init(&PORTC);                          // LCD is connected to PORTC
    Lcd_Cmd(LCD_CLEAR);
    Lcd_Out(1,1,"VOLTMETER");
    Delay_ms(2000);
//
// Configure A/D converter. AN0 is used in this project
//
    ADCON1 = 0x80;                             // Use AN0 and Vref=+5V
//
// Program loop
//
    for(;;)                                    // Endless loop
    {
      Lcd_Cmd(LCD_CLEAR);
      Vin = Adc_Read(0);                       // Read from channel 0 (AN0)
      Lcd_Out(1,1,"mV = ");                    // Display "mV = "
      Vin = 488*Vin;                           // Scale up the result
      Vdec = Vin / 100;                        // Decimal part
      Vfrac = Vin % 100;                       // Fractional part
      LongToStr(Vdec,op);                      // Convert Vdec to string in "op"
//
// Remove leading blanks
//
      j=0;
      for(i=0;i<=11;i++)
      {
        if(op[i] != ' ')                       // If a blank
        {
          lcd[j]=op[i];
          j++;
        }
      }
//
// Display result on LCD
//
      Lcd_Out(1,6,lcd);                        // Output to LCD
      Lcd_Out_Cp(".");                         // Display "."
      ch1 = Vfrac / 10;                        // Calculate fractional part
      ch2 = Vfrac % 10;                        // Calculate fractional part
      Lcd_Chr_Cp(48+ch1);                      // Display fractional part
      Lcd_Chr_Cp(48+ch2);                      // Display fractional part
      Delay_ms(1000);                          // Wait 1 second
    }

}
```

Figure 6.42: (Cont'd)

Figure 6.43: 4 × 4 keypad

Figure 6.44 shows the structure of the keypad used in this project which consists of sixteen switches formed in a 4 × 4 array and named numerals 0–9, Enter, "+", ".", "-", "*", and "/". Rows and columns of the keypad are connected to PORTB of a microcontroller which scans the keypad to detect when a switch is pressed. The operation of the keypad is as follows:

- A logic 1 is applied to the first column via RB0.

- Port pins RB4 to RB7 are read. If the data is nonzero, a switch is pressed. If RB4 is 1, key 1 is pressed, if RB5 is 1, key 4 is pressed, if RB6 is 1, key 9 is pressed, and so on.

- A logic 1 is applied to the second column via RB1.

- Again, port pins RB4 to RB7 are read. If the data is nonzero, a switch is pressed. If RB4 is 1, key 2 is pressed, if RB5 is 1, key 6 is pressed, if RB6 is 1, key 0 is pressed, and so on.

- This process is repeated for all four columns continuously.

In this project a simple integer calculator is designed. The calculator can add, subtract, multiply, and divide integer numbers and show the result on the LCD. The operation of the calculator is as follows: When power is applied to the system, the LCD displays text

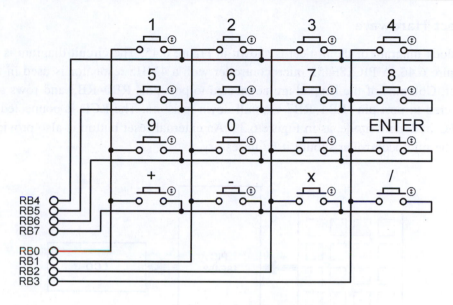

Figure 6.44: 4 × 4 keypad structure

"CALCULATOR" for 2 seconds. Then text "No1:" is displayed in the first row of the LCD and the user is expected to type the first number and then press the ENTER key. Then text "No2:" is displayed in the second row of the LCD, where the user enters the second number and presses the ENTER key. After this, the appropriate operation key should be pressed. The result is displayed on the LCD for five seconds and then the LCD is cleared, ready for the next calculation. The example that follows shows how numbers 12 and 20 can be added:

```
No1: 12 ENTER
No2: 20 ENTER
Op: +
Res = 32
```

In this project the keyboard is labeled as follows:

```
1    2    3    4
5    6    7    8
9    0         ENTER
+    −    X    /
```

One of the keys, between 0 and ENTER, is not used in this project.

Project Hardware

The block diagram of the project is shown in Figure 6.45. The circuit diagram is given in Figure 6.46. A PIC18F452 microcontroller with a 4MHz resonator is used in the project. Columns of the keypad are connected to port pins RB0–RB3 and rows are connected to port pins RB4–RB7 via pull-down resistors. The LCD is connected to PORTC in default mode, as in Figure 6.39. An external reset button is also provided to reset the microcontroller should it be necessary.

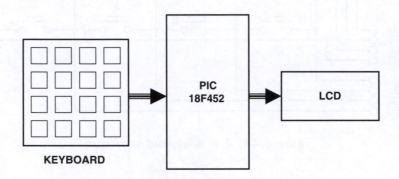

KEYBOARD

Figure 6.45: Block diagram of the project

Project PDL

The project PDL is shown in Figure 6.47. The program consist of two parts: function *getkeypad* and the main program. Function *getkeypad* receives a key from the keypad. Inside the main program the two numbers and the required operation are received from the keypad. The microcontroller performs the required operation and displays the result on the LCD.

Project Program

The program listing for the program KEYPAD.C is given in Figure 6.48. Each key is given a numeric value as follows:

```
0       1       2       3
4       5       6       7
8       9       10      11
12      13      14      15
```

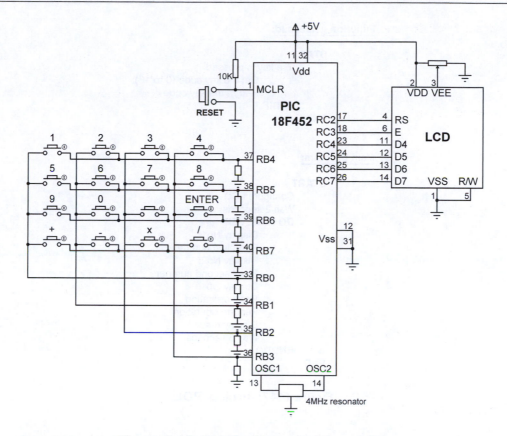

Figure 6.46: Circuit diagram of the project

The program consists of a function called *getkeypad*, which reads the pressed keys, and the main program. Variable *MyKey* stores the key value (0 to 15) pressed, variables *Op1* and *Op2* store respectively the first and second numbers entered by the user. All these variables are cleared to zero at the beginning of the program. A *while* loop is then formed to read the first number and store in variable *Op1*. This loop exits when the user presses the ENTER key. Similarly, the second number is read from the keyboard in a second *while* loop. Then the operation to be performed is read and stored in variable *MyKey*, and a *switch* statement is used to perform the required operation and store the result in variable *Calc*. The result is converted into a string array using function *LongToStr*. The leading blank characters are then removed as in Project 8. The program

Function getkeypad:

START
    IF a key is pressed
        Get the key code (0 to 15)
        Return the key code
    ENDIF
END

Main program:

START
    Configure LCD
    Wait 2 seconds
    DO FOREVER
        Display No1:
        Read first number
        Display No2:
        Read second number
        Display Op:
        Read operation
        Perform operation
        Display result
        Wait 5 seconds
    ENDDO
END

Figure 6.47: Project PDL

displays the result on the LCD, waits for five seconds, and then clears the screen and is ready for the next calculation. This process is repeated forever.

Function *getkeypad* receives a key from the keypad. We start by sending a 1 to column 1, and then we check all the rows. When a key is pressed, a logic 1 is detected in the corresponding row and the program jumps out of the *while* loop. Then a *for* loop is used to find the actual key pressed by the user as a number from 0 to 15.

It is important to realize that when a key is pressed or released, we get what is known as contact noise, where the key output pulses up and down momentarily, producing a number of logic 0 and 1 pulses at the output. Switch contact noise is usually removed either in hardware or by programming in a process called contact debouncing. In software the simplest way to remove the contact noise is to wait for about 20ms after a switch key is pressed or switch key is released. In Figure 6.46, contact debouncing is accomplished in function *getkeypad*.

```
/************************************************************
              CALCULATOR WITH KEYPAD AND LCD
              ================================
```

In this project a 4 x 4 keypad is connected to PORTB of a PIC18F452 microcontroller. Also an LCD is connected to PORTC. The project is a simple calculator which can perform integer arithmetic.

The keys are organized as follows:

```
0   1   2   3
4   5   6   7
8   9  10  11
12  13  14  15
```

The keys are labeled as follows:

```
1  2  3  4
5  6  7  8
9  0    Enter
+  −  *  /
```

```
Author:      Dogan Ibrahim
Date:        July 2007
File:        KEYPAD.C
************************************************************/
#define MASK 0xF0
#define Enter 11
#define Plus 12
#define Minus 13
#define Multiply 14
#define Divide 15

//
// This function gets a key from the keypad
//
unsigned char getkeypad()
{
    unsigned char i, Key = 0;

    PORTB = 0x01;                            // Start with column 1
    while((PORTB & MASK) == 0)               // While no key pressed
    {
        PORTB = (PORTB << 1);                // next column
        Key++;                               // column number
        if(Key == 4)
        {
            PORTB = 0x01;                    // Back to column 1
            Key = 0;
```

Figure 6.48: Program listing

(Continued)

```
        }
    }
    Delay_Ms(20);                                 // Switch debounce

    for(i = 0x10; i !=0; i <<=1)                   // Find the key pressed
    {
        if((PORTB & i) != 0)break;
        Key = Key + 4;
    }

    PORTB=0x0F;
    while((PORTB & MASK) != 0);                    // Wait until key released
    Delay_Ms(20);                                 // Switch debounce

    return (Key);                                 // Return key number
}


//
// Start of MAIN program
//
void main()
{
    unsigned char MyKey, i,j,lcd[5],op[12];
    unsigned long Calc, Op1, Op2;

    TRISC = 0;                                    // PORTC are outputs (LCD)
    TRISB = 0xF0;                                 // RB4-RB7 are inputs

//
// Configure LCD
//
    Lcd_Init(&PORTC);                            // LCD is connected to PORTC
    Lcd_Cmd(LCD_CLEAR);
    Lcd_Out(1,1,"CALCULATOR");                    // Display CALCULATOR
    Delay_ms(2000);                              // Wait 2 seconds
    Lcd_Cmd(LCD_CLEAR);                          // Clear display
//
// Program loop
//
    for(;;)                                      // Endless loop
    {
        MyKey = 0;
        Op1 = 0;
        Op2 = 0;

        Lcd_Out(1,1,"No1: ");                     // Display No1:
        while(1)                                 // Get first no
        {
```

Figure 6.48: (Cont'd)

```
        MyKey = getkeypad();
        if(MyKey == Enter)break;                    // If ENTER pressed
        MyKey++;
        if(MyKey == 10)MyKey = 0;                   // If 0 key pressed
        Lcd_Chr_Cp(MyKey + '0');
        Op1 = 10*Op1 + MyKey;                       // First number in Op1
    }

    Lcd_Out(2,1,"No2: ");                           // Display No2:
    while(1)                                        // Get second no
    {
        MyKey = getkeypad();
        if(MyKey == Enter)break;                    // If ENTER pressed
        MyKey++;
        if(MyKey == 10)MyKey = 0;                   // If 0 key pressed
        Lcd_Chr_Cp(MyKey + '0');
        Op2 = 10*Op2 + MyKey;                       // Second number in Op2
    }

    Lcd_Cmd(LCD_CLEAR);                             // Clear LCD
    Lcd_Out(1,1,"Op: ");                            // Display Op:

    MyKey = getkeypad();                            // Get operation
    Lcd_Cmd(LCD_CLEAR);
    Lcd_Out(1,1,"Res=");                            // Display Res=
    switch(MyKey)                                   // Perform the operation
    {
        case Plus:
            Calc = Op1 + Op2;                       // If ADD
            break;
        case Minus:
            Calc = Op1 - Op2;                       // If Subtract
            break;
        case Multiply:
            Calc = Op1 * Op2;                       // If Multiply
            break;
        case Divide:
            Calc = Op1 / Op2;                       // If Divide
            break;
    }

    LongToStr(Calc, op);                            // Convert to string in op
//
// Remove leading blanks
//
    j=0;
    for(i=0;i<=11;i++)
    {
        if(op[i] != ' ')                            // If a blank
        {
```

Figure 6.48: (Cont'd)

(Continued)

```
                lcd[j]=op[i];
                j++;
            }
        }

        Lcd_Out_Cp(lcd);                    // Display result
        Delay_ms(5000);                     // Wait 5 seconds
        Lcd_Cmd(LCD_CLEAR);
    }
}
```

Figure 6.48: (Cont'd)

Program Using a Built-in Keypad Function

In the program listing in Figure 6.48, a function called *getkeypad* has been developed to read a key from the keyboard. The mikroC language has built-in functions called *Keypad_Read* and *Keypad_Released* to read a key from a keypad when a key is pressed and when a key is released respectively. Figure 6.49 shows a modified program (KEYPAD2.C) listing using the *Keypad_Released* function to implement the preceding calculator project. The circuit diagram is the same as in Figure 6.46.

Before using the *Keypad_Released* function we have to call the *Keypad_Init* function to tell the microcontroller what the keypad is connected to. *Keypad_Released* detects when a key is pressed and then released. When released, the function returns a number between 1 and 16 corresponding to the key pressed. The remaining parts of the program are the same as in Figure 6.48.

PROJECT 6.10—Serial Communication–Based Calculator

Project Description

Serial communication is a simple means of sending data long distances quickly and reliably. The most common serial communication method is based on the RS232 standard, in which standard data is sent over a single line from a transmitting device to a receiving device in bit serial format at a prespecified speed, also known as the baud rate, or the number of bits sent each second. Typical baud rates are 4800, 9600, 19200, 38400, etc.

RS232 serial communication is a form of asynchronous data transmission where data is sent character by character. Each character is preceded with a start bit, seven or eight

```
/************************************************************
              CALCULATOR WITH KEYPAD AND LCD
              ================================
```

In this project a 4 x 4 keypad is connected to PORTB of a PIC18F452
microcontroller. Also an LCD is connected to PORTC. The project is a simple
calculator which can perform integer arithmetic.

The keys are labeled as follows:

```
1  2  3  4
5  6  7  8
9  0    Enter
+  -  *  /
```

In this program mikroC built-in functions are used.

```
Author:     Dogan Ibrahim
Date:       July 2007
File:       KEYPAD2.C
************************************************************/

#define Enter 12
#define Plus 13
#define Minus 14
#define Multiply 15
#define Divide 16

//
// Start of MAIN program
//
void main()
{
    unsigned char MyKey, i,j,lcd[5],op[12];
    unsigned long Calc, Op1, Op2;

    TRISC = 0;                          // PORTC are outputs (LCD)
//
// Configure LCD
//
    Lcd_Init(&PORTC);                   // LCD is connected to PORTC
    Lcd_Cmd(LCD_CLEAR);
    Lcd_Out(1,1,"CALCULATOR");          // Display CALCULATOR
    Delay_ms(2000);
    Lcd_Cmd(LCD_CLEAR);
//
// Configure KEYPAD
//
    Keypad_Init(&PORTB);                // Keypad on PORTB
```

Figure 6.49: Modified program listing

(Continued)

```
//
// Program loop
//
   for(;;)                                        // Endless loop
   {
     MyKey = 0;
     Op1 = 0;
     Op2 = 0;

     Lcd_Out(1,1,"No1: ");                        // Display No1:
     while(1)
     {
       do                                         // Get first number
          MyKey = Keypad_Released();
       while(!MyKey);
       if(MyKey == Enter)break;                   // If ENTER pressed
       if(MyKey == 10)MyKey = 0;                  // If 0 key pressed
       Lcd_Chr_Cp(MyKey + '0');
       Op1 = 10*Op1 + MyKey;
     }

     Lcd_Out(2,1,"No2: ");                        // Display No2:
     while(1)                                     // Get second no
     {
       do                                         // Get second number
          MyKey = Keypad_Released();
       while(!MyKey);
       if(MyKey == Enter)break;                   // If ENTER pressed
       if(MyKey == 10)MyKey = 0;                  // If 0 key pressed
       Lcd_Chr_Cp(MyKey + '0');
       Op2 = 10*Op2 + MyKey;
     }

     Lcd_Cmd(LCD_CLEAR);
     Lcd_Out(1,1,"Op: ");                         // Display Op:

     do
        MyKey = Keypad_Released();                // Get operation
     while(!MyKey);
     Lcd_Cmd(LCD_CLEAR);
     Lcd_Out(1,1,"Res=");                         // Display Res=
     switch(MyKey)                                // Perform the operation
     {
       case Plus:
           Calc = Op1 + Op2;                      // If ADD
           break;
       case Minus:
           Calc = Op1 - Op2;                      // If Subtract
           break;
       case Multiply:
```

Figure 6.49: (Cont'd)

```
            Calc = Op1 * Op2;                // If Multiply
            break;
        case Divide:
            Calc = Op1 / Op2;                // If Divide
            break;
        }

        LongToStr(Calc, op);                 // Convert to string
//
// Remove leading blanks
//
        j=0;
        for(i=0;i<=11;i++)
        {
          if(op[i] != ' ')                   // If a blank
          {
            lcd[j]=op[i];
            j++;
          }
        }

        Lcd_Out_Cp(lcd);                     // Display result
        Delay_ms(5000);                      // Wait 5 seconds
        Lcd_Cmd(LCD_CLEAR);
      }
    }
```

Figure 6.49: (Cont'd)

data bits, an optional parity bit, and one or more stop bits. The most common format is eight data bits, no parity bit, and one stop bit. The least significant data bit is transmitted first, and the most significant bit is transmitted last.

A logic high is defined at $-12V$, and a logic 0 is at $+12V$. Figure 6.50 shows how character "A" (ASCII binary pattern 0010 0001) is transmitted over a serial line. The line is normally idle at $-12V$. The start bit is first sent by the line going from high to low. Then eight data bits are sent, starting from the least significant bit. Finally, the stop bit is sent by raising the line from low to high.

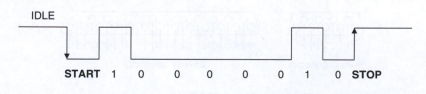

Figure 6.50: Sending character "A" in serial format

In a serial connection, a minimum of three lines is used for communication: transmit (TX), receive (RX), and ground (GND). Serial devices are connected to each other using two types of connectors: 9-way and 25-way. Table 6.11 shows the TX, RX, and GND pins of each type of connectors. The connectors used in RS232 serial communication are shown in Figure 6.51.

Table 6.11: Minimum required pins for serial communication

9-pin connector	
Pin	**Function**
2	Transmit (TX)
3	Receive (RX)
5	Ground (GND)

25-pin connector	
Pin	**Function**
2	Transmit (TX)
3	Receive (RX)
7	Ground (GND)

As just described, RS232 voltage levels are at ±12V. However, microcontroller input-output ports operate at 0 to +5V voltage levels, so the voltage levels must be translated before a microcontroller can be connected to a RS232 compatible device. Thus the output signal from the microcontroller has to be converted to ±12V, and the input from an RS232 device must be converted into 0 to +5V before it can be connected to a microcontroller. This voltage translation is normally done with special RS232 voltage

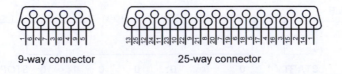

9-way connector   25-way connector

Figure 6.51: RS232 connectors

converter chips. One such popular chip is the MAX232, a dual converter chip having the pin configuration shown in Figure 6.52. The device requires four external 1μF capacitors for its operation.

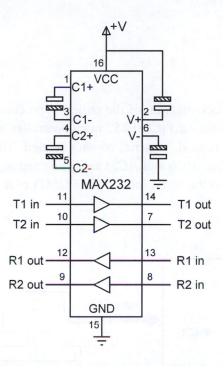

Figure 6.52: MAX232 pin configuration

In the PIC18 series of microcontrollers, serial communication can be handled either in hardware or in software. The hardware option is easy. PIC18 microcontrollers have built-in USART (universal synchronous asynchronous receiver transmitter) circuits providing special input-output pins for serial communication. For serial communication all the data transmission is handled by the USART, but the USART has to be configured before receiving and transmitting data. With the software option, all the serial bit timing is handled in software, and any input-output pin can be programmed and used for serial communication.

In this project a PC is connected to the microcontroller using an RS232 cable. The project operates as a simple integer calculator where data is sent to the microcontroller using the PC keyboard and displayed on the PC monitor.

A sample calculation is as follows:

```
CALCULATOR PROGRAM

Enter First Number: 12
Enter Second Number: 2
Enter Operation: +
Result = 14
```

Project Hardware

Figure 6.53 shows the block diagram of the project. The circuit diagram is given in
Figure 6.54. This project uses a PIC18F452 microcontroller with a 4MHz resonator,
and the built-in USART is used for serial communication. The serial communication
lines of the microcontroller (RC6 and RC7) are connected to a MAX232 voltage
translator chip and then to the serial input port (COM1) of a PC using a 9-pin
connector.

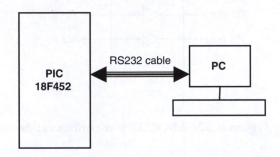

Figure 6.53: Block diagram of the project

Project PDL

The PDL of the project is shown in Figure 6.55. The project consists of a main program
and two functions called *Newline* and *Text_To_User*. Function *Newline* sends a
carriage-return and line-feed to the serial port. Function *Text_To_User* sends a text
message to USART. The main program receives two numbers and the operation to be
performed from the PC keyboard. The numbers are echoed on the PC monitor. The
result of the operation is also displayed on the monitor.

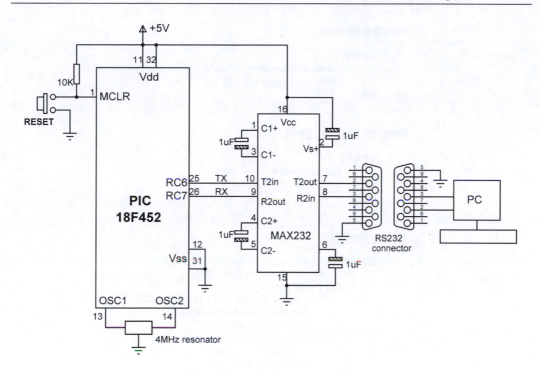

Figure 6.54: Circuit diagram of the project

Project Program

The program listing of the project is shown in Figure 6.56. The program consists of a main program and two functions called *Newline* and *Text_To_Usart*. Function *Newline* sends a carriage return and line feed to the USART to move the cursor to the next line. Function *Text_To_Usart* sends a text message to the USART.

At the beginning of the program various messages used in the program are defined as *msg1* to *msg5*. The USART is then initialized to 9600 baud using the mikroC library routine *Usart_Init*. Then the heading "CALCULATOR PROGRAM" is displayed on the PC monitor. The program reads the first number from the keyboard using the library function *Usart_Read*. Function *Usart_Data_Ready* checks when a new data byte is ready before reading it. Variable *Op1* stores the first number. Similarly, another loop is formed and the second number is read into variable *Op2*. The program then reads the operation to be performed (+ − * /). The required operation is performed inside a switch statement and the result is stored in variable *Calc*. The program then converts the result into string format by calling library function *LongToStr*. Leading blanks are

<u>**Function Newline:**</u>

START

Send carriage-return to USART

Send line-feed to USART

END

<u>**Function Text_To_Usart**</u>

START

Get text from the argument

Send text to USART

END

<u>**Main program:**</u>

START

Configure USART to 9600 Baud

DO FOREVER

Display "CALCULATOR PROGRAM"

Display "Enter First Number: "

Read first number

Display "Enter Second Number: "

Read second number

Display "Operation: "

Read operation

Perform operation

Display "Result= "

Display the result

ENDDO

END

Figure 6.55: Project PDL

removed from this string, and the final result is stored in character array *kbd* and sent to the USART to display on the PC keyboard.

Testing the Program

The program can be tested using a terminal emulator software such as HyperTerminal, which is distributed free of charge with Windows operating systems. The steps to test the program follow (these steps assume serial port COM2 is used):

- Connect the RS232 output from the microcontroller to the serial input port of a PC (e.g., COM2)

```
/*******************************************************************
                     CALCULATOR WITH PC INTERFACE
                     ================================
```

In this project a PC is connected to a PIC18F452 microcontroller. The project is a simple integer calculator. User enters the numbers using the PC keyboard. Results are displayed on the PC monitor.

The following operations can be performed:

   + – * /

This program uses the built in USART of the microcontroller. The USART is configured to operate with 9600 Baud rate.

The serial TX pin is RC6 and the serial RX pin is RC7.

```
Author:     Dogan Ibrahim
Date:       July 2007
File:       SERIAL1.C
*******************************************************************/

#define Enter 13
#define Plus '+'
#define Minus '–'
#define Multiply '*'
#define Divide '/'

//
// This function sends carriage-return and line-feed to USART
//
void Newline()
{
   Usart_Write(0x0D);                    // Send carriage-return
   Usart_Write(0x0A);                    // Send line-feed
}



//
// This function sends a text to USART
//
void Text_To_Usart(unsigned char *m)
{
   unsigned char i;

   i = 0;
   while(m[i] != 0)
   {                                     // Send TEXT to USART
     Usart_Write(m[i]);
     i++;
```

Figure 6.56: Program listing

(Continued)

```
    }
}


//
// Start of MAIN program
//
void main()
{
    unsigned char MyKey, i,j,kbd[5],op[12];
    unsigned long Calc, Op1, Op2,Key;
    unsigned char msg1[] = "    CALCULATOR PROGRAM";
    unsigned char msg2[] = "    Enter First Number: ";
    unsigned char msg3[]= "Enter Second Nummber: ";
    unsigned char msg4[] = "        Enter Operation: ";
    unsigned char msg5[] = "                Result = ";
//
// Configure the USART
//
    Usart_Init(9600);                          // Baud=9600
//
// Program loop
//
    for(;;)                                     // Endless loop
    {
      MyKey = 0;
      Op1 = 0;
      Op2 = 0;

      Newline();                                // Send newline
      Newline();                                // Send newline
      Text_To_Usart(msg1);                      // Send TEXT
      Newline();                                // Send newline
      Newline();                                // Send newline

//
// Get the first number
//
      Text_To_Usart(msg2);                      // Send TEXT to USART
      do                                        // Get first number
      {
        if(Usart_Data_Ready())                  // If a character ready
        {
          MyKey = Usart_Read();                 // Get a character
          if(MyKey == Enter)break;              // If ENTER key
          Usart_Write(MyKey);                   // Echo the character
          Key = MyKey - '0';
          Op1 = 10*Op1 + Key;                   // First number in Op1
        }
      }while(1);
```

Figure 6.56: (Cont'd)

```
        Newline();
    //
    // Get the second character
    //
        Text_To_Usart(msg3);                        // Send TEXT to USART
        do                                          // Get second number
        {
          if(Usart_Data_Ready())
          {
            MyKey = Usart_Read();                   // Get a character
            if(Mykey == Enter)break;                // If ENTER key
            Usart_Write(MyKey);                     // Echo the character
            Key = MyKey - '0';
            Op2 = 10*Op2 + Key;                     // Second number in Op2
          }
        }while(1);

        Newline();
    //
    // Get the operation
    //
        Text_To_Usart(msg4);
        do
        {
          if(Usart_Data_Ready())
          {
            MyKey = Usart_Read();                   // Get a character
            if(MyKey == Enter)break;                // If ENTER key
            Usart_Write(MyKey);                     // Echo the character
            Key = MyKey;
          }
        }while(1);

    //
    // Perform the operation
    //
        Newline();
        switch(Key)                                 // Calculate
        {
          case Plus:
              Calc = Op1 + Op2;                     // If ADD
              break;
          case Minus:
              Calc = Op1 - Op2;                     // If Subtract
              break;
          case Multiply:
              Calc = Op1 * Op2;                     // If Multiply
              break;
```

Figure 6.56: (Cont'd)

(Continued)

```
        case Divide:
            Calc = Op1 / Op2;                          // If Divide
            break;
    }

        LongToStr(Calc, op);                           // Convert to string
//
// Remove leading blanks
//
      j=0;
      for(i=0;i<=11;i++)
      {
       if(op[i] != ' ')                                // If a blank
       {
        kbd[j]=op[i];
          j++;
       }
      }

      Text_To_Usart(msg5);
      for(i=0; i<j;i++)Usart_Write(kbd[i]);            // Display result

    }
  }
```

Figure 6.56: (Cont'd)

- Start HyperTerminal terminal emulation software and give a name to the session

- Select *File -> New connection -> Connect using* and select COM2

- Select the baud rate as 9600, data bits as 8, no parity bits, and 1 stop bit

- Reset the microcontroller

An example output from the HyperTerminal screen is shown in Figure 6.57.

Using Software-Based Serial Communication

The preceding example made use of the microcontroller's USART and thus its special serial I/O pins. Serial communication can also be handled entirely in software, without using the USART. In this method, any pin of the microcontroller can be used for serial communication.

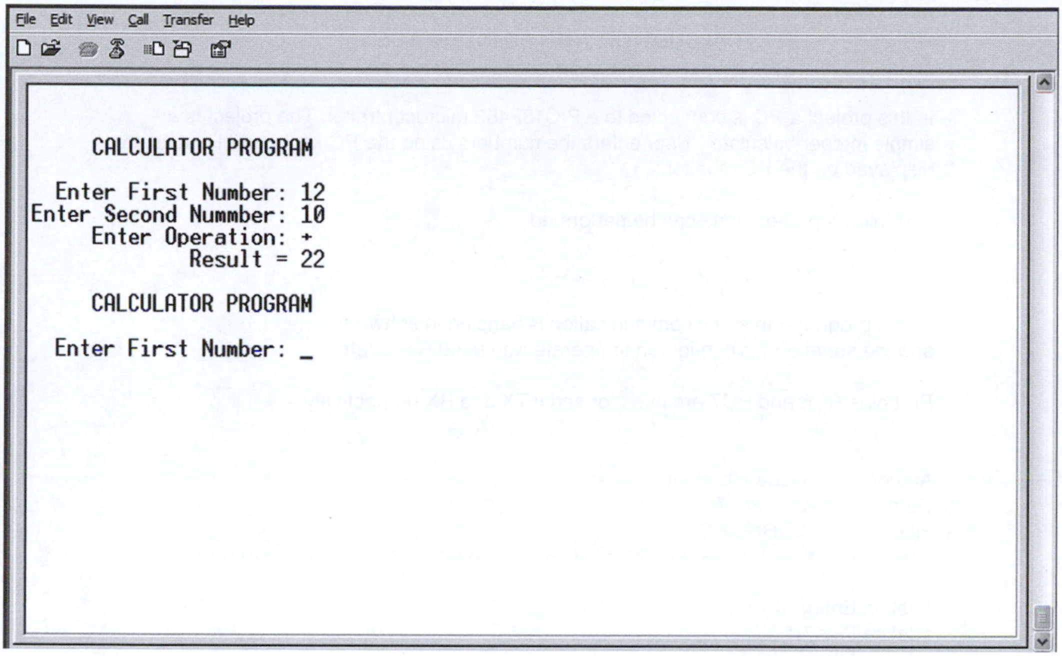

Figure 6.57: HyperTerminal screen

The calculator program given in Project 10 can be reprogrammed using the mikroC software serial communications library functions known as the *Software Uart Library*.

The modified program listing is given in Figure 6.58. The circuit diagram of the project is same as in Figure 6.54 (i.e., RC6 and RC7 are used for serial TX and RX respectively), although any other port pins can also be used. At the beginning of the program the serial I/O port is configured by calling function *Soft_Uart_Init*. The serial port name, the pins used for TX and RX, the baud rate, and the mode are specified. The mode tells the microcontroller whether or not the data is inverted. Setting mode to 1 inverts the data. When a MAX232 chip is used, the data should be noninverted (i.e., mode = 0).

Serial data is then output using function *Soft_Uart_Write*. Serial data is input using function *Soft_Uart_Read*. As the reading is a nonblocking function, it is necessary to check whether or not a data byte is available before attempting to read. This is done using the *error* argument of the function. The remaining parts of the program are the same.

```
/**********************************************************************
                    CALCULATOR WITH PC INTERFACE
                    =================================

In this project a PC is connected to a PIC18F452 microcontroller. The project is a
simple integer calculator. User enters the numbers using the PC keyboard. Results are
displayed on the PC monitor.

The following operations can be performed:

    + – * /

In this program the serial communication is handled in software
and the serial port is configured to operate with 9600 Baud rate.

Port pins RC6 and RC7 are used for serial TX and RX respectively.


Author:     Dogan Ibrahim
Date:       July 2007
File:       SERIAL2.C
**********************************************************************/

#define Enter 13
#define Plus '+'
#define Minus '–'
#define Multiply '*'
#define Divide '/'

//
// This function sends carriage-return and line-feed to USART
//
void Newline()
{
    Soft_Uart_Write(0x0D);                    // Send carriage-return
    Soft_Uart_Write(0x0A);                    // Send line-feed
}



//
// This function sends a text to serial port
//
void Text_To_Usart(unsigned char *m)
{
    unsigned char i;

    i = 0;
    while(m[i] != 0)
    {                                         // Send TEXT to serial port
        Soft_Uart_Write(m[i]);
```

Figure 6.58: Modified program

```
       i++;
     }
}


//
// Start of MAIN program
//
void main()
{
   unsigned char MyKey, i,j,error,kbd[5],op[12];
   unsigned long Calc, Op1, Op2,Key;
   unsigned char msg1[] = "    CALCULATOR PROGRAM";
   unsigned char msg2[] = " Enter First Number: ";
   unsigned char msg3[]=  "Enter Second Nummber: ";
   unsigned char msg4[] = "   Enter Operation: ";
   unsigned char msg5[] = "          Result = ";
//
// Configure the serial port
//
   Soft_Uart_Init(PORTC,7,6,2400,0);          // TX=RC6, RX=RC7, Baud=9600
//
// Program loop
//
   for(;;)                                     // Endless loop
   {
     MyKey = 0;
     Op1 = 0;
     Op2 = 0;

     Newline();                               // Send newline
     Newline();                               // Send newline
     Text_To_Usart(msg1);                     // Send TEXT
     Newline();                               // Send newline
     Newline();                               // Send newline


//
// Get the first number
//
     Text_To_Usart(msg2);                     // Send TEXT
     do                                       // Get first number
     {
       do                                     // If a character ready
          MyKey = Soft_Uart_Read(&error);     // Get a character
       while (error);
       if(MyKey == Enter)break;               // If ENTER key
       Soft_Uart_Write(MyKey);                // Echo the character
       Key = MyKey - '0';
       Op1 = 10*Op1 + Key;                    // First number in Op1
```

Figure 6.58: (Cont'd)

(Continued)

```
        }while(1);

        Newline();

//
// Get the second character
//
        Text_To_Usart(msg3);                    // Send TEXT
        do                                      // Get second number
        {
          do
            MyKey = Soft_Uart_Read(&error);     // Get a character
          while(error);
          if(Mykey == Enter)break;              // If ENTER key
          Soft_Uart_Write(MyKey);               // Echo the character
          Key = MyKey - '0';
          Op2 = 10*Op2 + Key;                   // Second number in Op2

        }while(1);

        Newline();
//
// Get the operation
//
        Text_To_Usart(msg4);
        do
        {
          do
            MyKey = Soft_Uart_Read(&error);     // Get a character
          while(error);
          if(MyKey == Enter)break;              // If ENTER key
          Soft_Uart_Write(MyKey);               // Echo the character
          Key = MyKey;

        }while(1);

//
// Perform the operation
//
        Newline();
        switch(Key)                             // Calculate
        {
          case Plus:
            Calc = Op1 + Op2;                   // If ADD
            break;
          case Minus:
            Calc = Op1 – Op2;                   // If Subtract
            break;
          case Multiply:
            Calc = Op1 * Op2;                   // If Multiply
```

Figure 6.58: (Cont'd)

```
                    break;
              case Divide:
                    Calc = Op1 / Op2;                    // If Divide
                    break;
          }

          LongToStr(Calc, op);                           // Convert to string
//
// Remove leading blanks
//
          j=0;
          for(i=0;i<=11;i++)
          {
           if(op[i] != ' ')                              // If a blank
           {
             kbd[j]=op[i];
             j++;
           }
          }

          Text_To_Usart(msg5);
          for(i=0; i<j;i++)Soft_Uart_Write(kbd[i]);      // Display result

       }
}
```

Figure 6.58: (Cont'd)

Advanced PIC18 Projects—SD Card Projects

In this and the remaining chapters we will look at the design of more complex PIC18 microcontroller–based projects. This chapter discusses the design of Secure Digital (SD) memory card–based projects. The remaining chapters of the book describe the basic theory and design of projects based on the popular USB bus and CAN bus protocols.

7.1 The SD Card

Before going into the design details of SD card–based projects, we should take a look at the basic principles and operation of SD card memory devices. Figure 7.1 shows a typical SD card.

The SD card is a flash memory storage device designed to provide high-capacity, nonvolatile, and rewritable storage in a small size. These devices are frequently used in many electronic consumer goods, such as cameras, computers, GPS systems, mobile phones, and PDAs. The memory capacity of the SD cards is increasing all the time. Currently they are available at capacities from 256MB to 8GB. The SD cards come in three sizes: standard, mini, and micro. Table 7.1 lists the main specifications of the most common standard SD and miniSD cards.

SD card specifications are maintained by the SD Card Association, which has over six hundred members. MiniSD and microSD cards are electrically compatible with the standard SD cards and can be inserted in special adapters and used as standard SD cards in standard card slots.

Figure 7.1: A typical SD card

Table 7.1: Standard SD and miniSD cards

	Standard SD	miniSD
Dimensions	32 × 24 × 2.1mm	21.5 × 20 × 1.4mm
Card weight	2.0 grams	1.0 grams
Operating voltage	2.7–3.6V	2.7–3.6V
Write protect	yes	no
Pins	9	11
Interface	SD or SPI	SD or SPI
Current consumption	<75mA (Write)	<40mA (Write)

SD card speeds are measured three different ways: in KB/s (kilobytes per second), in MB/s (megabytes per second), in an "x" rating similar to that of CD-ROMS where "x" is the speed corresponding to 150KB/s. The various "x" based speeds are:

- 4x: 600KB/s

- 16x: 2.4MB/s

- 40x: 6.0MB/s

- 66x: 10MB/s

In this chapter we are using the standard SD card only. The specifications of the smaller SD cards are the same and are not described further in this chapter.

SD cards can be interfaced to microcontrollers using two different protocols: SD card protocol and the SPI (Serial Peripheral Interface) protocol. The SPI protocol, being more widely used, is the one used in this chapter. The standard SD card has 9 pins with the pin layout shown in Figure 7.2. The pins have different functions depending on the interface protocol. Table 7.2 gives the function of each pin in both the SD and SPI modes of operation.

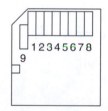

Figure 7.2: Standard SD card pin layout

Since the SD card projects described in this chapter are based on the SPI bus protocol, it is worth looking at the specifications of this bus before proceeding to the projects themselves.

7.1.1   The SPI Bus

The SPI (Serial Peripheral Interface) bus is a synchronous serial bus standard named by Motorola that operates in full duplex mode. Devices on a SPI bus operate in master-slave mode, where the master device initiates the data transfer, selects a slave, and provides a clock for the slaves. The selected slave responds and sends its data to the

Table 7.2: Standard SD card pin definitions

Pin	Name	SD description	SPI description
1	CD/DAT3/CS	Data line 3	Chip select
2	CMD/Datain	Command/response	Host to card command and data
3	VSS	Supply ground	Supply ground
4	VDD	Supply voltage	Supply voltage
5	CLK	Clock	Clock
6	VSS2	Supply voltage ground	Supply voltage ground
7	DAT0	Data line 0	Card to host data and status
8	DAT1	Data line 1	Reserved
9	DAT2	Data line 2	Reserved

master at each clock pulse. The SPI bus can operate with a single master device and one or more slave devices. This simple interface is also called a "four-wire" interface.

The signals in the SPI bus are named as follows:

- MOSI—master output, slave input
- MISO—master input, slave output
- SCLK—serial clock
- SS—slave select

These signals are also named as:

- DO—data out
- DI—data in
- CLK—clock
- CD—chip select

Figure 7.3 shows the basic connection between a master device and a slave device in SPI bus. The master sends out data on line MOSI and receives data on line MISO. The slave must be selected before data transfer can take place.

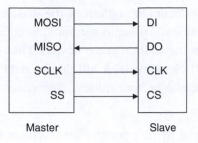

Figure 7.3: SPI master-slave connection

Figure 7.4 shows an instance where more than one slave device is connected to the SPI bus. Here, each slave is selected individually by the master, and although all the slaves receive the clock pulses, only the selected slave device responds. If an SPI device is not selected, its data output goes into a high-impedance state so it does not interfere with the currently selected device on the bus.

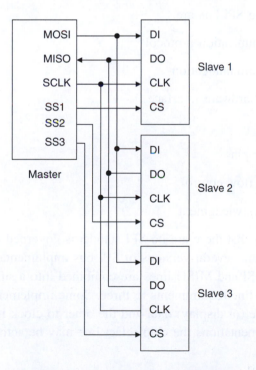

Figure 7.4: Multiple-slave SPI bus

Data transmission normally occurs in and out of the master and slave devices as the master supplies clock pulses. To begin a communication, the master first pulls the slave select line low for the desired slave device. Then the master issues clock pulses, and during each SPI clock cycle, a full duplex data transmission occurs. When there are no more data to be transmitted, the master stops toggling its clock output.

The SPI bus is currently used by microcontroller interface circuits to talk to a variety of devices such as:

- Memory devices (SD cards)

- Sensors

- Real-time clocks

- Communications devices

- Displays

The advantages of the SPI bus are:

- Simple communication protocol

- Full duplex communication

- Very simple hardware interface

Its disadvantages are:

- Requires four pins

- No hardware flow control

- No slave acknowledgment

It is worth remarking that there are no SPI standards governed by an international committee, so there are several versions of SPI bus implementation. In some applications, the MOSI and MISO lines are combined into a single data line, thereby reducing the line requirements to three. Some implementations have two clocks, one to capture (or display) data and the other to clock it into the device. Also, in some implementations the chip select line may be active-high rather than active-low.

7.1.2   Operation of the SD Card in SPI Mode

When the SD card is operated in SPI mode, only seven pins are used. Three (pins 3, 4, and 6) are used for the power supply, leaving four pins (pins 1, 2, 5, and 7) for the SPI mode of operation:

- Two power supply ground (pins 3 and 6)

- Power supply (pin 4)

- Chip select (pin 1)

- Data out (pin 7)

- Data in (pin 2)

- CLK (pin 5)

At power-up, the SD card defaults to the SD bus protocol. The card is switched to SPI mode if the Chip Select (CS) signal is asserted during reception of the reset command. When the card is in SPI mode, it only responds to SPI commands. The host may reset a card by switching the power supply off and then on again.

The mikroC compiler provides a library of commands for initializing, reading, and writing to SD cards. It is not necessary to know the internal structure of an SD card in detail before using one, since the library functions are available. However, a basic understanding of the internal structure of an SD card is helpful in making the best use of the card. In this section we will look briefly at the internal architecture and operation of SD cards.

An SD card has a set of registers that provide information about the status of the card. When the card is operated in SPI mode these are:

- Card identification register (CID)

- Card specific data register (CSD)

- SD configuration register (SCR)

- Operation control register (OCR)

The CID register consists of 16 bytes and contains the manufacturer ID, product name, product revision, card serial number, manufacturer date code, and a checksum byte. Table 7.3 shows the structure of the CID register.

Table 7.3: Structure of the CID register

Name	Type	Width	Comments
Manufacturer ID (MID)	Binary	1 byte	Manufacturer ID (e.g., 0×03 for SanDisk)
OEM/Application ID (OID)	ASCII	2 bytes	Identifies card OEM and/or card contents
Product Name (PNM)	ASCII	5 bytes	Product name
Product Revision (PRV)	BCD	1 byte	Two binary coded digits
Serial Number (PSN)	Binary	4 bytes	32 bits unsigned integer
Reserved		4 bits	Upper 4 bits
Manufacture Date Code (MDT)	BCD	12 bits	Manufacture date (offset from 2000)
CRC-7 Checksum	Binary	7 bits	Checksum
Not used	Binary	1 bit	Always 1

The CSD register consists of 16 bytes and contains card-specific data such as the card data transfer rate, read/write block lengths, read/write currents, erase sector size, file format, write protection flags, and checksum. Table 7.4 shows the structure of the CSD register.

The SCR register is 8 bytes long and contains information about the SD card's special features and capabilities, such as security support and data bus widths supported.

The OCR register is only 4 bytes long and stores the VDD voltage profile of the card. The OCR shows the voltage range in which the card data can be accessed.

All SD-card SPI commands are 6 bytes long, with the MSB transmitted first. Figure 7.5 shows the command format. The first byte is known as the command byte, and the remaining five bytes are called command arguments. Bit 6 of the command byte is set to 1 and the MSB bit is always 0. With the remaining six bits we have sixty-four possible commands, named CMD0 to CMD63. Some of the important commands are:

- CMD0   GO_IDLE_STATE (resets the SD card)

- CMD1   SEND_OP_COND (initializes the card)

- CMD9   SEND_CSD (gets CSD register data)

Table 7.4: Structure of the CSD register

Bytes	
Byte 0	0 0 XXXXXX
Byte 1	TAAC [7:0]
Byte 2	NSAC [7:0]
Byte 3	TRAN_SPEED [7:0]
Byte 4	CCC [11:4]
Byte 5	CCC [3:0] READ_BL_LEN [3:0]
Byte 6	READ_BL_PARTIAL WRITE_BLK_MISALIGN READ_BLK_MISALIGN DSR_IMP X X C_SIZE (11:10)
Byte 7	C_SIZE [9:2]
Byte 8	C_SIZE [1:0] VDD_R_CURR_MIN (2:0) VDD_R_CURR_MAX (2:0)
Byte 9	VDD_W_CURR_MIN (2:0) VDD_W_CURR_MAX (2:0) C_SIZE_MULT (2:1)
Byte 10	ERASE_BLK_EN SECTOR_SIZE (6:1)
Byte 11	SECTOR_SIZE (0) WP_GRP_SIZE (6:0)
Byte 12	WP_GRP_ENABLE X X R2W_FACTOR(2:0)
Byte 13	WRITE_BL_LEN (1:0) 0 X X X X X
Byte 14	FILE_FORMAT_GRP COPY PERM_WRITE_PROTECT TMP_WRITE_PROTECT FILE_FORMAT (1:0) X X
Byte 15	CRC (6:0) 1

Field definitions	
TAAC	data read access time 1 (e.g., 1.5ms)
NSAC	data read access time in CLK cycles
TRAN_SPEED	max data transfer rate
CCC	card command classes
READ_BL_LEN	max read data block length (e.g., 512 bytes)
READ_BL_PARTIAL	partial blocks for read allowed

(Continued)

Table 7.4: (Cont'd)

Field definitions	
WRITE_BLK_MISALIGN	write block misalignment
READ_BLK_MISALIGN	read block misalignment
DSR_IMP	DSR implemented
C_SIZE	device size
VDD_R_CURR_MIN	max read current at VDD min
VDD_R_CURR_MAX	max read current at VDD max
VDD_W_CURR_MIN	max write current at VDD min
VDD_W_CURR_MAX	max write current at VDD max
C_SIZE_MULT	device size multiplier
ERASE_BLK_EN	erase single block enable
SECTOR_SIZE	erase sector size
WP_GRP_SIZE	write protect group size
WP_GRP_ENABLE	write protect group enable
R2W_FACTOR	write speed factor
WRITE_BL_LEN	max write data block length (e.g., 512 bytes)
WRITE_BL_PARTIAL	partial blocks for write allowed
FILE_FORMAT_GRP	file format group
COPY	copy flag
PERM_WRITE_PROTECT	permanent write protect
TMP_WRITE_PROTECT	temporary write protect
FILE_FORMAT	file format

Byte 1			Byte 2 - 5		Byte 6	
7	6		31	0	7	0
0	1	Command	Command argument		CRC	1

Figure 7.5: SD card SPI command format

- CMD10   SEND_CID (gets CID register data)

- CMD16   SET_BLOCKLEN (selects a block length in bytes)

- CMD17   READ_SINGLE_BLOCK (reads a block of data)

- CMD24   WRITE_BLOCK (writes a block of data)

- CMD32   ERASE_WR_BLK_START_ADDR (sets the address of the first write block to be erased)

- CMD33   ERASE_WR_BLK_END_ADDR (sets the address of the last write block to be erased)

- CMD38   ERASE (erases all previously selected blocks)

In response to a command, the card sends a status byte known as R1. The MSB bit of this byte is always 0 and the other bits indicate the following error conditions:

- Card in idle state

- Erase reset

- Illegal command

- Communication CRC error

- Erase sequence error

- Address error

- Parameter error

Reading Data

The SD card in SPI mode supports single-block and multiple-block read operations. The host should set the block length. After a valid read command the card responds with a response token, followed by a data block and a CRC check. The block length can be between 1 and 512 bytes. The starting address can be any valid address in the address range of the card.

In multiple-block read operations, the card sends data blocks with each block having its own CRC check attached to the end of the block.

Writing Data

The SD card in SPI mode supports single- or multiple-block write operations. After receiving a valid write command from the host, the card responds with a response token and waits to receive a data block. A one-byte "start block" token is added to the beginning of every data block. After receiving the data block the card responds with a "data response" token, and the card is programmed as long as the data block is received with no errors.

In multiple-block write operations the host sends the data blocks one after the other, each preceded by a "start block" token. The card sends a response byte after receiving each data block.

Card Size Parameters   SD cards are available in various sizes. At the time of writing, SanDisk Corporation (www.sandisk.com) offered the models and capacities shown in Table 7.5. The company may now be offering models with 4GB or even greater capacity.

In addition to the normal storage area on the card, there is also a protected area pertaining to the secured copyright management. This area can be used by applications to save security-related data and can be accessed by the host using secured read/write commands. The card write protection mechanism does not affect this area. Table 7.6 shows the size of the protected area and the data area available to the user for reading and writing data. For example, a 1GB card has 20,480 blocks (one block is 512 bytes) of protected area and 1,983,744 blocks of user data area.

Table 7.5: SanDisk card models and capacities

Model	Capacities
SDSDB-16	16 MB
SDSDB-32	32 MB
SDSDJ-64	64 MB
SDSDJ-128	128 MB
SDSDJ-256	256 MB
SDSDJ-512	512 MB
SDSDJ-1024	1024 MB

Table 7.6: Protected area and data area sizes

Model	Protected area (blocks)	User area (blocks)
SDSDB-16	352	28,800
SDSDB-32	736	59,776
SDSDJ-64	1,376	121,856
SDSDJ-128	2,624	246,016
SDSDJ-256	5,376	494,080
SDSDJ-512	10,240	940,864
SDSDJ-1024	20,480	1,983,744

1 block = 512 bytes.

Data can be written to or read from any sector of the card using raw sector access methods. In general, SD card data is structured as a file system and two DOS-formatted partitions are placed on the card: the user area and the security protected area. The size of each area is shown in Table 7.7. For example, in a 1GB card, the size of the security protected area is 519 sectors (1 sector is 512 bytes), and the size of the user data area is 1,982,976 sectors.

Table 7.7: Size of the security protected area and the user area in a DOS-formatted card

Model	Protected area (sectors)	User area (sectors)
SDSDB-16	39	28,704
SDSDB-32	45	59,680
SDSDJ-64	57	121,760
SDSDJ-128	95	245,824
SDSDJ-256	155	493,824
SDSDJ-512	275	990,352
SDSDJ-1024	519	1,982,976

1 sector = 512 bytes.

A card can be inserted and removed from the bus without any damage. This is because all data transfer operations are protected by cyclic redundancy check (CRC) codes, and any bit changes caused by inserting or removing a card can easily be detected. SD cards typically operate with a supply voltage of 2.7V. The maximum allowed power supply voltage is 3.6V. If the card is to be operated from a standard 5.0V supply, a voltage regulator should be used to drop the voltage to 2.7V.

Using an SD card requires the card to be inserted into a special card holder with external contacts (see Figure 7.6) so connections are easily made to the required card pins.

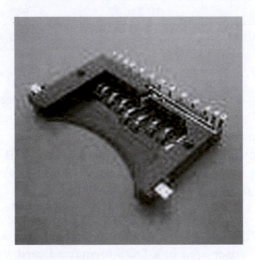

Figure 7.6: SD card holder

7.2   mikroC Language SD Card Library Functions

The mikroC language provides an extensive set of library functions to read and write data to SD cards (and also MultiMediaCards, MMC). Data can be written to or read from a given sector of the card, or the file system on the card can be used for more sophisticated applications.

The following library functions are provided:

- Mmc_Init   (initialize the card)

- Mmc_Read_Sector   (read one sector of data)

- Mmc_Write_Sector   (write one sector of data)

- Mmc_Read_Cid   (read CID register)

- Mmc_Read_Csd   (read CSD register)

- Mmc_Fat_Init   (initialize FAT)

- Mmc_Fat_QuickFormat   (format the card to FAT16)

- Mmc_Fat_Assign   (assign the file we will be working with)

- Mmc_Fat_Reset   (reset the file pointer; opens the currently assigned file for reading)

- Mmc_Fat_Rewrite   (reset the file pointer and clear assigned file; opens the assigned file for writing)

- Mmc_Fat_Append   (move file pointer to the end of assigned file so new data can be appended to the file)

- Mmc_Fat_Read   (read the byte the file pointer points to)

- Mmc_Fat_Write   (write a block of data to the assigned file)

- Mmc_Set_File_Date   (write system timestamp to a file)

- Mmc_Fat_Delete   (delete a file)

- Mmc_Fat_Get_File_Date   (read file timestamp)

- Mmc_Fat_Get_File_Size   (get file size in bytes)

- Mmc_Fat_Get_Swap_File   (create a swap file)

In the remainder of this chapter we will look at some SD-card and PIC18 microcontroller-based projects.

PROJECT 7.1—Read CID Register and Display on a PC Screen

In this project a SD card is interfaced to a PIC18F452-type microcontroller. The serial output port of the microcontroller is connected to the serial input port (e.g.,

COM1) of a PC. The microcontroller reads the contents of the card CID register and sends this data to the PC so it can be displayed on the PC screen.

Figure 7.7 shows the block diagram of the project.

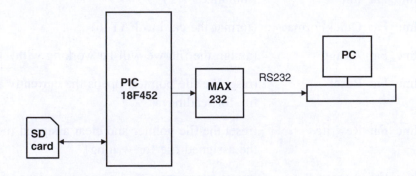

Figure 7.7: Block diagram of the project

The circuit diagram of the project is shown in Figure 7.8. The SD card is inserted into a card holder and then connected to PORTC of a PIC18F452 microcontroller through 2.2K and 3.3K resistors, using the following pins:

- Card CS   to PORTC pin RC2

- Card CLK to PORTC pin RC3

- Card DO to PORTC pin RC4

- Card DI   to PORTC pin RC5

According to the SD card specifications, when the card is operating with a supply voltage of VDD = 3.3V, the input-output pin voltage levels are as follows:

- Minimum produced output HIGH voltage, VOH = 2.475V

- Maximum produced output LOW voltage, VOL = 0.4125V

- Minimum required input HIGH voltage, VIH = 2.0625

- Maximum input HIGH voltage, VIH = 3.6V

- Maximum required input LOW voltage, VIL = 0.825V

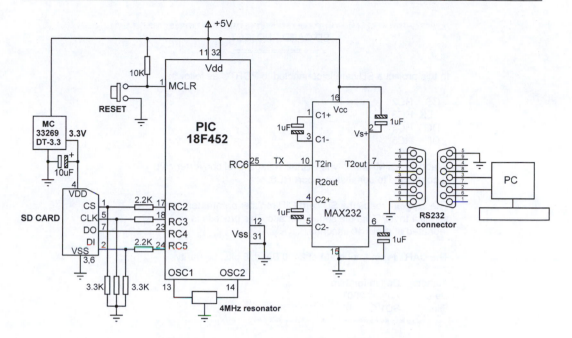

Figure 7.8: Circuit diagram of the project

Although the output produced by the card (2.475V) is sufficient to drive the input port of a PIC microcontroller, the logic HIGH output of the microcontroller (about 4.3V) is too high for the SD card inputs (maximum 3.6V). Therefore, a potential divider is set up at the three inputs of the SD card using 2.2K and 3.3K resistors. This limits the maximum voltage at the inputs of the SD card to about 2.5V:

$$\text{SD card input voltage} = 4.3\text{V} \times 3.3\text{K}/(2.2\text{K} + 3.3\text{K}) = 2.48\text{V}$$

Serial output port pin RC6 (TX) of the microcontroller is connected to a MAX232-type RS232 voltage level converter chip and then to a 9-way D-type connector so it can be connected to the serial input port of a PC.

The microcontroller is powered from a 5V supply which is obtained via a 7805-type 5V regulator with a 9V input. The 2.7V–3.6V supply required by the SD card is obtained via an MC33269DT-3.3 regulator with 3.3V output and is driven from the 5V input voltage.

The program listing of the project is given in Figure 7.9 (program SD1.C). At the beginning of the main program, character array *CID* is declared to have 16 bytes.

```
/************************************************************
                      SD CARD PROJECT
                      ================

In this project a SD card is connected to PORTC as follows:

        CS    RC2
        CLK   RC3
        DO    RC4
        DI    RC5

In addition, a MAX232 type RS232 voltage level converter chip
is connected to serial output port RC6.

The program reads the SD card CID register parameters and
sends it to a PC via the serial interface. This process is
repeated at every 10 seconds.

The UART is set to operate at 2400 Baud, 8 bits, no parity.

Author:    Dogan Ibrahim
Date:      August 2007
File:      SD1.C
*************************************************************/

//
// This function sends carriage-return and line-feed to USART
//
void Newline()
{
    Soft_Uart_Write(0x0D);                  // Send carriage-return
    Soft_Uart_Write(0x0A);                  // Send line-feed
}


//
// This function sends a space character to USART
//
void Space()
{
    Soft_Uart_Write(0x20);
}


//
// This function sends a text to serial port
//
void Text_To_Usart(unsigned char *m)
{
    unsigned char i;
```

Figure 7.9: Program listing

```
    i = 0;
    while(m[i] != 0)
    {                                                    // Send TEXT to serial port
      Soft_Uart_Write(m[i]);
      i++;
    }
}

//
// This function sends string to serial port. The string length is passed as an argument
//
void Str_To_Usart(unsigned char *m,unsigned char 1)
{
    unsigned char i;
    unsigned char txt[4];

    i=0;
    for(i=0; i<l; i++)
    {
      ByteToStr(m[i],txt);
      Text_To_Usart(txt);
      Space();
    }
}


//
// Start of MAIN program
//
void main()
{
    unsigned char error,CID[16];
    unsigned char msg[] = "    SD CARD CID REGISTER";

//
// Configure the serial port
//
    Soft_Uart_Init(PORTC,7,6,2400,0);                    // TX=RC6
//
// Initialise the SD card
//
    Spi_Init_Advanced(MASTER_OSC_DIV16,DATA_SAMPLE_MIDDLE,
            CLK_IDLE_LOW, LOW_2_HIGH);
//
// Initialise the SD bus
//
    while(Mmc_Init(&PORTC,2));
//
// Start of MAIN loop. Read the SD card CID register and send the data
```

Figure 7.9: (Cont'd)

```
// to serial port every 10 seconds
//
    for(;;)                                      // Endless loop
    {

      Text_To_Usart(msg);                        // Send TEXT
      Newline();                                     // Send newline
      Newline();                                     // Send newline
      error = Mmc_Read_Cid(CID);                     // Read CID register into CID
//
// Send the data to RS232 port
//
      Str_To_Usart(CID,16);                      // Send CID contents to UART
      Delay_Ms(10000);                           // Wait 10 seconds
      Newline();
      Newline();
    }
}
```

Figure 7.9: (Cont'd)

Variable *msg* is loaded with the message that is to be displayed when power is applied to the system. Then the UART is initialized at PORTC with a baud rate of 2400.

Before the SD card library functions are used, the function *Spi_Init_Advanced* must be called with the given arguments. Then the SD card bus is initialized by calling function *Mmc_Init*, where it is specified that the card is connected to PORTC. The program then enters an endless loop that repeats every ten seconds. Inside this loop the heading message is displayed followed by two new-line characters. The program then reads the contents of register CID by calling function *Mmc_Read_Cid* and stores the data in character array *CID*. The data is then sent to the serial port by calling function *Str_To_Usart*. At the end of the loop two new-line characters are displayed, the program waits for ten seconds, and the loop is repeated.

The operation of the project can be tested by connecting the device to a PC and starting the HyperTerminal terminal emulation program on the PC. Set the communications parameters to 2400 baud, 8 data bits, 1 stop bit, and no parity bit. An example output on the screen is shown in Figure 7.10.

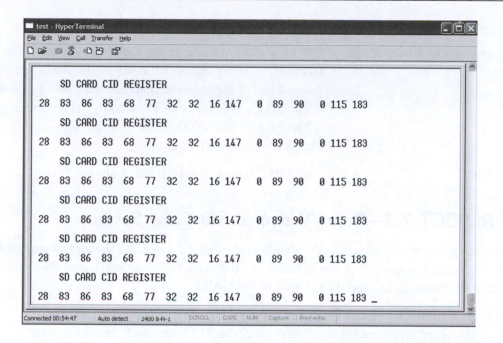

Figure 7.10: An example output from the project on HyperTerminal

The data returned by the card is:

28 83 86 83 68 77 32 32 16 147 0 89 90 0 115 183

Referring to Table 7.3, we can say the following about this card:

Manufacturer ID   =   28 decimal

OEM/Application ID =   SV

Product Name   =   SDM

Product Revision   =   1.0 (decimal 16 corresponds to binary "0001 0000" which is 10 in BCD; the revision number is as n.m, giving 1.0)

Serial Number   =   16 147 0 89 decimal

Reserved   = "0000" bits (4 bits only)

Manufacture Date Code = 073 (this 12-bit parameter has the binary value "0000 0111 0011" where the upper 4 bits are derived from the lower 4 bits of the reserved field and the lower 8 bits are decimal 115. This gives BCD value 073. The date is in YYM format since 2000. Thus, this card was manufactured in 2007, March).

CRC   = "1011100" binary (the LSB bit is always 1)

PROJECT 7.2—Read/Write to SD Card Sectors

The hardware of this project is the same as for Project 7.1 (i.e., as shown in Figure 7.8). In this project, sector 10 of the SD card is filled with "C" characters, and then this sector is read and the card data is sent to the UART.

The program listing of this project is given in Figure 7.11 (program SD2.C). Two character arrays called *data1* and *data2*, of 512 bytes each, are declared at the beginning of the program. Array *data1* is loaded with character "C," and the contents of this array are written to sector 10 of the SD card. Then the contents of sector 10 are read into character array *data2* and sent to the UART, displaying 512 "C" characters on the PC screen. Normally, only one array is used to read and write to the SD card. Two arrays are used here to make it clear that what is sent to the UART is the card data, not the contents of array *data1*.

PROJECT 7.3—Using the Card Filing System

The hardware of this project is the same as for Project 7.1 (i.e., as shown in Figure 7.8). In this project, a file called MYFILE55.TXT is created on the SD card. String "This is MYFILE.TXT" is written to the file initially. Then the string "This is the added data..." is appended to the file. The program then reads the contents of the file and sends the string "This is MYFILE.TXT. This is the added data..." to the UART, enabling the data to be displayed on the PC screen when HyperTerminal is run.

```
/****************************************************************
                    SD CARD PROJECT
                    ===============

In this project a SD card is connected to PORTC as follows:

    CS    RC2
    CLK   RC3
    DO    RC4
    DI    RC5

In addition, a MAX232 type RS232 voltage level converter chip
is connected to serial output port RC6.

The program loads sector 10 of the SD card with character "C".
The contents of sector 10 is then read and sent to the UART,
displaying 512 "C" characters on the PC display.

Author:    Dogan Ibrahim
Date:      August 2007
File:      SD2.C
****************************************************************/

unsigned char data1[512],data2[512];
unsigned int i;
unsigned short x;

void main()
{

//
// Configure the serial port
//
        Usart_Init(2400);
//
// Initialise the SD card
//  Spi_Init_Advanced(MASTER_OSC_DIV16,DATA_SAMPLE_MIDDLE,
            CLK_IDLE_LOW, LOW_2_HIGH);
//
// Initialise the SD bus
//
    while(Mmc_Init(&PORTC,2));
//
// Fill buffer with character "C"
//
    for(i=0; i<512; i++)data1[i] = 'C';
//
// Write to sector 10
//
    x = Mmc_Write_Sector(10, data1);
```

Figure 7.11: Program listing of the project

(Continued)

```
//
// Now read from sector 10 into data2 and send to UART
//
    x = Mmc_Read_Sector(10,data2);

    for(i=0; i<400; i++)Usart_Write(data2[i]);          // Send to UART

    for(;;);                                            // Wait here forever
}
```

Figure 7.11: (Cont'd)

The program listing of the project is given in Figure 7.12 (program SD3.C).
At the beginning of the program the UART is initialized to 2400 baud. Then the
SPI bus and the FAT file system are initialized as required by the library.
The program then creates file MYFILE55.TXT by calling library function
Mmc_Fat_Assign with the arguments as the *filename* and the creation flag 0×80,
which tells the function to create a new file if the file does not exist. The
filename should be in "filename.extension" format, though it is also possible to
specify an eight-digit filename and a three-digit extension with no "." between
them, as the "." will be inserted by the function. Other allowed values of the
creation flag are given in Table 7.8. Note that the SD card must have been
formatted in FAT16 before we can read or write to it. Most new cards are
already formatted, but we can also use the *Mmc_Fat_QuickFormat* function to
format a card.

The file is cleared (if it is not already empty) using function call *Mmc_Fat_Rewrite*,
and then the string "This is MYFILE.TXT" is written to the file by calling library
function *Mmc_Fat_Write*. Note that the size of the data to be written must be
specified as the second argument of this function call. Then *Mmc_Fat_Append* is
called and the second string "This is the added data…" is appended to the file.
Calling function *Mmc_Fat_Reset* sets the file pointer to the beginning of the
file and also returns the size of the file. Finally, a *for* loop is set up to read
each character from the file using the *Mmc_Fat_Read* function call, and
the characters read are sent to the UART with the *Usart_Write* function
call.

```
/*************************************************************
                   SD CARD PROJECT
                   ================

In this project a SD card is connected to PORTC as follows:

     CS    RC2
     CLK   RC3
     DO    RC4
     DI    RC5

In addition, a MAX232 type RS232 voltage level converter chip
is connected to serial output port RC6.

The program opens a file called MYFILE55.TXT on the SD card
and writes the string "This is MYFILE.TXT." to this file. Then
the string "This is the added data..." is appended to this file.
The program then sends the contents of this file to the UART.

Author:    Dogan Ibrahim
Date:      August 2007
File:      SD3.C
*************************************************************/

char filename[] = "MYFILE55TXT";
unsigned char txt[] = "This is the added data...";
unsigned short character;
unsigned long file_size,i;

void main()
{

//
// Configure the serial port
//
   Usart_Init(2400);
//
// Initialise the SPI bus
//
   Spi_Init_Advanced(MASTER_OSC_DIV16,DATA_SAMPLE_MIDDLE,
                     CLK_IDLE_LOW, LOW_2_HIGH);
//
// Initialise the SD card bus
//
   while(Mmc_Init(&PORTC,2));
//
// Initialise the FAT file system
//
   while(Mmc_Fat_Init(&PORTC,2));
//
```

Figure 7.12: Program listing of the project

(Continued)

```
// Create the file (if it doesn't exist)
//
   Mmc_Fat_Assign(&filename,0x80);
//
// Clear the file, start with new data
//
   Mmc_Fat_Rewrite();
//
// Write data to the file
//
   Mmc_Fat_Write("This is MYFILE.TXT.",19);
//
// Add more data to the end...
//
   Mmc_Fat_Append();
   Mmc_Fat_Write(txt,sizeof(txt));
//
// Now read the data and send to UART
//
   Mmc_Fat_Reset(&file_size);
   for(i=0; i<file_size; i++)
   {
      Mmc_Fat_Read(&character);
      Usart_Write(character);
   }

   for(;;);                                    // wait here forever

}
```

Figure 7.12: (Cont'd)

Table 7.8: *Mmc_Fat_Assign* file creation flags

Flag	Description
0×01	Read only
0×02	Hidden
0×04	System
0×08	Volume label
0×10	Subdirectory
0×20	Archive
0×40	Device (internal use only, never found on disk)
0×80	File creation flag. If file does not exist and this flag is set, a new file with the specified name will be created.

A snapshot of the screen with the HyperTerminal running is shown in Figure 7.13.

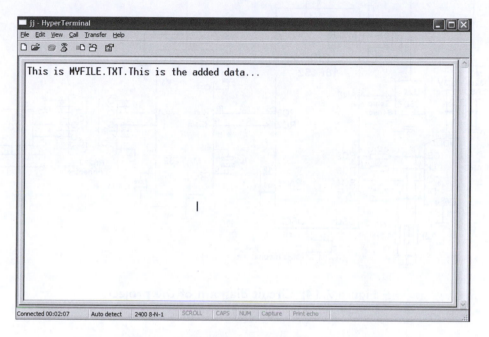

Figure 7.13: Snapshot of the screen

PROJECT 7.4—Temperature Logger

This project shows the design of a temperature data logger system. The ambient temperature is read every ten seconds and stored in a file on an SD card. The program is menu-based, and the user is given the option of:

- Sending the saved file contents to a PC

- Saving the temperature readings to a new file on an SD card

- Appending the temperature readings to an existing file on an SD card

The hardware of this project is similar to the one for Project 7.1 (i.e., as shown in Figure 7.8), but here, in addition, the serial input port pin (RC7) is connected to the RS232 connector so data can be received from the PC keyboard. In addition, a LM35DZ-type analog temperature sensor is connected to the microcontroller's analog input AN0 (pin 2). The new circuit diagram is shown in Figure 7.14.

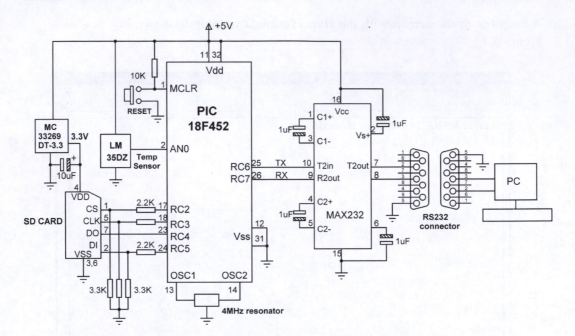

Figure 7.14: Circuit diagram of the project

The LM35 DZ is a three-pin analog temperature sensor that can measure with 1°C accuracy temperatures between 0°C and +100°C. One pin of the device is connected to the supply (+5V), another pin to the ground, and the third to the analog output. The output voltage of the sensor is directly proportional to the temperature (i.e., Vo = 10mV/°C). If, for example, the temperature is 10°C, the output voltage will be 100mV, and if the temperature is 35°C, the output voltage of the sensor will be 350mV.

When the program is started, the following menu is displayed on the PC screen:

TEMPERATURE DATA LOGGER

1. Send temperature data to the PC

2. Save temperature data in a new file

3. Append temperature data to an existing file

Choice?

The user then chooses one of the three options. When an option is completed, the program does not return to the menu. To display the menu again the system has to be restarted.

The program listing of the project is shown in Figure 7.15 (program SD4.C). In this project, a file called TEMPERTRTXT is created on the SD card to store the temperature readings (the library function call will insert the "." to make the filename "TEMPERTR.TXT"), if it does not already exist.

The following functions are created at the beginning of the program, before the main program:

Newline sends a carriage return and a line feed to the UART so the cursor moves to the next line.

Text_To_Usart receives a text string as its argument and sends it to the UART to display on the PC screen.

Get_Temperature starts the A/D conversion and receives the converted data into a variable called *Vin*. The voltage corresponding to this value is then calculated in millivolts and divided by 10 to find the actual measured temperature in °C. The decimal part of the found temperature is then converted into string form using function *LongToStr*. The leading spaces are removed from this string, and the resulting string is stored in character array *temperature*. Then the fractional parts of the measured temperature, a carriage return, and a line feed are added to this character array, which is later written to the SD card.

The following operations are performed inside the main program:

- Initialize the UART to 2400 baud

- Initialize the SPI bus

- Initialize the FAT file system

- Display menu on the PC screen

- Get a choice from the user (1, 2, or 3)

- If the choice = 1, assign the temperature file, read the temperature records, and display them on the PC screen

```
/***********************************************************
                 TEMPERATURE LOGGER PROJECT
                 ==============================

In this project a SD card is connected to PORTC as follows:

    CS    RC2
    CLK   RC3
    DO    RC4
    DI    RC5

In addition, a MAX232 type RS232 voltage level converter chip
is connected to serial ports RC6 and RC7. Also, a LM35DZ type
analog temperature sensor is connected to analog input AN0 of
the microcontroller.

The program is menu based. The user is given options of either
to send the saved temperature data to the PC, or to read and
save new data on the SD card, or to read temperature data and
append to the existing file. Temperature is read at every 10
seconds.

The temperature is stored in a file called "TEMPERTR.TXT"

Author:   Dogan Ibrahim
Date:     August 2007
File:     SD4.C
***********************************************************/

char filename[] = "TEMPERTRTXT";
unsigned short character;
unsigned long file_size,i,rec_size;
unsigned char ch1,ch2,flag,ret_status,choice;
unsigned char temperature[10],txt[12];

//
// This function sends carriage-return and line-feed to USART
//
void Newline()
{
    Usart_Write(0x0D);                    // Send carriage-return
    Usart_Write(0x0A);                    // Send line-feed
}


//
// This function sends a space character to USART
//
void Space()
{
```

Figure 7.15: Program listing of the project

```
      Usart_Write(0x20);
}


//
// This function sends a text to serial port
//
void Text_To_Usart(unsigned char *m)
{
    unsigned char i;

    i = 0;
    while(m[i] != 0)
    {                                        // Send TEXT to serial port
      Usart_Write(m[i]);
      i++;
    }
}


//
// This function reads the temperature from analog input AN0
//
void Get_Temperature()
{
    unsigned long Vin, Vdec,Vfrac;
    unsigned char op[12];
    unsigned char i,j;

    Vin = Adc_Read(0);                       // Read from channel 0 (AN0)
    Vin = 488*Vin;                           // Scale up the result
    Vin = Vin /10;                           // Convert to temperature in C
    Vdec = Vin / 100;                        // Decimal part
    Vfrac = Vin % 100;                       // Fractional part
    LongToStr(Vdec,op);                      // Convert Vdec to string in "op"
//
// Remove leading blanks
//
    j=0;
    for(i=0;i<=11;i++)
    {
      if(op[i] != ' ')                       // If a blank
      {
        temperature[j]=op[i];
        j++;
      }
    }

    temperature[j] = '.';                    // Add "."
    ch1 = Vfrac / 10;                        // fractional part
    ch2 = Vfrac % 10;
```

Figure 7.15: (Cont'd)

```
                j++;
                temperature[j] = 48+ch1;                                    // Add fractional part
                j++;
                temperature[j] = 48+ch2;
                j++;
                temperature[j] = 0x0D;                                      // Add carriage-return
                j++;
                temperature[j] = 0x0A;                                      // Add line-feed
                j++;
                temperature[j]='\0';
        }



//
// Start of MAIN program
//
void main()
{
    rec_size = 0;
//
// Configure A/D converter
//
    TRISA = 0xFF;
    ADCON1 = 0x80;                                                          // Use AN0, Vref = +5V
//
// Configure the serial port
//
    Usart_Init(2400);
//
// Initialise the SPI bus
//
    Spi_Init_Advanced(MASTER_OSC_DIV16,DATA_SAMPLE_MIDDLE,
                      CLK_IDLE_LOW, LOW_2_HIGH);
//
// Initialise the SD card bus
//
    while(Mmc_Init(&PORTC,2));
//
// Initialise the FAT file system
//
    while(Mmc_Fat_Init(&PORTC,2));
//
// Display the MENU and get user choice
//
    Newline();
    Text_To_Usart("TEMPERATURE DATA LOGGER");
    Newline();
    Newline();
    Text_To_Usart("1. Send temperature data to the PC");
```

Figure 7.15: (Cont'd)

```
            Newline();
            Text_To_Usart("2. Save temperature data in a new file");
            Newline();
            Text_To_Usart("3. Append temperature data to an existing file");
            Newline();
            Newline();
            Text_To_Usart("Choice ? ");

//
// Read a character from the PC keyboard
//
      flag = 0;
      do {
       if (Usart_Data_Ready())                        // If data received
       {
         choice = Usart_Read();                       // Read the received data
         Usart_Write(choice);                         // Echo received data
         flag = 1;
       }
      } while (!flag);
      Newline();
      Newline();

//
// Now process user choice
//
        switch(choice)
       {
           case '1':
                 ret_status = Mmc_Fat_Assign(&filename,1);
                 if(!ret_status)
                 {
                     Text_To_Usart("File does not exist..No saved data...");
                     Newline();
                     Text_To_Usart("Restart the program and save data to the file...");
                     Newline();
                     for(;;);
                 }
                 else
                 {
                   //
                   // Read the data and send to UART
                   //
                   Text_To_Usart("Sending saved data to the PC...");
                   Newline();
                   Mmc_Fat_Reset(&file_size);
                   for(i=0; i<file_size; i++)
                   {
                       Mmc_Fat_Read(&character);
                       Usart_Write(character);
```

Figure 7.15: (Cont'd)

```
          }
       Newline();
       text_To_Usart("End of data...");
       Newline();
       for(;;);
       }
  case '2':
       //
       // Start the A/D converter, get temperature readings every
       // 10 seconds, and then save in a NEW file
       //
       Text_To_Usart("Saving data in a NEW file...");
       Newline();
       Mmc_Fat_Assign(&filename,0x80);                    // Assign the file
       Mmc_Fat_Rewrite();                                 // Clear
       Mmc_Fat_Write("TEMPERATURE DATA - SAVED EVERY 10
                      SECONDS\r\n",43);
       //
       // Read the temperature from A/D converter, format and save
       //
       for(;;)
       {
         Mmc_Fat_Append();
         Get_Temperature();
         Mmc_Fat_Write(temperature,9);
         rec_size++;
         LongToStr(rec_size,txt);
         Newline();
         Text_To_Usart("Saving record:");
         Text_To_Usart(txt);
         Delay_ms(10000);
       }
       break;
  case '3':
       //
       // Start the A/D converter, get temperature readings every
       // 10 seconds, and then APPEND to the existing file
       //
       Text_To_Usart("Appending data to the existing file...");
       Newline();
       ret_status = Mmc_Fat_Assign(&filename,1);     // Assign the file
       if(!ret_status)
       {
         Text_To_Usart("File does not exist - can not append...");
         Newline();
         Text_To_Usart("Restart the program and choose option 2...");
         Newline();
         for(;;);
       }
       else
```

Figure 7.15: (Cont'd)

```
    {
    //
    // Read the temperature from A/D converter, format and save
    //
        for(;;)
        {
            Mmc_Fat_Append();
            Get_Temperature();
            Mmc_Fat_Write(temperature,9);
            rec_size++;
            LongToStr(rec_size,txt);
            Newline();
            Text_To_Usart("Appending new record:");
            Text_To_Usart(txt);
            Delay_ms(10000);
        }
    }
default:
        Text_To_Usart("Wrong choice...Restart the program and try again...");
        Newline();
        for(;;);
    }
}
```

Figure 7.15: (Cont'd)

- If the choice = 2, create a new temperature file, get new temperature readings every ten seconds, and store them in the file

- If the choice = 3, assign to the temperature file, get new temperature readings every ten seconds, and append them to the existing temperature file

- If the choice is not 1, 2, or 3, display an error message on the screen

The menu options are described here in more detail:

Option 1: The program attempts to assign the existing temperature file. If the file does not exist, the error messages "File does not exist...No saved data..." and "Restart the program and save data to the file..." are displayed on the screen, and the user is expected to restart the program. If, on the other hand, the temperature file already exists, then the message: "Sending saved data to the PC..." is displayed on the PC screen. Function *Mmc_Fat_Reset* is called to set the file pointer to the beginning of the file and also return the size of the file in bytes. Then a *for* loop is formed, temperature records are read from the card one byte at a time using function *Mmc_Fat_Read*, and these records are sent to the PC screen using function *Usart_Write*. At the end of the data the message "End of data…" is sent to the PC screen.

Option 2: In this option, the message "Saving data in a NEW file…" is sent to the PC screen, and a new file is created using function Mmc_Fat_Assign with the create flag set to 0×80. The message "TEMPERATURE DATA - SAVED EVERY 10 SECONDS" is written on the first line of the file using function *Mmc_Fat_Write*. Then, a *for* loop is formed, the SD card is set to file append mode by calling function *Mmc_Fat_Append*, and a new temperature reading is obtained by calling function *Get_Temperature*. The temperature is then written to the SD card. Also, the current record number appears on the PC screen to indicate that the program is actually working. This process is repeated after a ten-second delay.

Option 3: This option is very similar to Option 2, except that a new file is not created but rather the existing temperature file is opened in read mode. If the file does not exist, then an error message is displayed on the PC screen.

Default: If the user entry is a number other than 1, 2, or 3, then this option runs and displays the error message "Wrong choice…Restart the program and try again…" on the PC screen.

The project can be tested by connecting the output of the microcontroller to the serial port of a PC (e.g., COM1) and then running the HyperTerminal terminal emulation software. Set the communications parameters to 2400 baud, 8 data bits, 1 stop bit, and no parity bit. Figure 7.16 shows a snapshot of the PC screen when Option 2 is selected to save the temperature records in a new file. Notice that the current record numbers are displayed on the screen as they are written to the SD card.

Figure 7.17 shows a screen snapshot where Option 1 is selected to read the temperature records from the SD card and display them on the PC screen.

Figure 7.16: Saving temperature records on an SD card with Option 2

```
test - HyperTerminal
File  Edit  View  Call  Transfer  Help

 3. Append temperature data to an existing file

 Choice ? 1

 Sending saved data to the PC...
 TEMPERATURE DATA - SAVED EVERY 10 SECONDS
 27.81
 27.32
 30.25
 31.72
 32.69
 32.69
 30.25
 31.72
 32.69
 31.72
 32.20
 31.23
 30.25
 28.79
 28.30

 End of data...

Connected 02:55:31    Auto detect    2400 8-N-1    SCROLL   CAPS   NUM   Capture   Print echo
```

Figure 7.17: Displaying the records on the PC screen with Option 1

Finally, Figure 7.18 shows a screen snapshot when Option 3 is selected to append the temperature readings to the existing file.

```
  test - HyperTerminal
 File  Edit  View  Call  Transfer  Help

  TEMPERATURE DATA LOGGER

  1. Send temperature data to the PC
  2. Save temperature data in a new file
  3. Append temperature data to an existing file

  Choice ? 3

  Appending data to the existing file...

  Saving record:          1
  Saving record:          2
  Saving record:          3
  Saving record:          4
  Saving record:          5
  Saving record:          6
  Saving record:          7
  Saving record:          8
  Saving record:          9
  Saving record:         10
  Saving record:         11
  Saving record:         12
  Saving record:         13

 Connected 02:58:02    Auto detect   2400 8-N-1   SCROLL   CAPS   NUM   Capture   Print echo
```

Figure 7.18: Saving temperature records on an SD card with Option 3

Advanced PIC18 Projects—USB Bus Projects

The Universal Serial Bus (USB) is one of the most common interfaces used in electronic consumer products today, including PCs, cameras, GPS devices, MP3 players, modems, printers, and scanners, to name a few.

The USB was originally developed by Compaq, Microsoft, Intel, and NEC, and later by Hewlett-Packard, Lucent, and Philips as well. These companies eventually formed the nonprofit corporation USB Implementers Forum Inc. to organize the development and publication of USB specifications.

This chapter describes the basic principles of the USB bus and shows how to use USB-based applications with PIC microcontrollers. The USB bus is a complex protocol. A complete discussion of its design and use is beyond the scope of this chapter. Only the basic principles, enough to be able to use the USB bus, are outlined here. On the other hand, the functions offered by the mikroC language that simplify the design of USB-based microcontroller projects are described in some detail.

The USB is a high-speed serial interface that can also provide power to devices connected to it. A USB bus supports up to 127 devices (limited by the 7-bit address field—note that address 0 is not used as it has a special purpose) connected through a four-wire serial cable of up to three or even five meters in length. Many USB devices can be connected to the same bus with hubs, which can have 4, 8, or even 16 ports. A device can be plugged into a hub which is plugged into another hub, and so on. The maximum number of tiers permitted is six. According to the specification, the maximum distance of a device from its host is about thirty meters, accomplished by using five hubs. For longer-distance bus communications, other methods such as use of Ethernet are recommended.

The USB bus specification comes in two versions: the earlier version, USB1.1, supports 11Mbps, while the new version, USB 2.0, supports up to 480Mbps. The USB specification defines three data speeds:

- Low speed—1.5Mb/sec

- Full speed—12Mb/sec

- High speed—480Mb/sec

The maximum power available to an external device is limited to about 100mA at 5.0V.

USB is a four-wire interface implemented using a four-core shielded cable. Two types of connectors are specified and used: Type A and Type B. Figure 8.1 shows typical USB connectors. Figure 8.2 shows the pin-out of the USB connectors.

The signal wire colors are specified. The pins and wire colors of a Type A or Type B connector are given in Table 8.1.

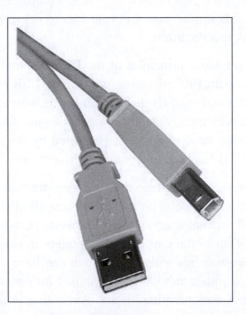

Figure 8.1: USB connectors

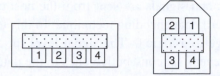

Figure 8.2: Pin-out of USB connectors

Table 8.1: USB connector pin assignments

Pin no.	Name	Color
1	+5.0V	Red
2	Data−	White
3	Data+	Green
4	Ground	Black

The specification also defines a mini-B connector, mainly used in smaller portable electronic devices such as cameras and other handheld devices. This connector has a fifth pin called ID, though this pin is not used. The pin assignment and wire colors of a mini-B connector are given in Table 8.2.

Two of the pins, Data+ and Data−, form a twisted pair and carry differential data signals and some single-ended data states.

Table 8.2: Mini USB pin assignments

Pin no.	Name	Color
1	+5.0V	Red
2	−Data	White
3	+Data	Green
4	Not used	–
5	Ground	Black

USB signals are bi-phase, and signals are sent from the host computer using the NRZI (non-return to zero inverted) data encoding technique. In this technique, the signal level is inverted for each change to a logic 0. The signal level for a logic 1 is not changed. A 0 bit is "stuffed" after every six consecutive ones in the data stream to make the data dynamic (this is called *bit stuffing* because the extra bit lengthens the data stream). Figure 8.3 shows how the NRZI is implemented.

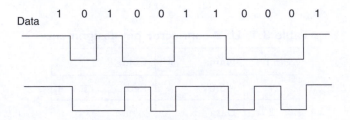

Figure 8.3: NRZI data

A packet of data transmitted by the host is sent to every device connected to the bus, traveling downward through the chain of hubs. All the devices receive the signal, but only one of them, the addressed one, accepts the data. Conversely, only one device at any time can transmit to the host, and the data travels upward through the chain of hubs until it reaches the host.

USB devices attached to the bus may be full-custom devices, requiring a full-custom device driver, or they may belong to a device class. Device classes enable the same device driver to be used for several devices having similar functionalities. For example, a printer device has the device class 0×07, and most printers use drivers of this type.

The most common device classes are given in Table 8.3. The USB human interface device (HID) class is of particular interest, as it is used in the projects in this chapter.

Some common USB terms are:

Endpoint: An endpoint is either a source or a sink of data. A single USB device can have a number of endpoints, the limit being sixteen IN and sixteen OUT endpoints.

Transaction: A transaction is a transfer of data on the bus.

Pipe: A pipe is a logical data connection between the host and an endpoint.

Table 8.3: USB device classes

Device class	Description	Example device
0×00	Reserved	—
0×01	USB audio device	Sound card
0×02	USB communications device	Modem, fax
0×03	USB human interface device	Keyboard, mouse
0×07	USB printer device	Printer
0×08	USB mass storage device	Memory card, flash drive
0×09	USB hub device	Hubs
0×0B	USB smart card reader device	Card reader
0×0E	USB video device	Webcam, scanner
0×E0	USB wireless device	Bluetooth

8.1   Speed Identification on the Bus

At the device end of the bus, a 1.5K pull-up resistor is connected from the D+ or D− line to 3.3V. On a full-speed bus, the resistor is connected from the D+ line to 3.3V, and on a low-speed bus the resistor is from D− line to 3.3V. When no device is plugged in, the host will see both data lines as low. Connecting a device to the bus will pull either the D+ or the D− line to logic high, and the host will know that a device is plugged into the bus. The speed of the device is determined by observing which line is pulled high.

8.2   USB States

Some of the USB bus states are:

Idle: The bus is in idle state when the pulled-up line is high and the other line is low. This is the state of the lines before and after a packet transmission.

Detached: When no device is connected to the bus, the host sees both lines as low.

Attached: When a device is connected to the bus, the host sees either D+ or D− go to logic high, which means a device has been plugged in.

J state: The same as idle state.

K state: The opposite of J state.

SE0: The single ended zero state, where both lines on the bus are pulled low.

SE1: The single ended one state, where both lines on the bus are high. SE1 is an illegal condition on the bus; it must never be in this state.

Reset: When the host wants to communicate with a device on the bus, it first sends a "reset" condition by pulling low both data lines (SE0 state) for at least 10ms.

EOP: The end of packet state, which is basically an SE0 state for 2 bit times, followed by a J state for 1 bit time.

Keep alive: The state achieved by EOP. Keep alive is sent at least once every millisecond to keep the device from suspending.

Suspend: Used to save power, suspend is implemented by not sending anything to a device for 3ms. A suspended device draws less than 0.5mA from the bus and must recognize reset and resume signals.

Resume: A suspended device is woken up by reversing the polarity of the signal on the data lines for at least 20ms, followed by a low-speed EOP signal.

8.3   USB Bus Communication

USB is a host-centric connectivity system where the host dictates the use of the USB bus. Each device on the bus is assigned a unique USB address, and no slave device can assert a signal on the bus until the host asks for it. When a new USB device is plugged into a bus, the USB host uses address 0 to ask basic information from the device. Then the host assigns it a unique USB address. After the host asks for and receives further information about the device, such as the name of the manufacturer, device capabilities, and product ID, two-way transactions on the bus can begin.

8.3.1   Packets

Data is transmitted on a USB bus in packets. A packet starts with a sync pattern to allow the receiver clock to synchronize with the data. The data bytes of the packet follow, ending with an end of packet signal.

A packet identifier (PID) byte immediately follows the sync field of every USB packet. A PID itself is 4 bits long, and the 4 bits are repeated in a complemented form. There are seventeen different PID values, as shown in Table 8.4. These include one reserved value and one that is used twice, with two different meanings.

There are four packet formats, based on which PID is at the start of the packet: token packets, data packets, handshake packets, and special packets.

Figure 8.4 shows the format of a token packet, which is used for OUT, IN, SOF (start of frame), and SETUP. The packet contains a 7-bit address, a 4-bit ENDP (endpoint number), a 5-bit CRC checksum, and an EOP (end of packet).

Table 8.4: PID values

PID type	PID name	Bits	Description
Token	OUT	1110 0001	Host to device transaction
	IN	0110 1001	Device to host transaction
	SOF	1010 0101	Start of frame
	SETUP	0010 1101	Setup command
Data	DATA0	1100 0011	Data packet PID even
	DATA1	0100 1011	Data packet PID odd
	DATA2	1000 0111	Data packet PID high speed
	MDATA	0000 1111	Data packet PID high speed
Handshake	ACK	1101 0010	Receiver accepts packet
	NAK	0101 1010	Receiver does not accept packet
	STALL	0001 1110	Stalled
	NYET	1001 0110	No response from receiver
Special	PRE	0011 1100	Host preample
	ERR	0011 1100	Split transaction error
	SPLIT	0111 1000	High-speed split transaction
	PING	1011 0100	High-speed flow control
	Reserved	1111 0000	Reserved

Sync	PID	ADDR	ENDP	CRC	EOP
	8 bits	7 bits	4 bits	5 bits	

Figure 8.4: Token packet

A data packet is used for DATA0, DATA1, DATA2, and MDATA data transactions. The packet format is shown in Figure 8.5 and consists of the PID, 0–1024 bytes of data, a 2-byte CRC checksum, and an EOP.

Sync	PID	Data	CRC	EOP
	1 byte	0–1024 bytes	2 bytes	

Figure 8.5: Data packet

Sync	PID	EOP
	1 byte	

Figure 8.6: Handshake packet

Figure 8.6 shows the format of a handshake packet, which is used for ACK, NAK, STALL, and NYET. ACK is used when a receiver acknowledges that it has received an error-free data packet. NAK is used when the receiving device cannot accept the packet. STALL indicates when the endpoint is halted, and NYET is used when there is no response from the receiver.

8.3.2   Data Flow Types

Data can be transferred on a USB bus in four ways: bulk transfer, interrupt transfer, isochronous transfer, and control transfer.

Bulk transfers are designed to transfer large amounts of data with error-free delivery and no guarantee of bandwidth. If an OUT endpoint is defined as using bulk transfers, then the host will transfer data to it using OUT transactions. Similarly, if an IN endpoint is defined as using bulk transfers, then the host will transfer data from it using IN transactions. In general, bulk transfers are used where a slow rate of transfer is not a problem. The maximum packet size in a bulk transfer is 8 to 64 packets at full speed, and 512 packets at high speed (bulk transfers are not allowed at low speeds).

Interrupt transfers are used to transfer small amounts of data with a high bandwidth where the data must be transferred as quickly as possible with no delay. Note that interrupt transfers have nothing to do with interrupts in computer systems. Interrupt packets can range in size from 1 to 8 bytes at low speed, from 1 to 64 bytes at full speed, and up to 1024 bytes at high speed.

Isochronous transfers have a guaranteed bandwidth, but error-free delivery is not guaranteed. This type of transfer is generally used in applications, such as audio data transfer, where speed is important but the loss or corruption of some data is not. An isochronous packet may contain 1023 bytes at full speed or up to 1024 bytes at high speed (isochronous transfers are not allowed at low speeds).

A *control transfer* is a bidirectional data transfer, using both IN and OUT endpoints. Control transfers are generally used for initial configuration of a device by the host. The maximum packet size is 8 bytes at low speed, 8 to 64 bytes at full speed, and 64 bytes at high speed. A control transfer is carried out in three stages: SETUP, DATA, and STATUS.

8.3.3 Enumeration

When a device is plugged into a USB bus, it becomes known to the host through a process called enumeration. The steps of enumeration are:

- When a device is plugged in, the host becomes aware of it because one of the data lines (D+ or D−) becomes logic high.

- The host sends a USB reset signal to the device to place the device in a known state. The reset device responds to address 0.

- The host sends a request on address 0 to the device to find out its maximum packet size using a *Get Descriptor* command.

- The device responds by sending a small portion of the device descriptor.

- The host sends a USB reset again.

- The host assigns a unique address to the device and sends a *Set Address* request to the device. After the request is completed, the device assumes the new address. At this point the host is free to reset any other newly plugged-in devices on the bus.

- The host sends a *Get Device Descriptor* request to retrieve the complete device descriptor, gathering information such as manufacturer, type of device, and maximum control packet size.

- The host sends a *Get Configuration Descriptors* request to receive the device's configuration data, such as power requirements and the types and number of interfaces supported.

- The host may request any additional descriptors from the device.

The initial communication between the host and the device is carried out using the control transfer type of data flow.

Initially, the device is addressed, but it is in an unconfigured state. After the host gathers enough information about the device, it loads a suitable device driver which configures the device by sending it a *Set Configuration* request. At this point the device has been configured, and it is ready to respond to device-specific requests (i.e., it can receive data from and send data to the host).

8.4   Descriptors

All USB devices have a hierarchy of descriptors that describe various features of the device: the manufacturer ID, the version of the device, the version of USB it supports, what the device is, its power requirements, the number and type of endpoints, and so forth.

The most common USB descriptors are:

- Device descriptors
- Configuration descriptors
- Interface descriptors
- HID descriptors
- Endpoint descriptors

The descriptors are in a hierarchical structure as shown in Figure 8.7. At the top of the hierarchy we have the device descriptor, then the configuration descriptors, followed by the interface descriptors, and finally the endpoint descriptors. The HID descriptor always follows the interface descriptor when the interface belongs to the HID class.

All descriptors have a common format. The first byte (*bLength*) specifies the length of the descriptor, while the second byte (*bDescriptorType*) indicates the descriptor type.

8.4.1   Device Descriptors

The device descriptor is the top-level set of information read from a device and the first item the host attempts to retrieve.

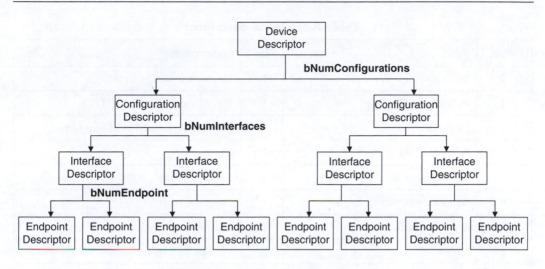

Figure 8.7: USB descriptor hierarchy

A USB device has only one device descriptor, since the device descriptor represents the entire device. It provides general information such as manufacturer, serial number, product number, the class of the device, and the number of configurations. Table 8.5 shows the format for a device descriptor with the meaning of each field.

bLength is the length of the device descriptor.

bDescriptorType is the descriptor type.

bcdUSB reports the highest version of USB the device supports in BCD format. The number is represented as $0\times JJMN$, where JJ is the major version number, M is the minor version number, and N is the subminor version number. For example, USB 1.1 is reported as 0×0110.

bDeviceClass, *bDeviceSubClass*, and *bDeviceProtocol* are assigned by the USB organization and are used by the system to find a class driver for the device.

bMaxPacketSize0 is the maximum input and output packet size for endpoint 0.

idVendor is assigned by the USB organization and is the vendor's ID.

idProduct is assigned by the manufacturer and is the product ID.

bcdDevice is the device release number and has the same format as the *bcdUSB*.

Table 8.5: Device descriptor

Offset	Field	Size	Description
0	bLength	1	Descriptor size in bytes
1	bDescriptorType	1	Device descriptor (0×01)
2	bcdUSB	2	Highest version of USB supported
4	bDeviceClass	1	Class code
5	bDeviceSubClass	1	Subclass code
6	bDeviceProtocol	1	Protocol code
7	bMaxPacketSize0	1	Maximum packet size
8	idVendor	2	Vendor ID
10	idProduct	2	Product ID
12	bcdDevice	2	Device release number
14	iManufacturer	1	Manufacturer string descriptor
15	iProduct	1	Index of product string descriptor
16	iSerialNumber	1	Index of serial number descriptor
17	bNumConfigurations	1	Number of possible configurations

iManufacturer, *iProduct*, and *iSerialNumber* are details about the manufacturer and the product. These fields have no requirement and can be set to zero.

bNumConfigurations is the number of configurations the device supports.

Table 8.6 shows an example device descriptor for a mouse device. The length of the descriptor is 18 bytes (*bLength* = 18), and the descriptor type is 0×01 (*bDescriptorType* = 0×01). The device supports USB 1.1 (*bcdUSB* = 0×0110). *bDeviceClass*, *bDeviceSubClass*, and *bDeviceProtocol* are set to zero to show that the class information is in the interface descriptor. *bMaxPacketSize0* is set to 8 to show that the maximum input and output packet size for endpoint 0 is 8 bytes. The next three bytes identify the device by the vendor ID, product ID, and device version number. The next three items define indexes to strings about the manufacturer, product, and the serial number. Finally, we notice that the mouse device has just one configuration (*bNumConfigurations* = 1).

Table 8.6: Example device descriptor

Offset	Field	Value	Description
0	bLength	18	Size is 18
1	bDescriptorType	0×01	Descriptor type
2	bcdUSB	0×0110	Highest USB supported = USB 1.1
4	bDeviceClass	0×00	Class information in interface descriptor
5	bDeviceSubClass	0×00	Class information in interface descriptor
6	bDeviceProtocol	0×00	Class information in interface descriptor
7	bMaxPacketSize0	8	Maximum packet size
8	idVendor	0×02A	XYZ Co Ltd.
10	idProduct	0×1001	Mouse
12	bcdDevice	0×0011	Device release number
14	iManufacturer	0×20	Index to manufacturer string
15	iProduct	0×21	Index of product string
16	iSerialNumber	0×22	Index of serial number string
17	bNumConfigurations	1	Number of possible configurations

8.4.2 Configuration Descriptors

The configuration descriptor provides information about the power requirements of the device and how many different interfaces it supports. There may be more than one configuration for a device.

Table 8.7 shows the format of the configuration descriptor with the meaning of each field.

bLength is the length of the device descriptor.

bDescriptorType is the descriptor type.

wTotalLength is the total combined size of this set of descriptors (i.e., total of configuration descriptor + interface descriptor + HID descriptor + endpoint descriptor). When the configuration descriptor is read by the host, it returns the entire configuration information, which includes all interface and endpoint descriptors.

Table 8.7: Configuration descriptor

Offset	Field	Size	Description
0	bLength	1	Descriptor size in bytes
1	bDescriptorType	1	Device descriptor (0×02)
2	wTotalLength	2	Total bytes returned
4	bNumInterfaces	1	Number of interfaces
5	bConfigurationValue	1	Value used to select configuration
6	iConfiguration	1	Index describing configuration string
7	bmAttributes	1	Power supply attributes
8	bMaxPower	2	Max power consumption in 2mA

bNumInterfaces is the number of interfaces present for this configuration.

bConfigurationValue is used by the host (in command *SetConfiguration*) to select the configuration.

iConfiguration is an index to a string descriptor describing the configuration in readable format.

bmAttributes describes the power requirements of the device. If the device is USB bus-powered, then bit D7 is set. If it is self-powered, it sets bit D6. Bit D5 specifies the remote wakeup of the device. Bits D7 and D0–D4 are reserved.

bMaxPower defines the maximum power the device will draw from the bus in 2mA units.

Table 8.8 shows an example configuration descriptor for a mouse device. The length of the descriptor is 9 bytes (*bLength* = 9), and the descriptor type is 0×02 (*bDescriptorType* = 0×02). The total combined size of the descriptors is 34 (*wTotalLength* = 34). The number of interfaces for the mouse device is 1 (*bNumInterfaces* = 1). *Host SetConfiguration* command must use the value 1 as an argument in *SetConfiguration()* to select this configuration. There is no string to describe this configuration. *bmAttributes* is set to 0×40 to indicate that the device is self-powered. *bMaxPower* is set to 10 to specify that the maximum current drawn by the device is 20mA.

Table 8.8: Example configuration descriptor

Offset	Field	Value	Description
0	bLength	9	Descriptor size is 9 bytes
1	bDescriptorType	0×02	Device descriptor is 0×02
2	wTotalLength	34	Total bytes returned is 34
4	bNumInterfaces	1	Number of interfaces is 1
5	bConfigurationValue	1	Value used to select configuration
6	iConfiguration	0×2A	Index describing configuration string
7	bmAttributes	0×40	Power supply attributes
8	bMaxPower	10	Max power consumption is 20mA

8.4.3   Interface Descriptors

The interface descriptors specify the class of the interface and the number of endpoints it uses. There may be more than one interface.

Table 8.9 shows the format of the interface descriptor with the meaning of each field.

Table 8.9: Interface descriptor

Offset	Field	Size	Description
0	bLength	1	Descriptor size in bytes
1	bDescriptorType	1	Device descriptor (0×04)
2	bInterfaceNumber	1	Number of interface
3	bAlternateSetting	1	Value to select alternate setting
4	bNumEndpoints	1	Number of endpoints
5	bInterfaceClass	1	Class code
6	bInterfaceSubClass	1	Subclass code
7	bInterfaceProtocol	1	Protocol code
8	iInterface	1	Index of string descriptor to interface

bLength is the length of the device descriptor.

bDescriptorType is the descriptor type.

bInterfaceNumber indicates the index of the interface descriptor.

bAlternateSetting can be used to specify alternate interfaces that can be selected by the host using command *Set Interface*.

bNumEndpoints indicates the number of endpoints used by the interface.

bInterfaceClass specifies the device class code (assigned by the USB organization).

bInterfaceSubClass specifies the device subclass code (assigned by the USB organization).

bInterfaceProtocol specifies the device protocol code (assigned by the USB organization).

iInterface is an index to a string descriptor of the interface.

Table 8.10 shows an example interface descriptor for a mouse device. The descriptor length is 9 bytes (*bLength* = 9) and the descriptor type is 0×04 (*bDescriptorType* = 0×04). The interface number used to reference this interface is 1 (*bInterfaceNumber* = 1).

Table 8.10: Example interface descriptor

Offset	Field	Value	Description
0	bLength	9	Descriptor size is 9 bytes
1	bDescriptorType	0×04	Device descriptor is 0×04
2	bInterfaceNumber	0	Number of interface
3	bAlternateSetting	0	Value to select alternate setting
4	bNumEndpoints	1	Number of endpoints is 1
5	bInterfaceClass	0×03	Class code is 0×03
6	bInterfaceSubClass	0×02	Subclass code is 0×02
7	bInterfaceProtocol	0×02	Protocol code is 0×02
8	iInterface	0	Index of string descriptor to interface

bAlternateSetting is set to 0 (i.e., no alternate interfaces). The number of endpoints used by this interface is 1 (excluding endpoint 0), and this is the endpoint used for the mouse to send its data. The device class code is 0×03 (*bInterfaceClass* = 0×03). This is an HID (human interface device) type class. The interface subclass is set to 0×02. The device protocol is 0×02 (mouse). There is no string to describe this interface (*iInterface* = 0).

8.4.4   HID Descriptors

An HID descriptor always follows an interface descriptor when the interface belongs to the HID class. Table 8.11 shows the format of the HID descriptor.

bLength is the length of the device descriptor.

bDescriptorType is the descriptor type.

bcdHID is the HID class specification.

bCountryCode specifies any special local changes.

bNumDescriptors specifes if there are any additional descriptors associated with this class.

bDescriptorType is the type of the additional descriptor specified in *bNumDescriptors*.

wDescriptorLength is the length of the additional descriptor in bytes.

Table 8.11: HID descriptor

Offset	Field	Size	Description
0	bLength	1	Descriptor size in bytes
1	bDescriptorType	1	HID (0×21)
2	bcdHID	2	HID class
4	bCountryCode	1	Special country dependent code
5	bNumDescriptors	1	Number of additional descriptors
6	bDescriptorType	1	Type of additional descriptor
7	wDescriptorLength	2	Length of additional descriptor

Table 8.12 shows an example HID descriptor for a mouse device. The length of the descriptor is 9 bytes (*bLength* = 9), and the descriptor type is 0×21 (*bDescriptorType* = 0×21). The HID class is set to 1.1 (*bcdHID* = 0×0110). The country code is set to zero (*bCountryCode* = 0), specifying that there is no special localization with this device. The number of descriptors is set to 1 (*bNumDescriptors* = 1) which specifies that there is one additional descriptor associated with this class. The type of the additional descriptor is REPORT (*bDescriptorType* = REPORT), and its length is 52 bytes (*wDescriptorLength* = 52).

Table 8.12: Example HID descriptor

Offset	Field	Value	Description
0	bLength	9	Descriptor size is 9 bytes
1	bDescriptorType	0×21	HID (0×21)
2	bcdHID	0×0110	Class version 1.1
4	bCountryCode	0	No special country dependent code
5	bNumDescriptors	1	Number of additional descriptors
6	bDescriptorType	REPORT	Type of additional descriptor
7	wDescriptorLength	5	Length of additional descriptor

8.4.5   Endpoint Descriptors

Table 8.13 shows the format of the endpoint descriptor.

bLength is the length of the device descriptor.

bDescriptorType is the descriptor type.

bEndpointAddress is the address of the endpoint.

bmAttributes specifies what type of endpoint it is.

wMaxPacketSize is the maximum packet size.

bInterval specifies how often the endpoint should be polled (in ms).

Table 8.14 shows an example endpoint descriptor for a mouse device. The length of the descriptor is 7 bytes (*bLength* = 7), and the descriptor type is 0×05 (*bDescriptorType*

Table 8.13: Endpoint descriptor

Offset	Field	Size	Description
0	bLength	1	Descriptor size in bytes
1	bDescriptorType	1	Endpoint (0×05)
2	bcdEndpointAddress	1	Endpoint address
4	bmAttributes	1	Type of endpoint
5	wMaxPacketSize	2	Max packet size
6	bInterval	1	Polling interval

Table 8.14: Example endpoint descriptor

Offset	Field	Size	Description
0	bLength	7	Descriptor size in bytes
1	bDescriptorType	0×05	Endpoint (0×05)
2	bcdEndpointAddress	0×50	Endpoint address
4	bmAttributes	0×03	Interrupt type endpoint
5	wMaxPacketSize	0×0002	Max packet size is 2
6	bInterval	0×14	Polling interval is 20ms

= 0×05). The endpoint address is 0×50 (*bEndpointAddress* = 0×50). The endpoint is to be used as an interrupt endpoint (*bmAttributes* = 0×03). The maximum packet size is set to 2 (*wMaxPacketSize* = 0×02) to indicate that packets longer than 2 bytes will not be sent from the endpoint. The endpoint should be polled at least once every 20ms (*bInterval* = 0×14).

8.5   PIC18 Microcontroller USB Bus Interface

Some of the PIC18 microcontrollers support USB interface directly. For example, the PIC18F4550 microcontroller contains a full-speed and low-speed compatible USB interface that allows communication between a host PC and the microcontroller. In the USB projects in this chapter we will use the PIC18F4550 microcontroller.

Figure 8.8 is an overview of the USB section of the PIC18F4550 microcontroller. PORTC pins RC4 (pin 23) and RC5 (pin 24) are used for USB interface. RC4 is the USB data D− pin, and RC5 is the USB data D+ pin. Internal pull-up resistors are provided which can be disabled (setting $UPUEN = 0$) if desired and external pull-up resistors can be used instead. For full-speed operation an internal or external resistor should be connected to data pin D+, and for low-speed operation an internal or external resistor should be connected to data pin D−.

Operation of the USB module is configured using three control registers, and a total of twenty-two registers are used to manage the actual USB transactions. Configuration

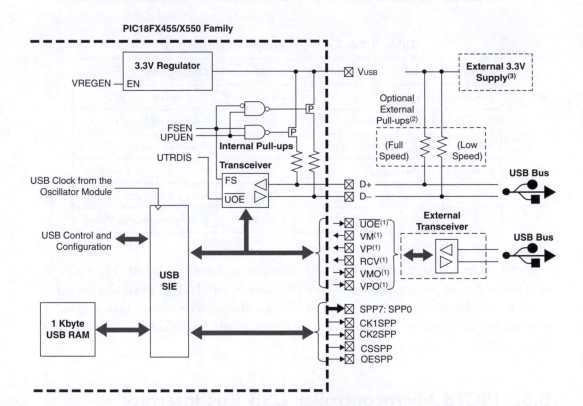

Note 1: This signal is only available if the internal transceiver is disabled (UTRDIS = 1).

2: The internal pull-up resistors should be disabled (UPUEN = 0) if external pull-up resistors are used.

3: Do not enable the internal regulator when using an external 3.3V supply.

Figure 8.8: PIC18F4550 microcontroller USB overview

of these registers is a highly complex task and is not covered in this book. Interested readers should refer to the PIC18F4550 data sheet and to books on USB internals. In this chapter we are using the mikroC language USB library functions to implement USB transactions. The details of these functions are given in the next section.

8.6   mikroC Language USB Bus Library Functions

The mikroC language supports a number of functions for USB HID-type communications. Each project based on the USB library should include a descriptor source file which contains vendor ID and name, product ID and name, report length, and other relevant information. To create a descriptor source file we can use mikroC's integrated USB HID terminal tool (see *Tools → HID Terminal*). The default name for descriptor file is *USBdsc.c*, but it can be renamed if required. The *USBdsc.c* file must be included in USB-based projects either via the mikroC IDE tool, or as an *#include* option in the program source file.

The mikroC language supports the following USB bus library functions when a PIC microcontroller with built-in USB is used (e.g., PIC18F4550), and port pins RC4 and RC5 are connected to the D+ and D− pins of the USB connector respectively:

Hid_Enable: This function enables USB communication and requires two arguments: the read-buffer address and the write-buffer address. It must be called before any other functions of the USB library, and it returns no data.

Hid_Read: This function receives data from the USB bus and stores it in the receive-buffer. It has no arguments but returns the number of characters received.

Hid_Write: This function sends data from the write-buffer to the USB bus. The name of the buffer (the same buffer used in the initialization) and the length of the data to be sent must be specified as arguments to the function. The function does not return any data.

Hid_Disable: This function disables the USB data transfer. It has no arguments and returns no data.

The USB interface of a PIC18F4550 microcontroller is shown in Figure 8.9. As the figure shows, the interface is very simple. In addition to the power supply and ground pins, it requires just two pins to be connected to the USB connector. The microcontroller receives power from the USB port.

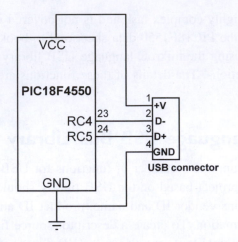

Figure 8.9: PIC18F4550 USB interface

PROJECT 8.1—USB-Based Microcontroller Output Port

This project describes the design of a USB-based microcontroller output port. A PIC18F4550 microcontroller is interfaced to a PC through a USB cable. A Visual Basic program runs on the PC and sends commands to the microcontroller through the USB bus, asking the microcontroller to set/reset the I/O bits of its PORTB.

The block diagram of the project is shown in Figure 8.10. The circuit diagram is given in Figure 8.11. The USB lines of the PIC18F4550 microcontroller are connected to a USB connector. The microcontroller is powered from the USB line (i.e., no external

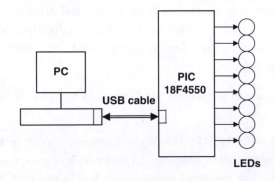

Figure 8.10: Block diagram of the project

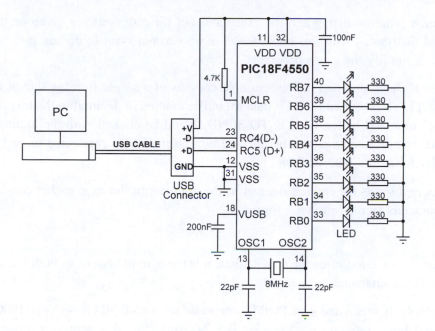

Figure 8.11: Circuit diagram of the project

power supply is required). This makes the design of USB-based products relatively cheap and very attractive in applications where the total power consumption is below 100mA. The microcontroller is operated from an 8MHz crystal.

The PORTB pins of the microcontroller are connected to LEDs so we can see the state changes as commands are sent from the PC. This makes testing the project very easy. Note that a capacitor (about 200nF) should be connected between the V_{USB} pin (pin 18) of the microcontroller and the ground for stability.

The project software consists of two parts: the PC software, and the microcontroller software. Both are described in this section.

The PC Software

The PC software is based on Visual Basic. It is assumed that the user has elementary knowledge of Visual Basic programming language. Instruction in programming using the Visual Basic language is beyond the scope of this book, and interested readers should refer to various books available on this topic.

The source program listing and the executables of the programs are given on the CDROM distributed with this book. Readers who do not want to do any programming can use or modify the given programs.

The Visual Basic program in this example consists of a single form as shown in Figure 8.12. The required PORTB data should be entered in decimal in the text box, and then the command button CLICK TO SEND should be clicked with the mouse. For example, entering decimal number 15 will turn on the LEDs connected to port pins RB0,RB1,RB2, and RB3 of PORTB.

The program sends the entered number to the microcontroller as a packet consisting of four characters in the following format:

$$P = nT$$

where character P indicates the start of data, n is the byte to be sent to PORTB, and T is the terminator character.

For example, if bits 3 and 4 of PORTB are to be set, i.e., PORTB = "00011000," then the Visual Basic program sends packet P = 24T (number 24 is sent as a single binary byte and not as two ASCII bytes) to the microcontroller over the USB link. The bottom part of the form displays the connection status.

The Visual Basic program used in this section is based on the USB utility known as EasyHID USB Wizard, developed by Mecanique, and can be downloaded free of charge

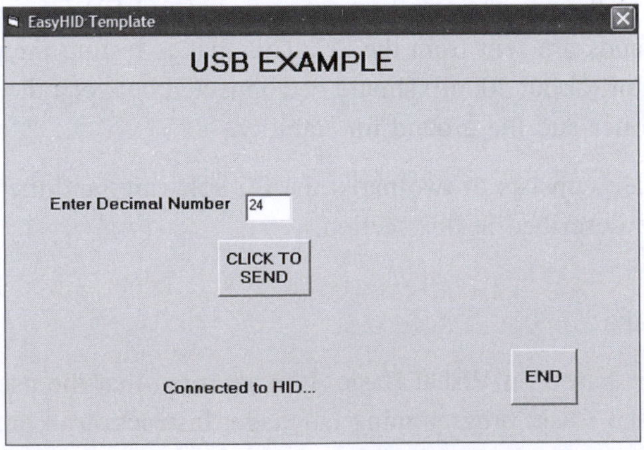

Figure 8.12: The PC Visual Basic form

from their web site (www.mecanique.co.uk). EasyHID is designed to work with USB 2.0, and there is no need to develop a driver, as the XP operating system is shipped with a HID-based USB driver. This utility generates Visual Basic, Visual C++, or Borland Delphi template codes for the PC end of a USB application using an HID-type device interface. In addition, the utility can generate USB template code for the PIC18F4550 and similar microcontrollers, based on the Proton Development Suite (www.crownhill.co.uk), Swordish PIC Basic, or PicBasic Pro (www.melabs.com) programming languages. The generated codes can be expanded with the user code to implement the required application.

The steps in generating a Visual Basic code template follow:

- Load the EasyHID zip file from the Mecanique web site by clicking on "Download EasyHID as a Standalone Application"

- Extract the files and install the application by double-clicking on SETUP.

- When the program has started, you should see a form as shown in Figure 8.13. Enter your data in the fields Company Name, Product Name, and the optional Serial Number.

Figure 8.13: EasyHID first form

- Enter your Vendor ID (VID) and Product ID (PID) as shown in the form in Figure 8.14. Vendor IDs are unique throughout the world and are issued by the USB implementers (www.usb.org) at a cost. Mecanique owns a Vendor ID and can issue you a set of Product IDs at low cost so your products can be shipped all over the world with unique VID and PID combinations. In this example, VID = 4660 and PID = 1 are selected for test purposes.

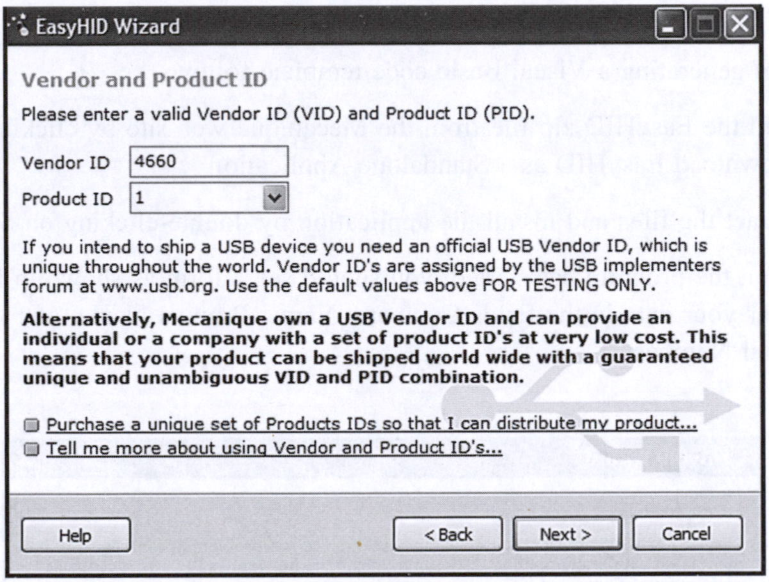

Figure 8.14: EasyHID VID and PID entry form

- Clicking *Next* displays the form shown in Figure 8.15. The important parameters here are the output and input buffer sizes, which specify the number of bytes to be sent and received respectively between the PC and the microcontroller during USB data transactions. In this example, 4 bytes are chosen for both fields (our output is in the format P = nT, which is 4 bytes).

- In the next form (see Figure 8.16), select a location for the generated files, choose the microcontroller compiler to be used (this field is not important, as we are only generating code for Visual Basic (i.e., the PC

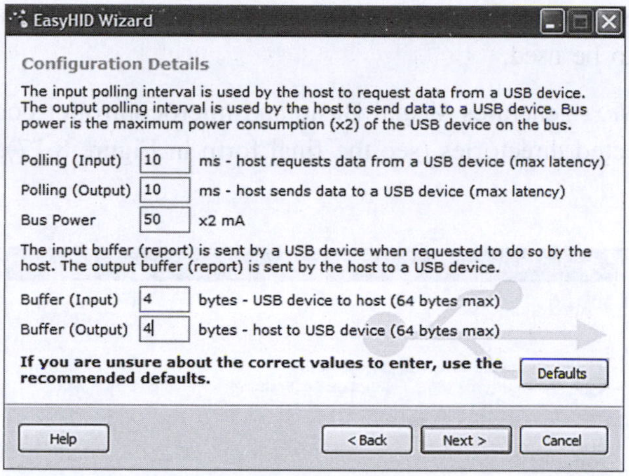

Figure 8.15: EasyHID input-output buffer selection

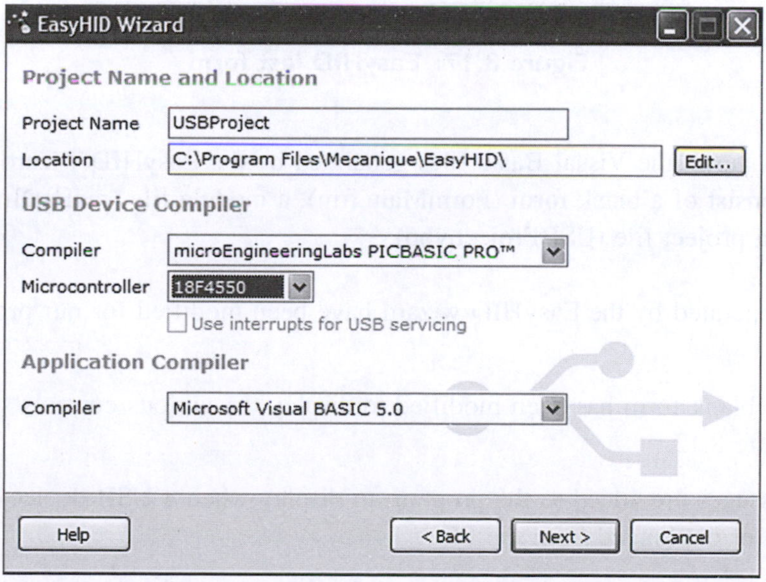

Figure 8.16: EasyHID output folder, microcontroller type, and host compiler selection

end), choose the microcontroller type, and finally select Visual Basic as the language to be used.

- Clicking *Next* generates Visual Basic and microcontroller code templates in the selected directories (see the final form in Figure 8.17).

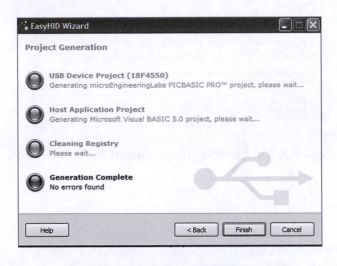

Figure 8.17: EasyHID last form

Figure 8.18 shows the Visual Basic files generated by the EasyHID wizard. The files basically consist of a blank form (FormMain.frm), a module file (mcHIDInterface.BAS), and a project file (USBProject.vbp).

The files generated by the EasyHID wizard have been modified for our project as follows:

- The blank form has been modified to display the various controls shown in Figure 8.12.

- Messages are added to the program to display when a USB device is plugged into or unplugged from the PC.

- A subroutine has been added to read the data entered by the user and then send this data to the microcontroller over the USB bus when the button CLICK TO SEND is clicked. This code is as follows:

```
Private Sub Command2_Click()

    BufferOut(0) = 0            ' first by is always the report ID
    BufferOut(1) = Asc("P")     ' first data item ("P")
    BufferOut(2) = Asc("=")     ' second data item ("=")
    BufferOut(3) = Val(txtno)   ' third data item (number to send)
    BufferOut(4) = Asc("T")     ' fourth data item ("T")

    ' write the data (don't forget, pass the whole array)...
    hidWriteEx VendorID, ProductID, BufferOut(0)
    lblstatus = "Data sent..."
End Sub
```

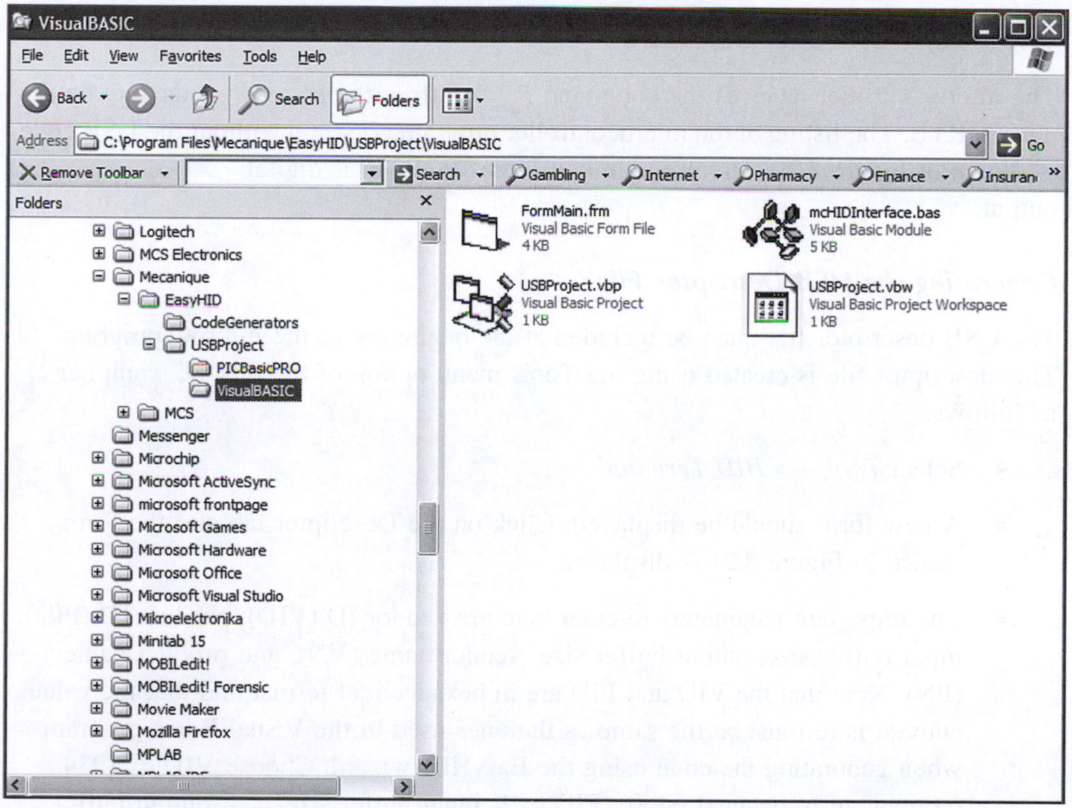

Figure 8.18: Files generated by the EasyHID wizard

BufferOut stores the data to be sent to the microcontroller over the USB bus. Notice that the first byte of this buffer is the report ID and must be set to 0. The actual data starts from address *BufferOut(1)* of the array and the data sent is in the format P = nT as described before. After the data is sent, the message "Data sent…" appears at the bottom part of the display.

Figure 8.19 shows the final listing of the Visual Basic program. The program is in two parts: the form USB1.FRM and the module USB1.BAS. The programs should be loaded and used in the Visual Basic development environment. An installable version of this program (in folder USB1) comes with the CDROM included with this book for those who do not have the Visual Basic development environment. This program should be installed as a normal Windows software installation.

The Microcontroller Software

The microcontroller receives the command *P = nT* from the PC and sends data byte *n* to PORTB. The listing of the microcontroller program (USB.C) without the USB code is shown in Figure 8.20. The program configures PORTB as digital output.

Generating the USB Descriptor File

The USB descriptor file must be included at the beginning of the mikroC program. This descriptor file is created using the Tools menu option of the mikroC compiler as follows:

- Select *Tools -> HID Terminal*

- A new form should be displayed. Click on the Descriptor tab and the form shown in Figure 8.21 is displayed.

- The important parameters to enter here are vendor ID (VID), product ID (PID), input buffer size, output buffer size, vendor name (VN), and product name (PN). Note that the VID and PID are in hexadecimal format and that the values entered here must be the same as the ones used in the Visual Basic program when generating the code using the EasyHID wizard. Choose VID = 1234 (equivalent to decimal 6460), PID = 1, input buffer size = 4, output buffer size = 4, and any names you like for the VN and PN fields.

- Check the mikroC compiler.

USB1.FRM

```
' vendor and product IDs
Private Const VendorID = 4660
Private Const ProductID = 1

' read and write buffers
Private Const BufferInSize = 8
Private Const BufferOutSize = 8
Dim BufferIn(0 To BufferInSize) As Byte
Dim BufferOut(0 To BufferOutSize) As Byte

Private Sub Command1_Click()
        Form_Unload (0)
        End
End Sub

Private Sub Command2_Click()
        BufferOut(0) = 0                ' first by is always the report ID
        BufferOut(1) = Asc("P")        ' first data item ("P")
        BufferOut(2) = Asc("=")        ' second data item ("-")
        BufferOut(3) = Val(txtno)      ' third data item (to send over USB)
        BufferOut(4) = Asc("T")        ' fourth data item ("T")

        ' write the data (don't forget, pass the whole array)...
        hidWriteEx VendorID, ProductID, BufferOut(0)
        lblstatus = "Data sent..."
End Sub

' ***********************************************************************
' when the form loads, connect to the HID controller - pass
' the form window handle so that you can receive notification
' events...
'************************************************************************
Private Sub Form_Load()
    ' do not remove!
        ConnectToHID (Me.hwnd)
        lblstatus = "Connected to HID..."
End Sub


'************************************************************************
' disconnect from the HID controller...
'************************************************************************
Private Sub Form_Unload(Cancel As Integer)
        DisconnectFromHID
End Sub


'************************************************************************
' a HID device has been plugged in...
```

Figure 8.19: **Visual Basic program for the PC end of USB link**

```
'***************************************************************************
Public Sub OnPlugged(ByVal pHandle As Long)
    If hidGetVendorID(pHandle) = VendorID And hidGetProductID(pHandle)=
ProductID Then
        lblstatus = "USB Plugged....."
        End If
End Sub


'***************************************************************************

' a HID device has been unplugged...
'***************************************************************************
Public Sub OnUnplugged(ByVal pHandle As Long)
    If hidGetVendorID(pHandle) = VendorID And hidGetProductID(pHandle) =
ProductID Then
        lblstatus = "USB Unplugged...."
    End If
End Sub


'***************************************************************************

' controller changed notification - called
' after ALL HID devices are plugged or unplugged
'***************************************************************************
Public Sub OnChanged()
    Dim DeviceHandle As Long

        ' get the handle of the device we are interested in, then set
        ' its read notify flag to true - this ensures you get a read
        ' notification message when there is some data to read...
        DeviceHandle = hidGetHandle(VendorID, ProductID)
        hidSetReadNotify DeviceHandle, True
End Sub


'***************************************************************************

' on read event...
'***************************************************************************
Public Sub OnRead(ByVal pHandle As Long)

    ' read the data (don't forget, pass the whole array)...
    If hidRead(pHandle, BufferIn(0)) Then
        ' ** YOUR CODE HERE **
        ' first byte is the report ID, e.g. BufferIn(0)
        ' the other bytes are the data from the microcontrolller...
    End If
End Sub
```

Figure 8.19: (Cont'd)

USB1.BAS

```
' this is the interface to the HID controller DLL - you should not
' normally need to change anything in this file.
'
' WinProc() calls your main form 'event' procedures - these are currently
' set to..
'
' MainForm.OnPlugged(ByVal pHandle as long)
' MainForm.OnUnplugged(ByVal pHandle as long)
' MainForm.OnChanged()
' MainForm.OnRead(ByVal pHandle as long)

Option Explicit

' HID interface API declarations...
Declare Function hidConnect Lib "mcHID.dll" Alias "Connect" (ByVal pHostWin As
Long) As Boolean
Declare Function hidDisconnect Lib "mcHID.dll" Alias "Disconnect" () As Boolean
Declare Function hidGetItem Lib "mcHID.dll" Alias "GetItem" (ByVal pIndex As
Long) As Long
Declare Function hidGetItemCount Lib "mcHID.dll" Alias "GetItemCount" () As
Long
Declare Function hidRead Lib "mcHID.dll" Alias "Read" (ByVal pHandle As Long,
ByRef pData As Byte) As Boolean
Declare Function hidWrite Lib "mcHID.dll" Alias "Write" (ByVal pHandle As Long,
ByRef pData As Byte) As Boolean
Declare Function hidReadEx Lib "mcHID.dll" Alias "ReadEx" (ByVal pVendorID As
Long, ByVal pProductID As Long, ByRef pData As Byte) As Boolean
Declare Function hidWriteEx Lib "mcHID.dll" Alias "WriteEx" (ByVal pVendorID
As Long, ByVal pProductID As Long, ByRef pData As Byte) As Boolean
Declare Function hidGetHandle Lib "mcHID.dll" Alias "GetHandle" (ByVal
pVendoID As Long, ByVal pProductID As Long) As Long
Declare Function hidGetVendorID Lib "mcHID.dll" Alias "GetVendorID" (ByVal
pHandle As Long) As Long
Declare Function hidGetProductID Lib "mcHID.dll" Alias "GetProductID" (ByVal
pHandle As Long) As Long
Declare Function hidGetVersion Lib "mcHID.dll" Alias "GetVersion" (ByVal
pHandle As Long) As Long
Declare Function hidGetVendorName Lib "mcHID.dll" Alias "GetVendorName"
(ByVal pHandle As Long, ByVal pText As String, ByVal pLen As Long) As Long
Declare Function hidGetProductName Lib "mcHID.dll" Alias "GetProductName"
(ByVal pHandle As Long, ByVal pText As String, ByVal pLen As Long) As Long
Declare Function hidGetSerialNumber Lib "mcHID.dll" Alias"GetSerialNumber"
(ByVal pHandle As Long, ByVal pText As String, ByVal pLen As Long) As Long
Declare Function hidGetInputReportLength Lib "mcHID.dll" Alias
"GetInputReportLength" (ByVal pHandle As Long) As Long
Declare Function hidGetOutputReportLength Lib "mcHID.dll" Alias
"GetOutputReportLength" (ByVal pHandle As Long) As Long
```

Figure 8.19: (Cont'd)

```
Declare Sub hidSetReadNotify Lib "mcHID.dll" Alias "SetReadNotify" (ByVal
pHandle As Long, ByVal pValue As Boolean)
Declare Function hidIsReadNotifyEnabled Lib "mcHID.dll" Alias
"IsReadNotifyEnabled" (ByVal pHandle As Long) As Boolean
Declare Function hidIsAvailable Lib "mcHID.dll" Alias "IsAvailable" (ByVal
pVendorID As Long, ByVal pProductID As Long) As Boolean

' windows API declarations - used to set up messaging...
Private Declare Function CallWindowProc Lib "user32" Alias "CallWindowProcA"
(ByVal lpPrevWndFunc As Long, ByVal hwnd As Long, ByVal Msg As Long,
ByVal wParam As Long, ByVal lParam As Long) As Long
Private Declare Function SetWindowLong Lib "user32" Alias "SetWindowLongA"
(ByVal hwnd As Long, ByVal nIndex As Long, ByVal dwNewLong As Long) As
Long

' windows API Constants
Private Const WM_APP = 32768
Private Const GWL_WNDPROC = -4

' HID message constants
Private Const WM_HID_EVENT = WM_APP + 200
Private Const NOTIFY_PLUGGED = 1
Private Const NOTIFY_UNPLUGGED = 2
Private Const NOTIFY_CHANGED = 3
Private Const NOTIFY_READ = 4

' local variables
Private FPrevWinProc As Long     ' Handle to previous window procedure
Private FWinHandle As Long       ' Handle to message window

' Set up a windows hook to receive notification
' messages from the HID controller DLL - then connect
' to the controller
Public Function ConnectToHID(ByVal pHostWin As Long) As Boolean
   FWinHandle = pHostWin
   ConnectToHID = hidConnect(FWinHandle)
   FPrevWinProc = SetWindowLong(FWinHandle, GWL_WNDPROC, AddressOf
WinProc)
End Function

' Unhook from the HID controller and disconnect...
Public Function DisconnectFromHID() As Boolean
   DisconnectFromHID = hidDisconnect
   SetWindowLong FWinHandle, GWL_WNDPROC, FPrevWinProc
End Function

' This is the procedure that intercepts the HID controller messages...
Private Function WinProc(ByVal pHWnd As Long, ByVal pMsg As Long,
ByVal wParam As Long, ByVal lParam As Long) As Long
   If pMsg = WM_HID_EVENT Then
```

Figure 8.19: (Cont'd)

```
    Select Case wParam

        ' HID device has been plugged message...
        Case Is = NOTIFY_PLUGGED
            MainForm.OnPlugged (lParam)

        ' HID device has been unplugged
        Case Is = NOTIFY_UNPLUGGED
            MainForm.OnUnplugged (lParam)

        ' controller has changed...
        Case Is = NOTIFY_CHANGED
            MainForm.OnChanged

        ' read event...
        Case Is = NOTIFY_READ
            MainForm.OnRead (lParam)
    End Select

    End If

    ' next...
    WinProc = CallWindowProc(FPrevWinProc, pHWnd, pMsg, wParam, lParam)

End Function
```

Figure 8.19: (Cont'd)

- Clicking the CREATE button will ask for a folder name and then create descriptor file USBdsc in this folder. Rename this file to have extension ".C" (i.e., the full file name should be USBdsc.C) and then copy it to the following folder (other required mikroC files are already in this folder, so it makes sense to copy USBdsc.C here as well).

```
C:\Program Files\Mikroelektronika\mikroC\Examples\EasyPic4
\extra_examples\HID-library\USBdsc.c
```

Do not modify the contents of file USBdsc.C. A listing of this file is given on the CDROM.

The microcontroller program listing with the USB code included is shown in Figure 8.22 (program USB1.C). At the beginning of the program the USB descriptor file USBdsc.C is included. The operation of the USB link requires the microcontroller to keep the connection alive by sending keep-alive messages to the PC every several milliseconds. This is achieved by setting up a timer interrupt service routine using

```
/***************************************************************************
                  USB BASED MICROCONTROLLER OUTPUT PORT
                  =======================================
```

In this project a PIC18F4550 type microcontroller is connected to a PC through the USB link.

A Visual Basic program runs on the PC where the user enters the bits to be set or cleared on PORTB of the microcontroller. The PC sends a command to the microcontroller requesting it to set or reset the required bits of the microcontroller PORTB.

The command sent by the PC to the microcontroller is in the following format:

```
      P=nT
```

where n is the byte the microcontroller is requested to send to PORTB of the microcontroller.

```
Author:       Dogan Ibrahim
Date:         September 2007
File:         USB.C
***************************************************************************/

void main()
{
    ADCON1 = 0xFF;                        // Set PORTB to digital I/O
    TRISB = 0;                            // Set PORTB to outputs
    PORTB = 0;                            // Clear all outputs
}
```

Figure 8.20: Microcontroller program without the USB code

TIMER 0. Inside the timer interrupt service routine the mikroC USB function *HID_InterruptProc* is called. Timer TMR0L is reloaded and timer interrupts are re-enabled just before returning from the interrupt service routine.

Inside the main program PORTB is defined as digital I/O and TRISB is cleared to 0 so all PORTB pins are outputs. All the interrupt registers are then set to their power-on-reset values for safety. The timer interrupts are then set up. The timer is operated in 8-bit mode with a prescaler of 256. Although the crystal clock frequency is 8MHZ, the CPU is operated with a 48MHz clock, as described later. Selecting a timer value of TMR0L = 100 with a 48MHz clock (CPU clock period of $0.083\mu s$) gives timer interrupt intervals of:

$$(256 - 100) * 256 * 0.083\mu s$$

or, about 3.3ms. Thus, the keep-alive messages are sent every 3.3ms.

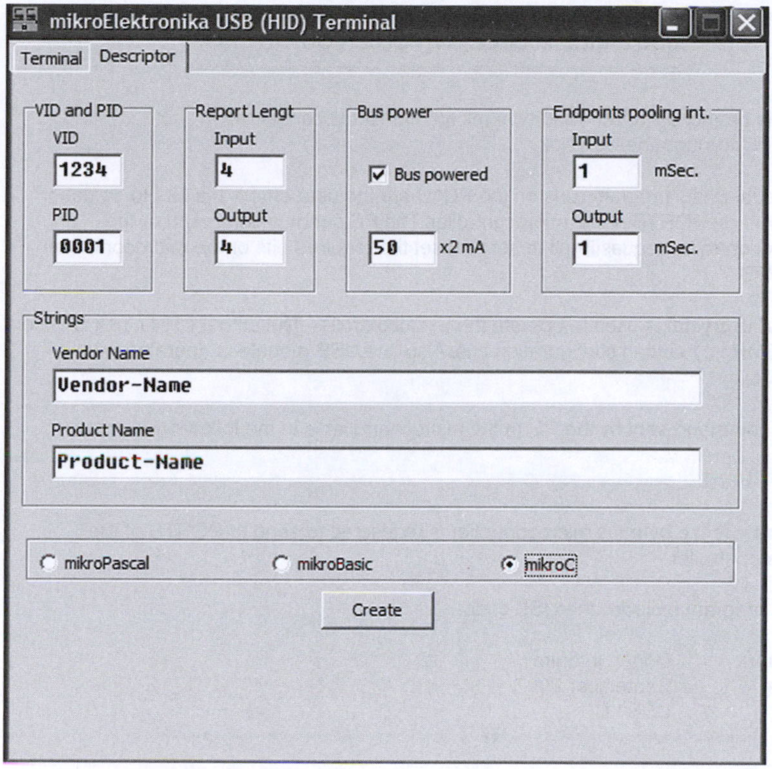

Figure 8.21: Creating the USBdsc descriptor file

The USB port is then enabled by calling function *Hid_Enable*. The program then enters an indefinite loop and reads data from the USB port with *Hid_Read*. When 4 bytes are received at the correct format (i.e., byte 0 = "P," byte 1 = "=", and byte 3 = "T") then the data byte is read from byte 2 and sent to PORTB of the microcontroller.

It is important to note that when data is received using the *Hid_Read* function, the function returns the number of bytes received. In addition, the first byte received is the first actual data byte and not the report ID.

Microcontroller Clock

The USB module of the PIC18F4550 microcontroller requires a 48MHz clock. In addition, the microcontroller CPU requires a clock that can range from 0 to 48MHz. In this project the CPU clock is set to be 48MHz.

There are several ways to provide the required clock pulses.

```
/*****************************************************************************
                USB BASED MICROCONTROLLER OUTPUT PORT
                ==========================================
```

In this project a PIC18F4550 type microcontroller is connected
to a PC through the USB link.

A Visual Basic program runs on the PC where the user enters the bits to be set or
cleared on PORTB of the microcontroller. The PC sends a command to the
microcontroller requesting it to set or reset the required bits of the microcontroller
PORTB.

A 8MHz crystal is used to operate the microcontroller. The actual CPU clock is raised
to 48MHz by setting configuration bits. Also, the USB module is operated with
48MHz.

The command sent by the PC to the microcontroller is in the following format:

         P=nT

where n is the byte the microcontroller is requested to send to PORTB of the
microcontroller.

This program includes the USB code.

```
Author:        Dogan Ibrahim
Date:          September 2007
File:          USB1.C
*****************************************************************************/

#include "C:\Program
Files\Mikroelektronika\mikroC\Examples\EasyPic4\extra_examples\HID-
library\USBdsc.c"

unsigned char Read_buffer[64];
unsigned char Write_buffer[64];
unsigned char num;
//
// Timer interrupt service routine
//
void interrupt()
{
  HID_InterruptProc();                              // Keep alive
  TMR0L = 100;                                      // Re-load TMR0L
  INTCON.TMR0IF = 0;                                // Re-enable TMR0 interrupts
}


//
// Start of MAIN program
```

Figure 8.22: Microcontroller program with USB code

```
//
void main()
{
    ADCON1 = 0xFF;                          // Set PORTB to digital I/O
    TRISB = 0;                              // Set PORTB to outputs
    PORTB = 0;                              // Clear all outputs
//
// Set interrupt registers to power-on defaults
// Disable all interrupts
//
    INTCON=0;
    INTCON2=0xF5;
    INTCON3=0xC0;
    RCON.IPEN=0;
    PIE1=0;
    PIE2=0;
    PIR1=0;
    PIR2=0;
//
// Configure TIMER 0 for 3.3ms interrupts. Set prescaler to 256
// and load TMR0L to 100 so that the time interval for timer
// interrupts at 48MHz is 256*(256-100)*0.083 = 3.3ms
//
// The timer is in 8-bit mode by default
//
    T0CON  = 0x47;                          // Prescaler = 256
    TMR0L = 100;                            // Timer count is 256-156 = 100
    INTCON.TMR0IE = 1;                      // Enable T0IE
    T0CON.TMR0ON = 1;                       // Turn Timer 0 ON
    INTCON = 0xE0;                          // Enable interrupts

//
// Enable USB port
//
    Hid_Enable(&Read_buffer, &Write_buffer);
    Delay_ms(1000);
    Delay_ms(1000);
//
// Read from the USB port. Number of bytes read is in num
//

    for(;;)                                 // do forever
{
    num=0;
    while(num != 4)                         // Get 4 characters
    {num = Hid_Read();
    }
    if(Read_buffer[0] == 'P' && Read_buffer[1] == '=' && Read_buffer[3] == 'T')
    {
        PORTB = Read_buffer[2];
    }
}
    Hid_Disable();

}
```

Figure 8.22: (Cont'd)

Figure 8.23 shows part of the PIC18F4550 clock circuit. The circuit consists of a 1:1 – 1:12 PLL prescaler and multiplexer, a 4:96MHz PLL, a 1:2 – 1:6 PLL postscaler, and a 1:1 – 1:4 oscillator postscaler. Assuming the crystal frequency is 8MHz and we want to operate the microcontroller with a 48MHz clock, and also remembering that a 48MHz clock is required for the USB module, we should make the following choices in the Edit Project option of the mikroC IDE:

- Set *PLL_DIV2_1L* so the 8MHz clock is divided by 2 to produce 4MHZ at the output of the PLL prescaler multiplexer. The output of the 4:96MHZ PLL is now 96MHz. This is further divided by 2 to give 48MHz at the input of multiplexer USBDIV.

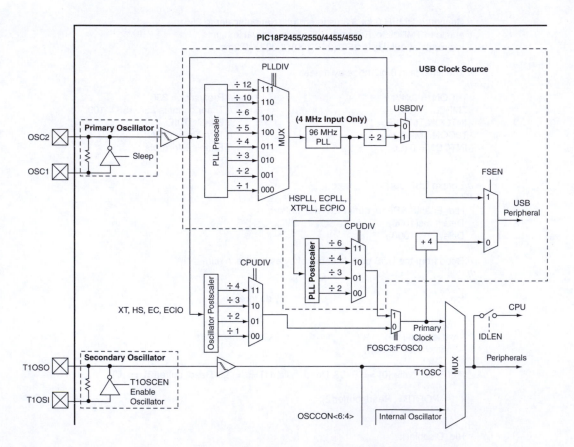

Figure 8.23: PIC18F4550 microcontroller clock

- Check *_USBDIV_2_1L* to provide a 48MHz clock to USB module and to select ÷2 for the PLL postscaler.

- Check *CPUDIV_OSC1_PLL2_1L* to select PLL as the clock source.

- Check *_FOSC_HSPLL_HS_1H* to select a 48MHz clock for the CPU.

- Set the CPU clock to 48MHz in mikroC IDE (using Edit Project).

The clock bits selected for the 48MHz USB operation with a 48MHz CPU clock are shown in Figure 8.24.

Setting other configuration bits in addition to the clock bits is recommended. The following list gives all the bits that should be set in the Edit Project option of the IDE (most of these settings are the power-on-reset values of the bits):

```
PLLDIV_2_1L
CPUDIV_OSC1_PLL2_1L
USBDIV_2_1L

FOSC_HSPLL_HS_1H
FCMEM_OFF_1H
IESO_OFF_1H

PWRT_ON_2L
BOR_ON_2L
BORV_43_2L
VREGEN_ON_2L

WDT_OFF_2H
WDTPS_256_2H

MCLRE_ON_3H
LPT1OSC_OFF_3H
PBADEN_OFF_3H
CCP2MX_ON_3H

STVREN_ON_4L
LVP_OFF_4L
ICPRT_OFF_4L
XINST_OFF_4L
DEBUG_OFF_4L
```

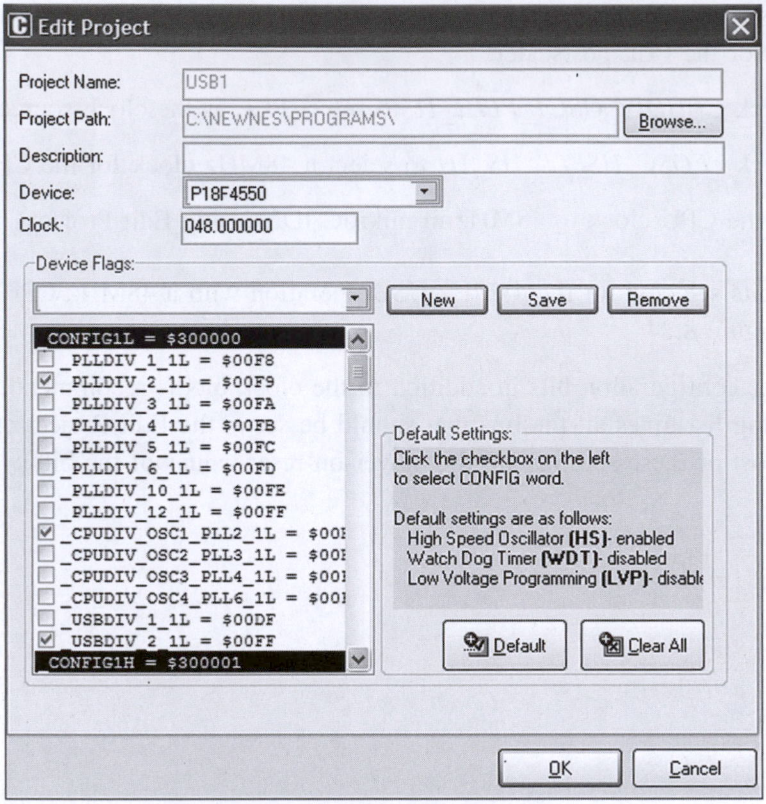

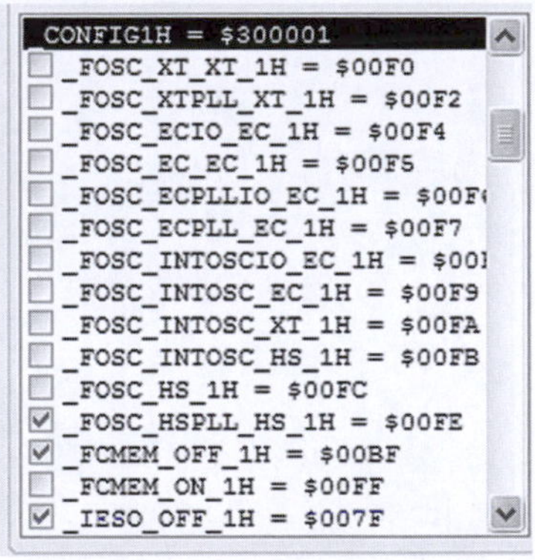

Figure 8.24: Selecting clock bits for USB operation

Testing the Project

Testing the project is relatively easy. The steps are:

- Construct the hardware

- Load the program (Figure 8.22) into the PIC18F4550 microcontroller

- Copy or run the PC-based Visual Basic program

When the microcontroller is connected to one of the USB ports of the PC, a message should be visible at the bottom right-hand corner of the screen similar to the one in Figure 8.25. This message shows that the new USB HID device has been plugged in and is recognized by the PC.

(i) Found New Hardware ☒
USB Human Interface Device

Figure 8.25: USB connection message

In addition, the device manager display should show an HID-compliant device and a USB human interface device as in Figure 8.26. The properties of these drivers can be displayed to make sure the VIP is 0×1234 and the PID is 1.

Enter data into the Visual Basic form and click the CLICK TO SEND button. The corresponding microcontroller LEDs should turn on. For example, entering 3 should turn on LEDs 0 and 1.

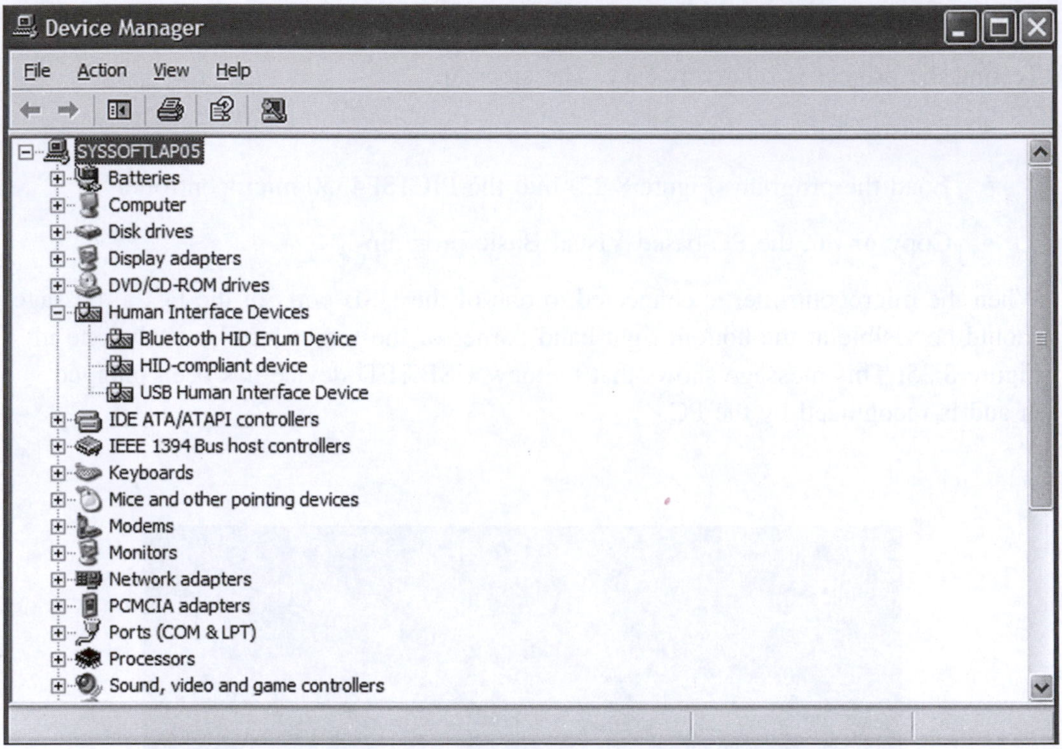

Figure 8.26: Device manager display showing the USB devices

Using a USB Protocol Analyzer

If for any reason the project is not working, a USB protocol analyzer can be used to check the data transactions on the USB bus. There are many USB protocol analyzers on the market. Some expensive professional ones are hardware-based and require the purchase of special hardware. Most low-cost USB protocol analyzers are software-based. Two such tools are described here briefly.

UVCView

UVCView is a free Microsoft product that runs on a PC and displays the descriptors of a USB device after it is plugged in. Figure 8.27 shows the UVCView display after the

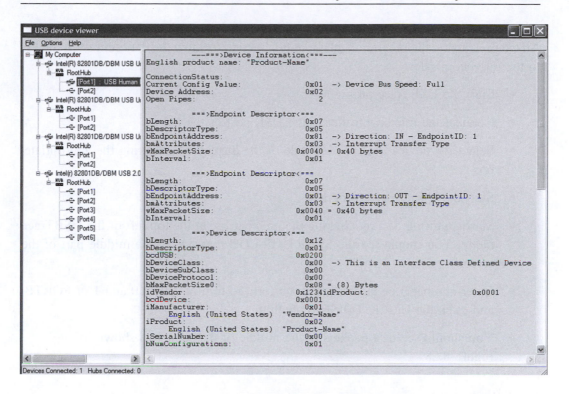

Figure 8.27: UVCView display of the project

microcontroller is plugged into the PC. The left side of the display shows the USB ports
available in the system. Clicking on a device in this part of the display shows descriptor
details of the device in the middle of the screen. In Figure 8.27 the descriptors of
our device are shown. The UVCView display is useful when various fields of the
device descriptors must be checked.

USBTrace

USBTrace is a software USB protocol analyzer developed by SysNucleus (www.
sysnucleus.com) and runs on a PC. The software monitors the USB ports of the PC it is
running on and displays all the transactions on the bus. This software can be an
invaluable tool when all the transactions on the line must be monitored and logged.

A limited-time demo version of USBTrace is available on the manufacturer's web site. An example using the program is given in this section to show the data sent from the PC to the microcontroller:

- Start the USBTrace program.

- Connect the microcontroller to the USB port of the PC.

- Select the device from the left side of the display by checking the appropriate box.

- Start the Visual Basic program.

- Start capturing data by clicking the green arrow at the top left of the USBTrace menu. You should see the START OF LOG message in the middle part of the screen

- Enter number 3 on the Visual Basic form to turn on LEDs 0 and 1 of PORTB, and click the CLICK TO SEND button.

- You should see data packets in the middle of the screen as shown in Figure 8.28.

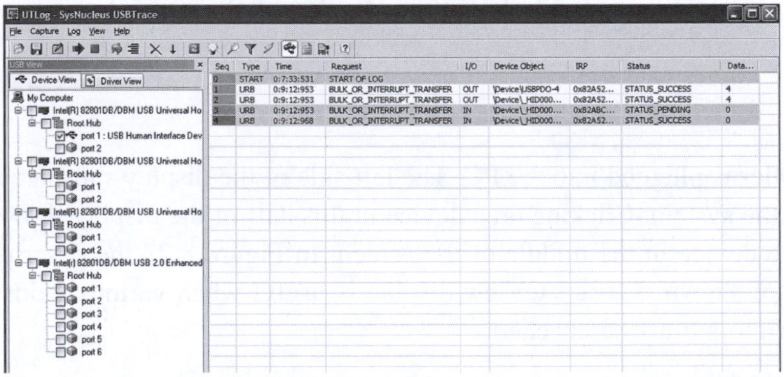

Figure 8.28: Transactions on the bus when CLICK TO SEND is clicked

- Move the cursor over the first packet. This is the packet sent from the PC to the microcontroller (OUT packet). A pop-up window will appear, and information about this packet will be displayed, with the data sent appearing in hexadecimal

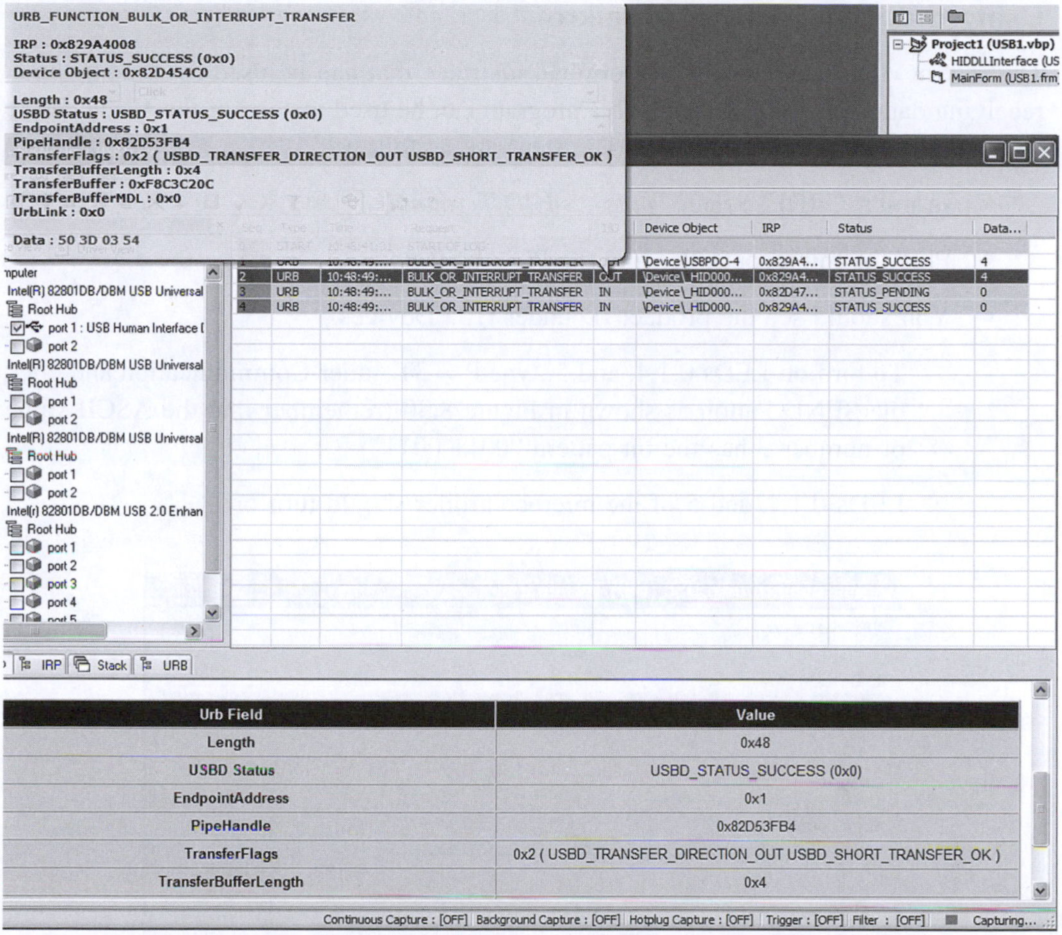

Figure 8.29: Displaying contents of the packet

at the bottom of the display, as shown in Figure 8.29. Note that the data consists of the following 4 bytes:

```
50 3D 03 54
P = 3T
```

which correspond to the ASCII string P = 3T. This is the actual packet sent from the PC to the microcontroller.

USBTrace can also display the device descriptors in detail, as shown in the lower part of the screen in Figure 8.29.

Using the HID Terminal of mikroC

The mikroC IDE provides a USB terminal interface that can be used for sending and receiving data over the USB bus. This program can be used instead of the Visual Basic program to test the USB interface. The steps are as follows:

- In mikroC IDE, Select *Tools -> HID Terminal*

- Plug the microcontroller into the PC's USB port

- You should see the product ID under HID Devices:

  ○ To turn on LEDs 0,1,4, and 5, type P = 3T under Communication and click the SEND button as shown in Figure 8.30 (remember that the ASCII value of number 3 has the bit pattern "0011 0011")

  ○ LEDs 0,1,4, and 5 of the microcontroller should turn on

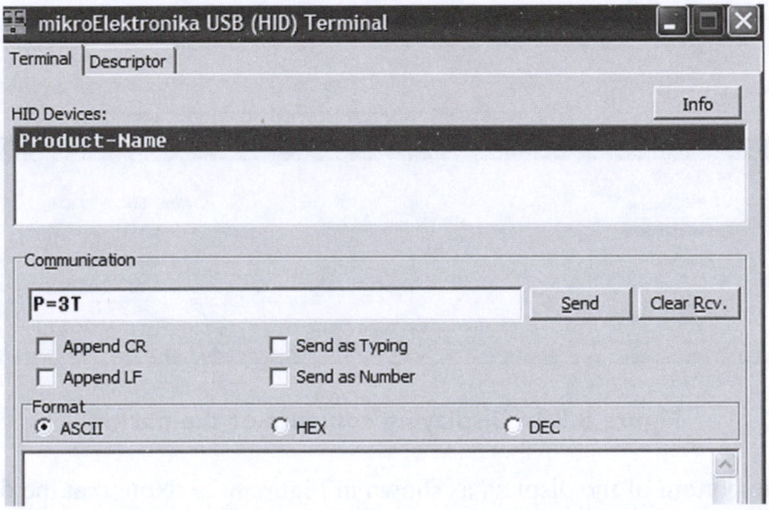

Figure 8.30: Using the HID terminal to send data to a USB device

PROJECT 8.2—USB-Based Microcontroller Input/Output

This project is very similar to Project 8.1, except that it includes two-way communication, while in Project 8.1 data to be output on PORTB was sent to the microcontroller. In addition, PORTB data is received from the microcontroller and displayed on the PC.

The PC sends two commands to the microcontroller:

- Command $P = nT$ requests the microcontroller to send data byte n to PORTB.

- Command $P = ??$ requests the microcontroller to read its PORTB data and send it as a byte to the PC. The PC then displays this data on the screen. The microcontroller sends its data in the familiar format $P = nT$.

The hardware of this project is the same as the hardware for the previous project, shown in Figure 8.11, where eight LEDs are connected to PORTB of a PIC18F4550 microcontroller which is operated from a 8MHz crystal.

A single form is used in this project, and Figure 8.31 shows the format of this form. The upper part of the form is the same as in Project 8.1, i.e., sending data to PORTB of the microcontroller. A text box and a command button named CLICK TO RECEIVE are also placed on the form. When the button is pressed, the PC sends command $P = ??$ to the microcontroller. The microcontroller reads its PORTB data and sends it in the format $P = nT$ to the PC where it is displayed in the text box.

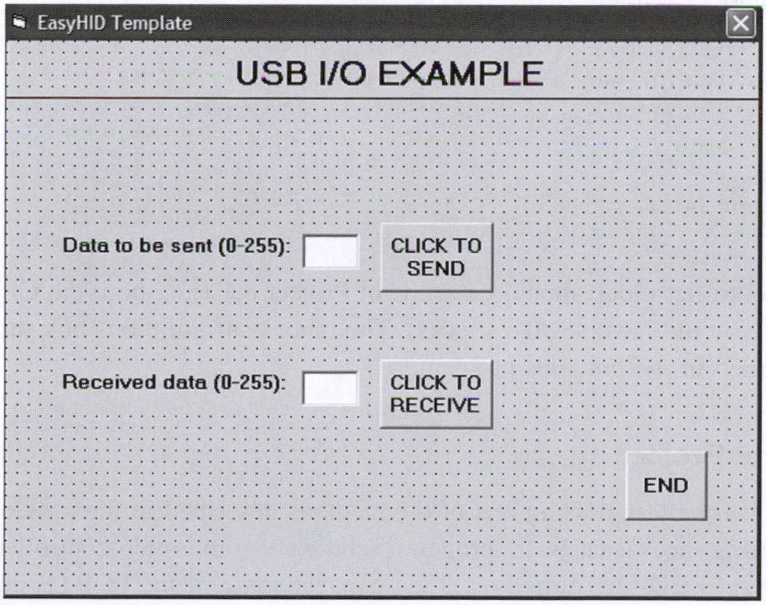

Figure 8.31: Visual Basic form of the project

Figure 8.32 shows the mikroC program of the project. The program is named
USB2.C and is very similar to the one for the previous project. But here, in
addition, when the command *P = ??* is received from the PC, the microcontroller
reads PORTB data and sends it to the PC in the format using the mikroC function
Hid_Write.

The program checks the format of the received command. For P = ?? type commands,
PORTB is configured as inputs, PORTB data is read into *Write_buffer[2]*, and
Write_buffer is sent to the PC, where *Write_buffer[0]* = "P," *Write_buffer[1]* = "=",
and *Write_buffer[3]* = "T" as follows:

```
if(Read_buffer[0] == 'P' && Read_buffer[1] == '=' &&
Read_buffer[2] == '?' && Read_Buffer[3] == '?')
{
TRISB = 0×FF;
Write_buffer[0] = 'P'; Write_buffer[1] = '='; Write_buffer[2] =
  PORTB; Write_buffer[3] = 'T';
Hid_Write(&Write_buffer,4);
}
```

For P = nT type commands, PORTB is configured as outputs and *Read_buffer[2]* is
sent to PORTB as follows:

```
if(Read_buffer[0] == 'P' && Read_buffer[1] == '=' &&
Read_buffer[3] == 'T')
{
TRISB = 0;
PORTB = Read_buffer[2];
}
```

The microcontroller clock should be set as in Project 8.1 (i.e., both the CPU and the
USB module should have 48MHz clocks). The other configurations bits should also be
set as described in the previous problem.

Testing the Project

The project can be tested using one of the methods described in the previous project.
If you are using the Visual Basic program, send data to the microcontroller and make
sure the correct LEDs are turned on. Then connect some of the PORTB pins to
logic 0 and click the CLICK TO RECEIVE button. The microcontroller will read its
PORTB data and send it to the PC, where it will be displayed on the PC screen.

```
/**************************************************************************
          USB BASED MICROCONTROLLER INPUT/OUTPUT PORT
          ===============================================
```

In this project a PIC18F4550 type microcontroller is connected
to a PC through the USB link.

A Visual Basic program runs on the PC where the user enters the
bits to be set or cleared on PORTB of the microcontroller. The
PC sends a command to the microcontroller requesting it to set
or reset the required bits of the microcontroller PORTB. In addition,
the PORTB data can be requested from the microcontroller and displayed
on the PC.

The microcontroller is operated from a 8MHz crystal, but the CPU
clock frequency is increased to 48MHz. Also, the USB module operates
with 48MHz.

The commands are:

From PC to microcontroller: P=nT (Send data byte n to PORTB)
                            P=?? (Give me PORTB data)

From microcontroller to PC: P=nT (Here is my PORTB data)

```
Author:      Dogan Ibrahim
Date:        September 2007
File:        USB2.C
 **************************************************************************/

#include "C:\Program
Files\Mikroelektronika\mikroC\Examples\EasyPic4\extra_examples\HID-
library\USBdsc.c"

unsigned char Read_buffer[64];
unsigned char Write_buffer[64];
unsigned char num,i;
//
// Timer interrupt service routine
//
void interrupt()
{
  HID_InterruptProc();                    // Keep alive
  TMR0L = 100;                            // Reload TMR0L
  INTCON.TMR0IF = 0;                      // Re-enable TMR0 interrupts
}


//
// Start of MAIN program
```

Figure 8.32: mikroC program listing of the project

(Continued)

```
//
void main()
{

    ADCON1 = 0xFF;                      // Set PORTB to digital I/O
    TRISB = 0;                          // Set PORTB to outputs
    PORTB = 0;                          // PORTB all 0s to start with

//
// Set interrupt registers to power-on defaults
// Disable all interrupts
//
    INTCON=0;
    INTCON2=0xF5;
    INTCON3=0xC0;
    RCON.IPEN=0;
    PIE1=0;
    PIE2=0;
    PIR1=0;
    PIR2=0;
//
// Configure TIMER 0 for 20ms interrupts. Set prescaler to 256
// and load TMR0L to 156 so that the time interval for timer
// interrupts at 8MHz is 256*156*0.5 = 20ms
//
// The timer is in 8-bit mode by default
//
    T0CON  = 0x47;                      // Prescaler = 256
    TMR0L  = 100;                       // Timer count is 256-156 = 100
    INTCON.TMR0IE = 1;                  // Enable T0IE
    T0CON.TMR0ON = 1;                   // Turn Timer 0 ON
    INTCON = 0xE0;                      // Enable interrupts

//
// Enable USB port
//
    Hid_Enable(&Read_buffer, &Write_buffer);
    Delay_ms(1000);
    Delay_ms(1000);
//
// Read from the USB port. Number of bytes read is in num
//

    for(;;)                            // do forever
{
    num=0;
    while(num != 4)
    {num = Hid_Read();
    }
```

Figure 8.32: (Cont'd)

```
        if(Read_buffer[0] == 'P' && Read_buffer[1] == '=' &&
          Read_buffer[2] == '?' && Read_Buffer[3] == '?')
        {
          TRISB = 0xFF;
          Write_buffer[0] = 'P';   Write_buffer[1] = '=';
          Write_buffer[2] = PORTB; Write_buffer[3] = 'T';
          Hid_Write(&Write_buffer,4);
        }
        else
        {
          if(Read_buffer[0] == 'P' && Read_buffer[1] == '=' &&
            Read_buffer[3] == 'T')
          {
            TRISB = 0;
            PORTB = Read_buffer[2];
          }
        }
      }
    Hid_Disable();

}
```

Figure 8.32: (Cont'd)

The project can also be tested using the HID terminal of mikroC IDE. The steps are:

- Start the HID terminal.

- Send a command to the microcontroller to turn on the LEDs (e.g., $P = 1T$) and make sure the correct LEDs are turned on (in this case, LEDs 0, 4, and 5 should turn on, corresponding to the data pattern "0011 0001").

- Connect bits 2 and 3 of PORTB to logic 1 and the other six bits to ground.

- Send command $P = ??$ to the microcontroller.

- The PC will display the number 12, corresponding to bit pattern "0000 1100".

The Visual Basic program listing of the project is given in Figure 8.33. Only the main program is given here, as the library declarations are the same as in Figure 8.19. The program jumps to subroutine *OnRead* when data arrives at the USB bus. The format of this data is checked to be in the format P = nT, and if the format is correct, the received data byte is displayed in the text box.

An installable version of the Visual Basic PC program is available in folder USB2 on the CDROM included with this book.

```
' vendor and product IDs
Private Const VendorID = 4660
Private Const ProductID = 1

' read and write buffers
Private Const BufferInSize = 8
Private Const BufferOutSize = 8
Dim BufferIn(0 To BufferInSize) As Byte
Dim BufferOut(0 To BufferOutSize) As Byte

Private Sub Command1_Click()
  Form_Unload (0)
End
End Sub

Private Sub Command2_Click()
   BufferOut(0) = 0                        ' first byte is always the report ID
   BufferOut(1) = Asc("P")                 ' first data item ("P")
   BufferOut(2) = Asc("=")                 ' second data item ("=")
   BufferOut(3) = Val(txtno)               ' third data item (data)
   BufferOut(4) = Asc("T")                 ' fourth data item ("T")

   ' write the data (don't forget, pass the whole array)...
   hidWriteEx VendorID, ProductID, BufferOut(0)
   lblstatus = "Data sent..."

End Sub
' *************************************************************************

' Send command P=?? to the microcontroller to request its PORTB data
' *************************************************************************

*
Private Sub Command3_Click()
   BufferOut(0) = 0                        ' first byte is always the report ID
   BufferOut(1) = Asc("P")                 ' first data item ("P")
   BufferOut(2) = Asc("=")                 ' second data item ("=")
   BufferOut(3) = Asc("?")                 ' third data item ("?")
   BufferOut(4) = Asc("?")                 ' fourth data item ("?")

   ' write the data (don't forget, pass the whole array)...
   hidWriteEx VendorID, ProductID, BufferOut(0)
   lblstatus = "Data requested..."

End Sub

' *************************************************************************
' when the form loads, connect to the HID controller - pass
' the form window handle so that you can receive notification
' events...
' *************************************************************************
Private Sub Form_Load()
```

Figure 8.33: Visual Basic program listing of the project

```
' do not remove!
ConnectToHID (Me.hwnd)
lblstatus = "Connected to HID..."
End Sub


'*******************************************************************

' disconnect from the HID controller...
'*******************************************************************
Private Sub Form_Unload(Cancel As Integer)
DisconnectFromHID
End Sub


'*******************************************************************

' a HID device has been plugged in...
'*******************************************************************
Public Sub OnPlugged(ByVal pHandle As Long)
If hidGetVendorID(pHandle) = VendorID And hidGetProductID(pHandle) =
ProductID Then
lblstatus = "USB Plugged....."
End If
End Sub


'*******************************************************************

' a HID device has been unplugged...
'*******************************************************************
Public Sub OnUnplugged(ByVal pHandle As Long)
If hidGetVendorID(pHandle) = VendorID And hidGetProductID(pHandle) =
ProductID Then
lblstatus = "USB Unplugged...."
End If
End Sub


'*******************************************************************

' controller changed notification - called
' after ALL HID devices are plugged or unplugged
'*******************************************************************
Public Sub OnChanged()
Dim DeviceHandle As Long

' get the handle of the device we are interested in, then set
' its read notify flag to true - this ensures you get a read
' notification message when there is some data to read...
DeviceHandle = hidGetHandle(VendorID, ProductID)
hidSetReadNotify DeviceHandle, True
End Sub


'*******************************************************************

' on read event...
'*******************************************************************
Public Sub OnRead(ByVal pHandle As Long)
```

Figure 8.33: (Cont'd)

```
' read the data (don't forget, pass the whole array)...
If hidRead(pHandle, BufferIn(0)) Then
   ' The data is received in the format: P=nT where the first byte
   ' is the report ID. i.e. BufferIn(0)=reportID, BufferIn(0)="P" and so on
   ' Check to make sure that received data is in correct format
   If (BufferIn(1) = Asc("P") And BufferIn(2) = Asc("=") And
      BufferIn(4) = Asc("T")) Then
      txtreceived = Str$(BufferIn(3))
      lblstatus = "Data received..."
   End If
End If
End Sub
```

Figure 8.33: (Cont'd)

PROJECT 8.3—USB-Based Ambient Pressure Display on the PC

In this project, an ambient atmospheric pressure sensor is connected to a PIC18F4550 microcontroller, and the measured pressure is sent and displayed on a PC every second using a USB link.

An MPX4115A-type pressure sensor is used in this project. This sensor generates an analog voltage proportional to the ambient pressure. The device is available in either a 6-pin or an 8-pin package.

The pin configuration of a 6-pin sensor is:

Pin	Description
1	Output voltage
2	Ground
3	+5V supply
4–6	not used

and for an 8-pin sensor:

Pin	Description
1	not used
2	+5V supply
3	Ground
4	Output voltage
5–8	not used

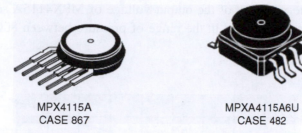

MPX4115A
CASE 867

MPXA4115A6U
CASE 482

Figure 8.34: MPX4115A pressure sensors

Figure 8.34 shows pictures of this sensor with both types of pin configurations.

The output voltage of the sensor is determined by:

$$V = 5.0 * (0.009 * kPa - 0.095) \qquad (8.1)$$

or

$$kPa = \frac{\frac{V}{5.0} + 0.095}{0.009} \qquad (8.2)$$

where

kPa = atmospheric pressure (kilopascals)

V   = output voltage of the sensor (V)

The atmospheric pressure measurements are usually shown in millibars. At sea level and at 15°C the atmospheric pressure is 1013.3 millibars. In Equation (8.2) the pressure is given in kPa. To convert kPa to millibars we have to multiply Equation (8.2) by 10 to give:

$$mb = 10 \times \frac{\frac{V}{5.0} + 0.095}{0.009} \qquad (8.3)$$

or

$$mb = \frac{2.0V + 0.95}{0.009} \qquad (8.4)$$

Figure 8.35 shows the variation of the output voltage of MPX4115A sensor as the pressure varies. We are interested in the range of pressure between 800 and 1100 millibars.

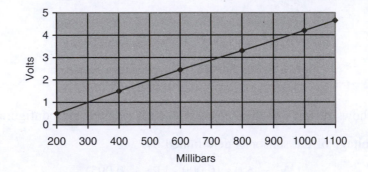

Figure 8.35: Variation of sensor output voltage with pressure

The steps to calculate the pressure in millibars are:

- Read the output voltage of the pressure sensor using one of the A/D channels of the microcontroller

- Use Equation (8.4) to convert the voltage into pressure in millibars

The block diagram of the project is shown in Figure 8.36.

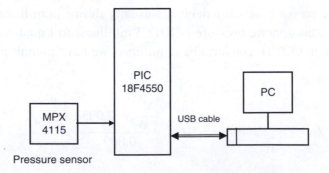

Figure 8.36: Block diagram of the project

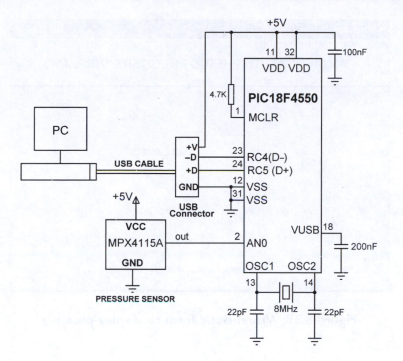

Figure 8.37: Circuit diagram of the project

The circuit diagram of the project is shown in Figure 8.37. The sensor output is connected to analog input AN0 of the microcontroller. As in Project 8.2, the USB connector is connected to port pins RC4 and RC5 and the microcontroller is operated from an 8MHz crystal.

The program on the PC is based on Visual Basic, as in the previous projects. A single form is used, as shown in Figure 8.38, to display the pressure in millibars every second.

The microcontroller program listing (named PRESSURE.C) of the project is given in Figure 8.39. At the beginning of the main program the PORTA pins are defined as analog inputs by clearing ADCON1 to 0 and setting port pins as inputs. Then the interrupt registers are set to their default power-on values. Timer interrupt TMR0 is set

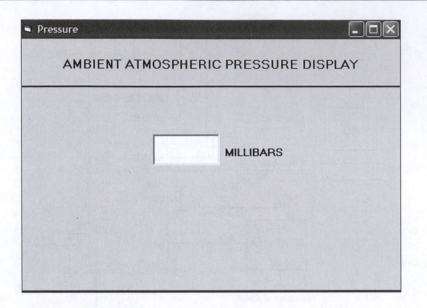

Figure 8.38: Visual Basic form to display pressure

to generate an interrupt every 3.3ms to keep the USB bus alive. The USB port of the microcontroller is then enabled, and ADCON2 is initialized by setting the A/D clock frequency to Fosc/64.

An endless loop is formed using a *for* statement. Inside this loop the pressure sensor data is read into variable *Vin* and then converted into physical voltage in millivolts and stored in variable *mV*. The atmospheric pressure is then calculated using Equation (8.4) and stored in variable *Pint* as a long integer. The mikroC function *LongToStr* converts this integer into a string in array *op*. Any leading spaces are removed from this array, and the resulting pressure is stored in a character array called *Pressure*. The mikroC USB function *Hid_Write* is then called to send the pressure data to the USB bus as 4-character data. The program then waits for one second, and the above process is repeated forever.

An 8MHz crystal is used to provide clock pulses to the microcontroller. The microcontroller CPU clock and the USB module are operated at 48MHz, and the clock and configuration register settings are as in the other projects in this chapter.

```
/************************************************************************
              USB BASED ATMOSPHERIC PRESSURE DISPLAY ON PC
              ===============================================
```

In this project a PIC18F4550 type microcontroller is connected
to a PC through the USB link.

In addition, a MPX4115A type pressure sensor IC is connected to analog port AN0 of
the microcontroller. The microcontroller reads the atmospheric presure and sends it to
the PC every second. The PC displays the pressure on the screen.

A Visual Basic program runs on the PC which reads the pressure from the USB port
and then displays it on a form.

The microcontroller is operated from a 8MHz crystal, but the CPU clock frequency is
increased to 48MHz. Also, the USB module operates with 48MHz.

The pressure is sent to the PC in millibars as a 4 digit integer
number.

```
Author:       Dogan Ibrahim
Date:         September 2007
File:         PRESSURE.C
************************************************************************/

#include "C:\Program
Files\Mikroelektronika\mikroC\Examples\EasyPic4\extra_examples\HID-
library\USBdsc.c"

unsigned char num,i,j;
unsigned long Vin, Pint;
unsigned char op[12], Pressure[4], Read_buffer[4];
float mV,V,Pmb;

//
// Timer interrupt service routine
//
void interrupt()
{
  HID_InterruptProc();                      // Keep alive
  TMR0L = 100;                              // Reload TMR0L
  INTCON.TMR0IF = 0;                        // Re-enable TMR0 interrupts
}


//
// Start of MAIN program
//
void main()
{
```

Figure 8.39: Microcontroller program of the project

(Continued)

```
        ADCON1 = 0;                          // Set inputs as analog, Ref=+5V
        TRISA = 0xFF;                         // Set PORT A as inputs
//
// Set interrupt registers to power-on defaults
// Disable all interrupts
//
    INTCON=0;
    INTCON2=0xF5;
    INTCON3=0xC0;
    RCON.IPEN=0;
    PIE1=0;
    PIE2=0;
    PIR1=0;
    PIR2=0;
//
// Configure TIMER 0 for 3.3ms interrupts. Set prescaler to 256
// and load TMR0L to 156 so that the time interval for timer
// interrupts at 48MHz is 256*156*0.083 = 3.3ms
//
// The timer is in 8-bit mode by default
//
    T0CON  = 0x47;                           // Prescaler = 256
    TMR0L  = 100;                            // Timer count is 256-156 = 100
    INTCON.TMR0IE = 1;                       // Enable T0IE
    T0CON.TMR0ON = 1;                        // Turn Timer 0 ON
    INTCON = 0xE0;                           // Enable interrupts


//
// Enable USB port
//
    Hid_Enable(&Read_buffer, &Pressure);
    Delay_ms(1000);
    Delay_ms(1000);


//
// Configure A/D converter. AN0 is used in this project
//
    ADCON2 = 0xA6;                           // A/D clock = Fosc/64, 8TAD
//
// Endless loop. Read pressure from the A/D converter,
// convert into millibars and send to the PC over the
// USB port every second
//
    for(;;)                                  // do forever
    {
        Vin = Adc_Read(0);                   // Read from channel 0 (AN0)
        mV = (Vin * 5000.0) / 1024.0;        // In mv=Vin x 5000/1024
        V = mV /1000.0;                      // Pressure in Volts
        Pmb = (2.0*V + 0.95) / 0.009;        // Pressure in mb
```

Figure 8.39: (Cont'd)

```
        Pint = (int)Pmb;                          // As an integer number
        LongToStr(Pint,op);                       // Convert to string in "op"
//
// Remove leading blanks
//
        for(j=0; j<4; j++)Pressure[j]=' ';

        j=0;
        for(i=0;i<=11;i++)
        {
          if(op[i] != ' ')                        // If a blank
          {
            Pressure[j]=op[i];
            j++;
          }
        }
//
// Send pressure (in array Pressure) to the PC
//
        Hid_Write(&Pressure,4);                   // Send to USB as 4 characters
        Delay_ms(1000);                           // Wait 1 second
     }
     Hid_Disable();

}
```

Figure 8.39: (Cont'd)

The PC program, based on Visual Basic, is called PRESSURE. Subroutine *OnRead* receives the data arriving at the USB port of the PC and then displays it on the screen form. The program does not send any data to the USB bus. The program listing (except the global variable declarations) is given in Figure 8.40.

Figure 8.41 shows a typical output from the Visual Basic program, displaying the atmospheric pressure.

An installable version of the Visual Basic program is provided on the CDROM that comes with this book, in folder PRESSURE.

```
' vendor and product IDs
Private Const VendorID = 4660
Private Const ProductID = 1

' read and write buffers
Private Const BufferInSize = 8
Private Const BufferOutSize = 8
Dim BufferIn(0 To BufferInSize) As Byte
Dim BufferOut(0 To BufferOutSize) As Byte

Private Sub Command1_Click()
  Form_Unload (0)
End
End Sub


' *********************************************************************


' when the form loads, connect to the HID controller - pass
' the form window handle so that you can receive notification
' events...
' *********************************************************************
Private Sub Form_Load()
  ' do not remove!
  ConnectToHID (Me.hwnd)
  lblstatus = "Connected to HID..."
End Sub


' *********************************************************************


' disconnect from the HID controller...
' *********************************************************************
Private Sub Form_Unload(Cancel As Integer)
  DisconnectFromHID
End Sub


' *********************************************************************


' a HID device has been plugged in...
' *********************************************************************
Public Sub OnPlugged(ByVal pHandle As Long)
  If hidGetVendorID(pHandle) = VendorID And hidGetProductID(pHandle) =
ProductID Then
  lblstatus = "USB Plugged....."
  End If
End Sub
```

Figure 8.40: Visual Basic program of the project

```
'************************************************************************
' a HID device has been unplugged...
'************************************************************************
Public Sub OnUnplugged(ByVal pHandle As Long)
    If hidGetVendorID(pHandle) = VendorID And hidGetProductID(pHandle) =
ProductID Then
    lblstatus = "USB Unplugged...."
    End If
End Sub


'************************************************************************
' controller changed notification - called
' after ALL HID devices are plugged or unplugged
'************************************************************************
Public Sub OnChanged()
    Dim DeviceHandle As Long

    ' get the handle of the device we are interested in, then set
    ' its read notify flag to true - this ensures you get a read
    ' notification message when there is some data to read...
    DeviceHandle = hidGetHandle(VendorID, ProductID)
    hidSetReadNotify DeviceHandle, True
End Sub


'************************************************************************
' on read event...
'************************************************************************
Public Sub OnRead(ByVal pHandle As Long)
    Dim pressure As String

    If hidRead(pHandle, BufferIn(0)) Then
        ' The first byte is the report ID. i.e. BufferIn(0)=reportID
        pressure = Chr(BufferIn(1)) & Chr(BufferIn(2)) & Chr(BufferIn(3)) &
Chr(BufferIn(4))
        txtno = pressure
    End If
End Sub
```

Figure 8.40: (Cont'd)

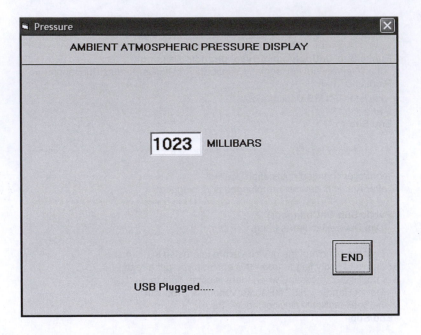

Figure 8.41: Typical output from the Visual Basic program

Advanced PIC18 Projects—CAN Bus Projects

The Controller Area Network (CAN) is a serial bus communications protocol developed by Bosch (an electrical equipment manufacturer in Germany) in the early 1980s. Thereafter, CAN was standardized as ISO-11898 and ISO-11519, establishing itself as the standard protocol for in-vehicle networking in the auto industry. In the early days of the automotive industry, localized stand-alone controllers had been used to manage various actuators and electromechanical subsystems. By networking the electronics in vehicles with CAN, however, they could be controlled from a central point, the engine control unit (ECU), thus increasing functionality, adding modularity, and making diagnostic processes more efficient.

Early CAN development was mainly supported by the vehicle industry, as it was used in passenger cars, boats, trucks, and other types of vehicles. Today the CAN protocol is used in many other fields in applications that call for networked embedded control, including industrial automation, medical applications, building automation, weaving machines, and production machinery. CAN offers an efficient communication protocol between sensors, actuators, controllers, and other nodes in real-time applications, and is known for its simplicity, reliability, and high performance.

The CAN protocol is based on a bus topology, and only two wires are needed for communication over a CAN bus. The bus has a multimaster structure where each device on the bus can send or receive data. Only one device can send data at any time while all the others listen. If two or more devices attempt to send data at the same time, the one with the highest priority is allowed to send its data while the others return to receive mode.

As shown in Figure 9.1, in a typical vehicle application there is usually more than one CAN bus, and they operate at different speeds. Slower devices, such as door control, climate control, and driver information modules, can be connected to a slow speed bus. Devices that require faster response, such as the ABS antilock braking system, the transmission control module, and the electronic throttle module, are connected to a faster CAN bus.

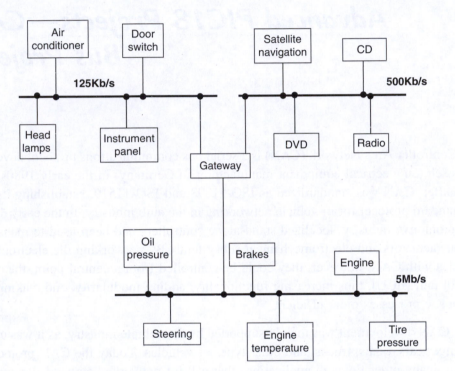

Figure 9.1: Typical CAN bus application in a vehicle

The automotive industry's use of CAN has caused mass production of CAN controllers. Current estimate is that 400 million CAN modules are sold every year, and CAN controllers are integrated on many microcontrollers, including PIC microcontrollers, and are available at low cost.

Figure 9.2 shows a CAN bus with three nodes. The CAN protocol is based on CSMA/CD+AMP (Carrier-Sense Multiple Access/Collision Detection with Arbitration on Message Priority) protocol, which is similar to the protocol used in Ethernet LAN. When Ethernet detects a collision, the sending nodes simply stop transmitting and wait

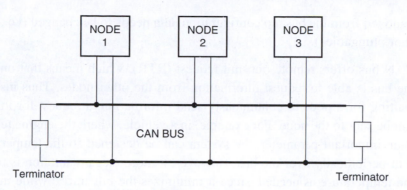

Figure 9.2: Example CAN bus

a random amount of time before trying to send again. CAN protocol, however, solves the collision problem using the principle of arbitration, where only the higheest priority node is given the right to send its data.

There are basically two types of CAN protocols: 2.0A and 2.0B. CAN 2.0A is the earlier standard with 11 bits of identifier, while CAN 2.0B is the new extended standard with 29 bits of identifier. 2.0B controllers are completely backward-compatible with 2.0A controllers and can receive and transmit messages in either format.

There are two types of 2.0A controllers. The first is capable of sending and receiving 2.0A messages only, and reception of a 2.0B message will flag an error. The second type of 2.0A controller (known as 2.0B passive) sends and receives 2.0A messages but will also acknowledge receipt of 2.0B messages and then ignore them.

Some of the CAN protocol features are:

- CAN bus is multimaster. When the bus is free, any device attached to the bus can start sending a message.

- CAN bus protocol is flexible. The devices connected to the bus have no addresses, which means messages are not transmitted from one node to another based on addresses. Instead, all nodes in the system receive every message transmitted on the bus, and it is up to each node to decide whether the received message should be kept or discarded. A single message can be destined for a particular node or for many nodes, depending on how the system is designed. Another advantage of having no addresses is that when a device is added to or removed from the bus, no configuration data needs to be changed (i.e., the bus is "hot pluggable").

- CAN bus offers remote transmit request (RTR), which means that one node on the bus is able to request information from the other nodes. Thus instead of waiting for a node to continuously send information, a request for information can be sent to the node. For example, in a vehicle, where the engine temperature is an important parameter, the system can be designed so the temperature is sent periodically over the bus. However, a more elegant solution is to request the temperature as needed, since it minimizes the bus traffic while maintaining the network's integrity.

- CAN bus communication speed is not fixed. Any communication speed can be set for the devices attached to a bus.

- All devices on the bus can detect an error. The device that has detected an error immediately notifies all other devices.

- Multiple devices can be connected to the bus at the same time, and there are no logical limits to the number of devices that can be connected. In practice, the number of units that can be attached to a bus is limited by the bus's delay time and electrical load.

The data on CAN bus is differential and can be in two states: dominant and recessive. Figure 9.3 shows the state of voltages on the bus. The bus defines a logic bit 0 as a dominant bit and a logic bit 1 as a recessive bit. When there is arbitration on the bus, a

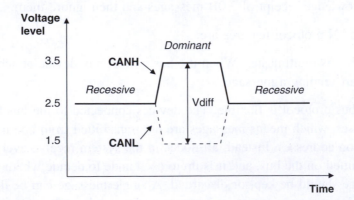

Figure 9.3: CAN logic states

dominant bit state always wins out over a recessive bit state. In the recessive state, the differential voltage CANH and CANL is less than the minimum threshold (i.e., less than 0.5V receiver input and less than 1.5V transmitter output). In the dominant state, the differential voltage CANH and CANL is greater than the minimum threshold.

The ISO-11898 CAN bus specifies that a device on that bus must be able to drive a forty-meter cable at 1Mb/s. A much longer bus length can usually be achieved by lowering the bus speed. Figure 9.4 shows the variation of bus length with the communication speed. For example, with a bus length of one thousand meters we can have a maximum speed of 40Kb/s.

Figure 9.4: CAN bus speed and bus length

A CAN bus is terminated to minimize signal reflections on the bus. The ISO-11898 requires that the bus has a characteristic impedance of 120 ohms. The bus can be terminated by one of the following methods:

- Standard termination

- Split termination

- Biased split termination

In *standard termination*, the most common termination method, a 120-ohm resistor is used at each end of the bus, as shown in Figure 9.5(a). In *split termination*, the ends of the bus are split and a single 60-ohm resistor is used as shown in Figure 9.5(b). Split termination allows for reduced emission, and this method is gaining popularity. *Biased split termination* is similar to split termination except that a voltage divider

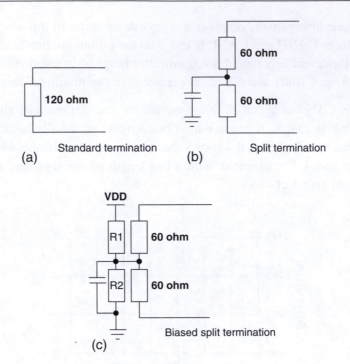

Figure 9.5: Bus termination methods

circuit and a capacitor are used at either end of the bus. This method increases the EMC performance of the bus (Figure 9.5(c)).

Many network protocols are described using the seven-layer Open Systems Interconnection (OSI) model. The CAN protocol includes the data link layer, and the physical layer of the OSI reference model (see Figure 9.6). The data link layer (DLL) consists of the Logical Link Control (LLC) and Medium Access Control (MAC). LLC manages the overload notification, acceptance filtering, and recovery management. MAC manages the data encapsulation, frame coding, error detection, and serialization/deserialization of the data. The physical layer consists of the physical signaling layer (PSL), physical medium attachment (PMA), and the medium dependent interface (MDI). PSL manages the bit encoding/decoding and bit timing. PMA manages the driver/receiver characteristics, and MDI is the connections and wires.

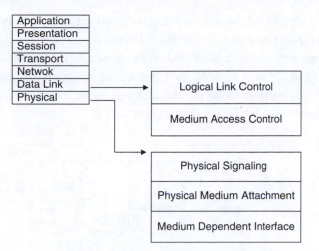

Figure 9.6: CAN and the OSI model

There are basically four message frames in CAN: data, remote, error, and overload. The data and remote frames need to be set by the user. The other two are set by the CAN hardware.

9.1 Data Frame

The data frame is in two formats: standard (having an 11-bit ID) and extended (having a 29-bit ID). The data frame is used by the transmitting device to send data to the receiving device, and the data frame is the most important frame handled by the user. Figure 9.7 shows the data frame's structure. A standard data frame starts with the start of frame (SOF) bit, which is followed by an 11-bit identifier and the remote transmission request (RTR) bit. The identifier and the RTR form the 12-bit arbitration field. The control field is 6 bits wide and indicates how many bytes of data are in the data field. The data field can be 0 to 8 bytes. The data field is followed by the

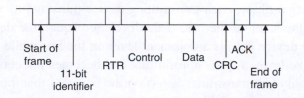

Figure 9.7: Standard data frame

CRC field, which checks whether or not the received bit sequence is corrupted. The ACK field is 2 bits and is used by the transmitter to receive acknowledgment of a valid frame from any receiver. The end of the message is indicated by a 7-bit end of frame (EOF) field. In an extended data frame, the arbitration field is 32 bits wide (29-bit identifier +1-bit IDE to define the message as an extended data frame +1-bit SRR which is unused +1-bit RTR) (see Figure 9.8).

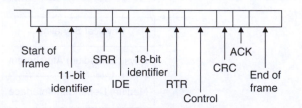

Figure 9.8: Extended data frame

The data frame consists of the following fields:

9.1.1   Start of Frame (SOF)

The start of frame field indicates the beginning of a data frame and is common to both standard and extended formats.

9.1.2   Arbitration Field

Arbitration is used to resolve bus conflicts that occur when several devices at once start sending messages on the bus. The arbitration field indicates the priority of a frame, and it is different in the standard and extended formats. In the standard format there are 11 bits, and up to 2032 IDs can be set. The extended format ID consists of 11 base IDs plus 18 extended IDs. Up to 2032×2^{18} discrete IDs can be set.

During the arbitration phase, each transmitting device transmits its identifier and compares it with the level on the bus. If the levels are equal, the device continues to transmit. If the device detects a dominant level on the bus while it is trying to transmit a recessive level, it quits transmitting and becomes a receiving device. After arbitration only one transmitter is left on the bus, and this transmitter continues to send its control field, data field, and other data.

The process of arbitration is illustrated in Figure 9.9 by an example consisting of three nodes having identifiers:

Node 1: 11100110011   Node 2: 11100111111   Node 3: 11100110001

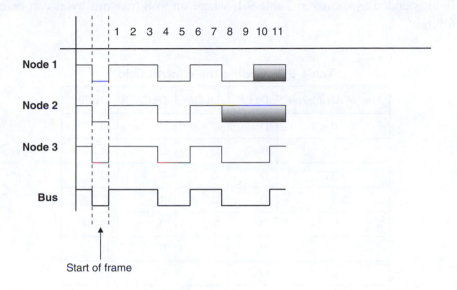

Figure 9.9: Example CAN bus arbitration

Assuming the recessive level corresponds to 1 and the dominant level to 0, the arbitration is performed as follows:

- All the nodes start transmitting simultaneously, first sending SOF bits.

- Then they send their identifier bits. The 8th bit of Node 2 is in the recessive state, while the corresponding bits of Nodes 1 and 3 are in the dominant state. Therefore Node 2 stops transmitting and returns to receive mode. The receiving phase is indicated by a gray field.

- The 10th bit of Node 1 is in the recessive state, while the same bit of Node 3 is in dominant state. Thus Node 1 stops transmitting and returns to receive mode.

- The bus is now left to Node 3, which can send its control and data fields freely.

Notably, the devices on the bus have no addresses. Instead, all the devices pick up all the data on the bus, and every node must filter out the messages it does not want.

9.1.3 Control Field

The control field is 6 bits wide, consisting of 2 reserved bits and 4 data length code (DLC) bits, and indicates the number of data bytes in the message being transmitted. This field is coded as shown in Table 9.1, where up to 8 transmit bytes can be coded with 6 bits.

Table 9.1: Coding the control field

No. of data bytes	DLC3	DLC2	DLC1	DLC0
0	D	D	D	D
1	D	D	D	R
2	D	D	R	D
3	D	D	R	R
4	D	R	D	D
5	D	R	D	R
6	D	R	R	D
7	D	R	R	R
8	R	D or R	D or R	D or R

D: Dominant level, R: Recessive level.

9.1.4 Data Field

The data field carries the actual content of the message. The data size can vary from 0 to 8 bytes. The data is transmitted with the MSB first.

9.1.5 CRC Field

The CRC field, consisting of a 15-bit CRC sequence and a 1-bit CRC delimiter, is used to check the frame for a transmission error. The CRC calculation includes the start of frame, arbitration field, control field, and data field. The calculated CRC and the received CRC sequence are compared, and if they do not match, an error is assumed.

9.1.6   ACK Field

The ACK field indicates that the frame has been received normally. This field consists of 2 bits, one for ACK slot and one for ACK delimiter.

9.2   Remote Frame

The remote frame is used by the receiving unit to request transmission of a message from the transmitting unit. It consists of six fields (see Figure 9.10): start of frame, arbitration field, control field, CRC field, ACK field, and end of frame field. A remote frame is the same as a data frame except that it lacks a data field.

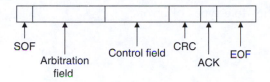

SOF

Arbitration
field

Control field

CRC

ACK

EOF

Figure 9.10: Remote frame

9.3   Error Frame

Error frames are generated and transmitted by the CAN hardware and are used to indicate when an error has occurred during transmission. An error frame consists of an error flag and an error delimiter. There are two types of error flags: active, which consists of 6 dominant bits, and passive, which consists of 6 recessive bits. The error delimiter consists of 8 recessive bits.

9.4   Overload Frame

The overload frame is used by the receiving unit to indicate that it is not yet ready to receive frames. This frame consists of an overload flag and an overload delimiter. The overload flag consists of 6 dominant bits and has the same structure as the active error flag of the error frame. The overload delimiter consists of 8 recessive bits and has the same structure as the error delimiter of the error frame.

9.5   Bit Stuffing

The CAN bus makes use of bit stuffing, a technique to periodically synchronize transmit-receive operations to prevent timing errors between receive nodes. After 5 consecutive bits with the same level, one bit of inverted data is added to the sequence. If, during sending of a data frame or remote frame, the same level occurs in 5 consecutive bits anywhere from the start of frame to the CRC sequence, an inverted bit is inserted in the next (i.e., the sixth) bit. If, during receiving of a data frame or remote frame, the same level occurs in 5 consecutive bits anywhere from the start of frame to CRC sequence, the next (sixth) bit is deleted from the received frame. If the deleted sixth bit is at the same level as the fifth bit, an error (stuffing error) is detected.

9.6   Types of Errors

The CAN bus identifies five types of errors:

- Bit error

- CRC error

- Form error

- ACK error

- Stuffing error

Bit errors are detected when the output level and the data level on the bus do not match. Both transmit and receive units can detect bit errors. *CRC errors* are detected only by receiving units. CRC errors are detected if the calculated CRC from the received message and the received CRC do not match. *Form errors* are detected by the transmitting or receiving units when an illegal frame format is detected. *ACK errors* are detected only by the transmitting units if the ACK field is found recessive. *Stuffing errors* are detected when the same level of data is detected for 6 consecutive bits in any field that should have been bit-stuffed. This error can be detected by both the transmitting and receiving units.

9.7   Nominal Bit Timing

The CAN bus nominal bit rate is defined as the number of bits transmitted every second without resynchronization. The inverse of the nominal bit rate is the nominal bit time. All devices on the CAN bus must use the same bit rate, even though each device can have its own different clock frequency. One message bit consists of four nonoverlapping time segments:

- Synchronization segment (Sync_Seg)

- Propagation time segment (Prop_Seg)

- Phase buffer segment 1 (Phase_Seg1)

- Phase buffer segment 2 (Phase_Seg2)

The *Sync_Seg* segment is used to synchronize various nodes on the bus, and an edge is expected to lie within this segment. The *Prop_Seg* segment compensates for physical delay times within the network. The *Phase_Seg1* and *Phase_Seg2* segments compensate for edge phase errors. These segments can be lengthened or shortened by synchronization. The sample point is the point in time where the actual bit value is located and occurs at the end of *Phase_Seg1*. A CAN controller can be configured to sample three times and use a majority function to determine the actual bit value.

Each segment is divided into units known as time quantum, or T_Q. A desired bit timing can be set by adjusting the number of T_Q's that comprise one message bit and the number of T_Q's that comprise each segment in it. The T_Q is a fixed unit derived from the oscillator period, and the time quantum of each segment can vary from 1 to 8. The lengths of the various time segments are:

- Sync_Seg is 1 time quantum long

- Prop_Seg is programmable as 1 to 8 time quanta long

- Phase_Seg1 is programmable as 1 to 8 time quanta long

- Phase_Seg2 is programmable as 2 to 8 time quanta long

By setting the bit timing, a sampling point can be set so multiple units on the bus can sample messages with the same timing.

The nominal bit time is programmable from a minimum of 8 time quanta to a maximum of 25 time quanta. By definition, the minimum nominal bit time is 1µs, corresponding to a maximum 1Mb/s rate. The nominal bit time (T_{BIT}) is given by:

$$T_{BIT} = T_Q * (Sync\_Seg + Prop\_Seg + Phase\_Seg1 + Phase\_Seg2) \qquad (9.1)$$

and the nominal bit rate (NMR) is

$$NBR = 1/T_{BIT} \qquad (9.2)$$

The time quantum is derived from the oscillator frequency and the programmable baud rate prescaler, with integer values from 1 to 64. The time quantum can be expressed as:

$$T_Q = 2 * (BRP + 1)/F_{OSC} \qquad (9.3)$$

where T_Q is in μs, F_{OSC} is in MHz, and BRP is the baud rate prescaler (0 to 63).

Equation (9.2) can be written as

$$T_Q = 2 * (BRP + 1) * T_{OSC} \qquad (9.4)$$

where T_{OSC} is in μs.

An example of the calculation of a nominal bit rate follows.

Example 9.1

Assuming a clock frequency of 20MHz, a baud rate prescaler value of 1, and a nominal bit time of $T_{BIT} = 8 * T_Q$, determine the nominal bit rate.

Solution 9.1

Using equation (9.3),

$$T_Q = 2 * (1 + 1)/20 = 0.2 μs$$

also

$$T_{BIT} = 8 * T_Q = 8 * 0.2 = 1.6 μs$$

From Equation (9.2),

$$NBR = 1/T_{BIT} = 1/1.6 μs = 625,000 bites/s \text{ or } 625Kb/s$$

In order to compensate for phase shifts between the oscillator frequencies of nodes on a bus, each CAN controller must synchronize to the relevant signal edge of the received signal. Two types of synchronization are defined: hard synchronization and resynchronization. Hard synchronization is used only at the beginning of a message frame, when each CAN node aligns the *Sync_Seg* of its current bit time to the recessive or dominant edge of the transmitted start of frame. According to the rules of synchronization, if a hard synchronization occurs, there will not be a resynchronization within that bit time.

With resynchronization, *Phase_Seg1* may be lengthened or *Phase_Seg2* may be shortened. The amount of change in the phase buffer segments has an upper bound given by the synchronization jump width (SJW). The SJW is programmable between 1 and 4, and its value is added to *Phase_Seg1* or subtracted from *Phase_Seg2*.

9.8   PIC Microcontroller CAN Interface

In general, any type of PIC microcontroller can be used in CAN bus–based projects, but some PIC microcontrollers (e.g., PIC18F258) have built-in CAN modules, which can simplify the design of CAN bus–based systems. Microcontrollers with no built-in CAN modules can also be used in CAN bus applications, but additional hardware and software are required, making the design costly and also more complex.

Figure 9.11 shows the block diagram of a PIC microcontroller–based CAN bus application, using a PIC16 or PIC12-type microcontroller (e.g., PIC16F84) with no

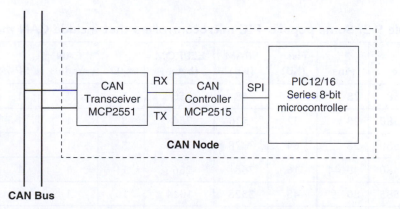

Figure 9.11: CAN node with any PIC microcontroller

built-in CAN module. The microcontroller is connected to the CAN bus using an external MCP2515 CAN controller chip and an MCP2551 CAN bus transceiver chip. This configuration is suitable for a quick upgrade to an existing design using any PIC microcontroller.

For new CAN bus–based designs it is easier to use a PIC microcontroller with a built-in CAN module. As shown in Figure 9.12, such devices include built-in CAN controller hardware on the chip. All that is required to make a CAN node is to add a CAN transceiver chip. Table 9.2 lists some of the PIC microcontrollers that include a CAN module.

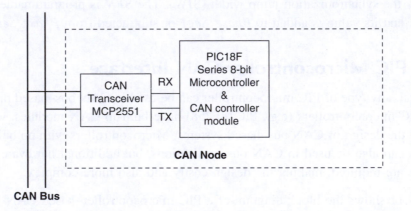

Figure 9.12: CAN node with integrated CAN module

Table 9.2: Some popular PIC microcontrollers that include CAN modules

Device	Pins	Flash (KB)	SRAM (KB)	EEPROM (bytes)	A/D	CAN module	SPI	UART
18F258	28	16	768	256	5	1	1	1
18F2580	28	32	1536	256	8	1	1	1
18F2680	28	64	3328	1024	8	1	1	1
18F4480	40/44	16	768	256	11	1	1	1
18F8585	80	48	3328	1024	16	1	1	1
18F8680	80	64	3328	1024	16	1	1	1

9.9   PIC18F258 Microcontroller

Later in this chapter the PIC18F258 microcontroller is used in a CAN bus–based project. This section describes this microcontroller and its operating principles with respect to its built-in CAN bus. The principles here are in general applicable to other PIC microcontrollers with CAN modules.

The PIC18F258 is a high performance 8-bit microcontroller with integrated CAN module. The device has the following features:

- 32K flash program memory

- 1536 bytes RAM data memory

- 256 bytes EEPROM memory

- 22 I/O ports

- 5-channel 10-bit A/D converters

- Three timers/counters

- Three external interrupt pins

- High-current (25mA) sink/source

- Capture/compare/PWM module

- SPI/I^{2}C module

- CAN 2.0A/B module

- Power-on reset and power-on timer

- Watchdog timer

- Priority level interrupts

- DC to 40MHz clock input

- 8 × 8 hardware multiplier

- Wide operating voltage (2.0V to 5.5V)

- Power-saving sleep mode

The features of the PIC18F258 microcontroller's CAN module are as follows:

- Compatible with CAN 1.2, CAN 2.0A, and CAN 2.0B
- Supports standard and extended data frames
- Programmable bit rate up to 1Mbit/s
- Double-buffered receiver
- Three transmit buffers
- Two receive buffers
- Programmable clock source
- Six acceptance filters
- Two acceptance filter masks
- Loop-back mode for self-testing
- Low-power sleep mode
- Interrupt capabilities

The CAN module uses port pins RB3/CANRX and RB2/CANTX for CAN bus receive and transmit functions respectively. These pins are connected to the CAN bus via an MCP2551-type CAN bus transceiver chip.

The PIC18F258 microcontroller supports the following frame types:

- Standard data frame
- Extended data frame
- Remote frame
- Error frame
- Overload frame
- Interframe space

A node uses filters to decide whether or not to accept a received message. Message filtering is applied to the whole identifier field, and mask registers are used to specify which bits in the identifier the filters should examine.

The CAN module in the PIC18F258 microcontroller has six modes of operation:

- Configuration mode

- Disable mode

- Normal operation mode

- Listen-only mode

- Loop-back mode

- Error recognition mode

9.9.1   Configuration Mode

The CAN module is initialized in configuration mode. The module is not allowed to enter configuration mode while a transmission is taking place. In configuration mode the module will neither transmit nor receive, the error counters are cleared, and the interrupt flags remain unchanged.

9.9.2   Disable Mode

In disable mode, the module will neither transmit nor receive. In this mode the internal clock is stopped unless the module is active. If the module is active, it will wait for 11 recessive bits on the CAN bus, detect that condition as an IDLE bus, and then accept the module disable command. The WAKIF interrupt (wake-up interrupt) is the only CAN module interrupt that is active in disable mode.

9.9.3   Normal Operation Mode

The normal operation mode is the CAN module's standard operating mode. In this mode, the module monitors all bus messages and generates acknowledge bits, error frames, etc. This is the only mode that can transmit messages.

9.9.4   Listen-only Mode

The listen-only mode allows the CAN module to receive messages, including messages with errors. It can be used to monitor bus activities or to detect the baud rate on the bus. For automatic baud rate detection, at least two other nodes must be communicating with each other. The baud rate can be determined by testing different values until valid messages are received. The listen-only mode cannot transmit messages.

9.9.5   Loop-Back Mode

In the loop-back mode, messages can be directed from internal transmit buffers to receive buffers without actually transmitting messages on the CAN bus. This mode is useful during system developing and testing.

9.9.6   Error Recognition Mode

The error recognition mode is used to ignore all errors and receive all messages. In this mode, all messages, valid or invalid are received and copied to the receive buffer.

9.9.7   CAN Message Transmission

The PIC18F258 microcontroller implements three dedicated transmit buffers: TXB0, TXB1, and TXB2. Pending transmittable messages are in a priority queue. Before the SOF is sent, the priorities of all buffers queued for transmission are compared. The transmit buffer with the highest priority is sent first. If two buffers have the same priority, the one with the higher buffer number is sent first. There are four levels of priority.

9.9.8   CAN Message Reception

Reception of a message is a more complex process. The PIC18F258 microcontroller includes two receive buffers, RXB0 and RXB1, with multiple acceptance filters for each (see Figure 9.13). All received messages are assembled in the message assembly buffer (MAB). Once a message is received, regardless of the type of identifier and the number of data bytes, the entire message is copied into the MAB.

Received messages have priorities. RXB0 is the higher priority buffer, and it has two message acceptance filters, RXF0 and RXF1. RXB1 is the lower priority buffer and has four acceptance filters: RXF2, RXF3, RXF4, and RXF5. Two programmable acceptance filter masks, RXM0 and RXM1, are also available, one for each receive buffer.

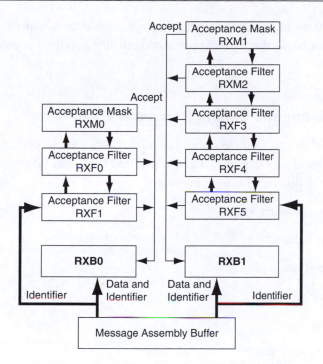

Figure 9.13: Receive buffer block diagram

The CAN module uses message acceptance filters and masks to determine if a message in the MAB should be loaded into a receive buffer. Once a valid message is received by the MAB, the identifier field of the message is compared to the filter values. If there is a match, that message is loaded into the appropriate receive buffer. The filter masks determine which bits in the identifier are examined with the filters. The truth table in Table 9.3 shows how each bit in the identifier is compared against

Table 9.3: Filter/mask truth table

Mask bit n	Filter bit n	Message identifier bit n001	Accept or reject bit n
0	×	×	Accept
1	0	0	Accept
1	0	1	Reject
1	1	0	Reject
1	1	1	Accept

the masks and filters to determine if the message should be accepted. If a mask bit is set to 0, that bit in the identifier is automatically accepted regardless of the filter bit.

9.9.9   Calculating the Timing Parameters

Setting the nodes' timing parameters is essential for the bus to operate reliably. Given the microcontroller clock frequency and the required CAN bus bit rate, we can calculate the values of the following timing parameters:

- Baud rate prescaler value

- Prop_Seg value

- Phase_Seg1 value

- Phase_Seg2 value

- SJW value

Correct timing requires that

- Prop_Seg + Phase_Seg1 $\geq$ Phase_Seg2

- Phase_Seg2 $\geq$ SJW

The following example illustrates the calculation of these timing parameters.

Example 9.2

Assuming the microcontroller oscillator clock rate is 20MHz and the required CAN bit rate is 125KHz, calculate the timing parameters.

Solution 9.2

With a 20MHz clock rate, the clock period is 50ns. Choosing a baud rate prescaler value of 4, from Equation (9.4), $T_Q = 2 * (BRP + 1) * T_{OSC}$, gives a time quantum of $T_Q = 500ns$. To obtain a nominal bit rate of 125KHz, the nominal bit time must be:

$$T_{BIT} = 1/0.125MHz = 8\mu s, \text{ or } 16T_Q$$

The *Sync_Segment* is $1T_Q$. Choosing $2T_Q$ for the *Prop_Seg*, and $7T_Q$ for *Phase_Seg1* leaves $6T_Q$ for *Phase_Seg2* and places the sampling point at $10T_Q$ at the end of *Phase_Seg1*.

By the rules described earlier, the SJW can be the maximum allowed (i.e., 4). However, a large SJW is only necessary when the clock generation of different nodes is not stable or accurate (e.g., if ceramic resonators are used). Typically, a SJW of 1 is enough. In summary, the required timing parameters are:

```
Baud rate prescaler (BRP) = 4
Sync_Seg                  = 1
Prop_Seg                  = 2
Phase_Seg1                = 7
Phase_Seg2                = 6
SJW                       = 1
```

The sampling point is at $10T_Q$ which corresponds to 62.5% of the total bit time.

There are several tools available for free on the Internet for calculating CAN bus timing parameters. One such tool is the CAN Baud Rate Calculator, developed by Artic Consultants Ltd (http://www.articconsultants.co.uk). An example using this tool follows.

Example 9.3

Assuming the microcontroller oscillator clock rate is 20MHz and the required CAN bit rate is 125KHz, calculate the timing parameters using the CAN Baud Rate Calculator.

Solution 9.3

Figure 9.14 shows the output of the CAN Baud Rate Calculator program. The device type is selected as PIC18Fxxx8, the oscillator frequency is entered as 20MHz, and the CAN bus baud rate is entered as 125KHz.

Clicking the Calculate Settings button calculates and displays the recommended timing parameters. In general, there is more than one solution, and different solutions are given in the Calculated Solutions field's drop-down menu.

In choosing Solution 2 from the drop-down menu, the following timing parameters are recommended by the program:

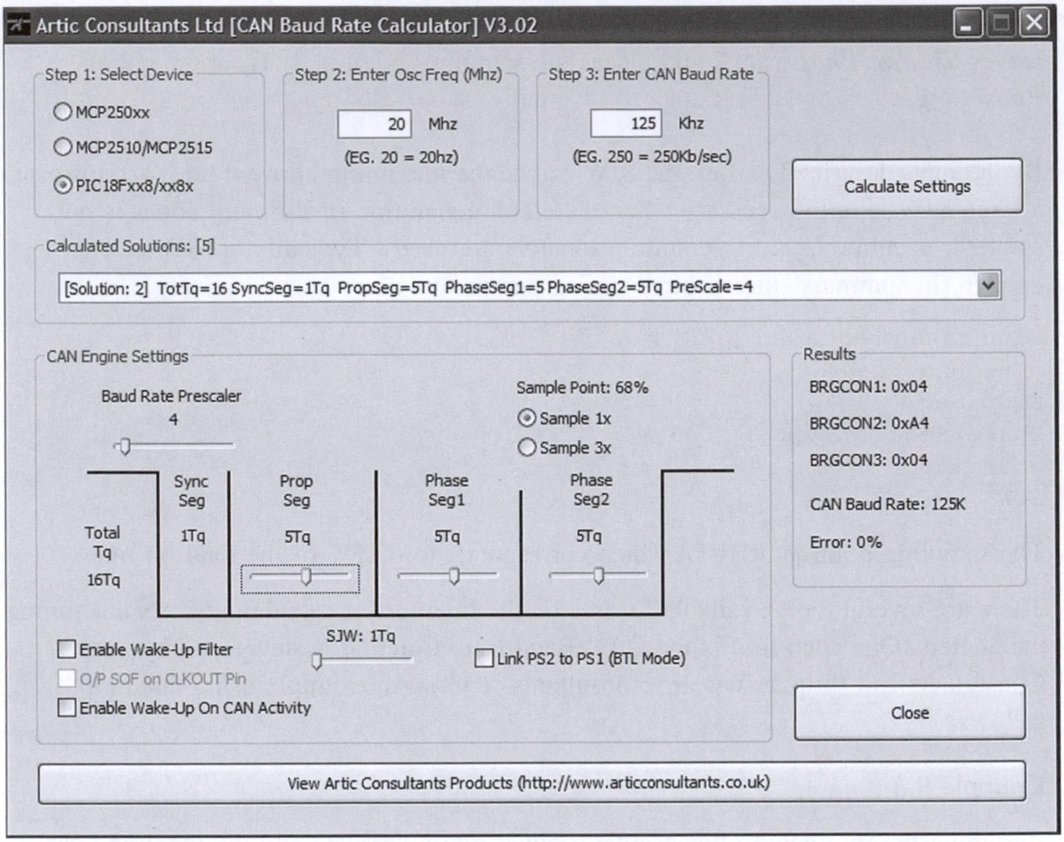

Figure 9.14: Output of the CAN Baud Rate Calculator program

```
Baud rate prescaler (BRP) = 4
Sync_Seg                  = 1
Prop_Seg                  = 5
Phase_Seg1                = 5
Phase_Seg2                = 5
SJW                       = 1
Sample point              = 68%
Error                     = 0%
```

9.10   mikroC CAN Functions

The mikroC language provides two libraries for CAN bus applications: the library for PIC microcontrollers with built-in CAN modules and the library based on using a SPI bus for PIC microcontrollers having no built-in CAN modules. In this section we will discuss only the library functions available for PIC microcontrollers with built-in CAN modules. Similar functions are available for the PIC microcontrollers with no built-in CAN modules.

The mikroC CAN functions are supported only by PIC18XXX8 microcontrollers with MCP2551 or similar CAN transceivers. Both standard (11 identifier bits) and extended format (29 identifier bits) messages are supported.

The following mikroC functions are provided:

- CANSetOperationMode
- CANGetOperationMode
- CANInitialize
- CANSetBaudRAte
- CANSetMask
- CANSetFilter
- CANRead
- CANWrite

9.10.1 CANSetOperationMode

The *CANSetOperationMode* function sets the CAN operation mode. The function prototype is:

```
void CANSetOperationMode(char mode, char wait_flag)
```

The parameter *wait_flag* is either 0 or 0 × FF. If it is set to 0 × FF, the function blocks and will not return until the requested mode is set. If it is set to 0, the function returns as a nonblocking call.

The mode can be one of the following:

- CAN_MODE_NORMAL   Normal mode of operation
- CAN_MODE_SLEEP   Sleep mode of operation
- CAN_MODE_LOOP   Loop-back mode of operation

- CAN_MODE_LISTEN   Listen-only mode of operation
- CAN_MODE_CONFIG   Configuration mode of operation

9.10.2 CANGetOperationMode

The *CANGetOperationMode* function returns the current CAN operation mode. The function prototype is:

```
char CANGetOperationMode(void)
```

9.10.3 CANInitialize

The *CANInitialize* function initializes the CAN module. All mask registers are cleared to 0 to allow all messages. Upon execution of this function, the normal mode is set. The function prototype is:

```
void CANInitialize(char SJW, char BRP, char PHSEG1, char PHSEG2,
char PROPEG, char CAN_CONFIG_FLAGS)
```

where

     SJW   is the synchronization jump width

     BRP   is the baud rate prescaler

     PHSEG1   is the Phase_Seg1 timing parameter

     PHSEG2   is the Phase_Seg2 timing parameter

     PROPSEG   is the Prop_Seg

CAN_CONFIG_FLAGS can be one of the following configuration flags:

- CAN_CONFIG_DEFAULT   Default flags
- CAN_CONFIG_PHSEG2_PRG_ON   Use supplied PHSEG2 value
- CAN_CONFIG_PHSEG2_PRG_OFF   Use maximum of PHSEG1 or information processing time (IPT), whichever is greater
- CAN_CONFIG_LINE_FILTER_ON   Use CAN bus line filter for wake-up
- CAN_CONFIG_FILTER_OFF   Do not use CAN bus line filter

- CAN_CONFIG_SAMPLE_ONCE   Sample bus once at sample point

- CAN_CONFIG_SAMPLE_THRICE   Sample bus three times prior to sample point

- CAN_CONFIG_STD_MSG   Accept only standard identifier messages

- CAN_CONFIG_XTD_MSG   Accept only extended identifier messages

- CAN_CONFIG_DBL_BUFFER_ON   Use double buffering to receive data

- CAN_CONFIG_DBL_BUFFER_OFF   Do not use double buffering

- CAN_CONFIG_ALL_MSG   Accept all messages including invalid ones

- CAN_CONFIG_VALID_XTD_MSG   Accept only valid extended identifier messages

- CAN_CONFIG_VALID_STD_MSG   Accept only valid standard identifier messages

- CAN_CONFIG_ALL_VALID_MSG   Accept all valid messages

These configuration values can be bitwise AND'ed to form complex configuration values.

9.10.4 CANSetBaudRate

The *CANSetBaudRate* function is used to set the CAN bus baud rate. The function prototype is:

```
void CANSetBaudRate(char SJW, char BRP, char PHSEG1, char PHSEG2,
char PROPSEG, char CAN_CONFIG_FLAGS)
```

The arguments of the function are as in function *CANInitialize*.

9.10.5 CANSetMask

The *CANSetMask* function sets the mask for filtering messages. The function prototype is:

```
void CANSetMask(char CAN_MASK, long value, char
CAN_CONFIGFLAGS)
```

CAN_MASK can be one of the following:

- CAN_MASK_B1   Receive buffer 1 mask value

- CAN_MASK_B2   Receive buffer 2 mask value

value is the mask register value. CAN_CONFIG_FLAGS can be either
CAN_CONFIG_XTD (extended message), or CAN_CONFIG_STD (standard
message).

9.10.6   CANSetFilter

The *CANSetFilter* function sets filter values. The function prototype is:

```
void CANSetFilter(char CAN_FILTER, long value, char
CAN_CONFIG_FLAGS)
```

CAN_FILTER can be one of the following:

- CAN_FILTER_B1_F1   Filter 1 for buffer 1

- CAN_FILTER_B1_F2   Filter 2 for buffer 1

- CAN_FILTER_B2_F1   Filter 1 for buffer 2

- CAN_FILTER_B2_F2   Filter 2 for buffer 2

- CAN_FILTER_B2_F3   Filter 3 for buffer 2

- CAN_FILTER_B2_F4   Filter 4 for buffer 2

CAN_CONFIG_FLAGS can be either CAN_CONFIG_XTD (extended message) or
CAN_CONFIG_STD (standard message).

9.10.7   CANRead

The *CANRead* function is used to read messages from the CAN bus. If no message is
available, 0 is returned. The function prototype is:

```
char CANRead(long *id, char *data, char *datalen, char
*CAN_RX_MSG_FLAGS)
```

id is the CAN message identifier. Only 11 or 29 bits may be used depending on message type (standard or extended). *data* is an array of bytes up to 8 where the received data is stored. *datalen* is the length of the received data (1 to 8).

CAN_RX_MSG_FLAGS can be one of the following:

• CAN_RX_FILTER_1	Receive buffer filter 1 accepted this message
• CAN_RX_FILTER_2	Receive buffer filter 2 accepted this message
• CAN_RX_FILTER_3	Receive buffer filter 3 accepted this message
• CAN_RX_FILTER_4	Receive buffer filter 4 accepted this message
• CAN_RX_FILTER_5	Receive buffer filter 5 accepted this message
• CAN_RX_FILTER_6	Receive buffer filter 6 accepted this message
• CAN_RX_OVERFLOW	Receive buffer overflow occurred
• CAN_RX_INVALID_MSG	Invalid message received
• CAN_RX_XTD_FRAME	Extended identifier message received
• CAN_RX_RTR_FRAME	RTR frame message received
• CAN_RX_DBL_BUFFERED	This message was double buffered

These flags can be bitwise AND'ed if desired.

9.10.8   CANWrite

The *CANWrite* function is used to send a message to the CAN bus. A zero is returned if message can not be queued (buffer full). The function prototype is:

```
char CANWrite(long id, char *data, char datalen, char
CAN_TX_MSG_FLAGS)
```

id is the CAN message identifier. Only 11 or 29 bits may be used depending on message type (standard or extended). *data* is an array of bytes up to 8 where the data to be sent is stored. *datalen* is the length of the data (1 to 8).

CAN_TX_MSG_FLAGS can be one of the following:

- • CAN_TX_PRIORITY_0   Transmit priority 0
- • CAN_TX_PRIORITY_1   Transmit priority 1

- CAN_TX_PRIORITY_2   Transmit priority 2

- CAN_TX_PRIORITY_3   Transmit priority 3

- CAN_TX_STD_FRAME   Standard identifier message

- CAN_TX_XTD_FRAME   Extended identifier message

- CAN_TX_NO_RTR_FRAME   Non RTR message

- CAN_TX_RTR_FRAME   RTR message

These flags can be bitwise AND'ed if desired.

9.11   CAN Bus Programming

To operate the PIC18F258 microcontroller on the CAN bus, perform the following steps:

- Configure the CAN bus I/O port directions (RB2 and RB3)

- Initialize the CAN module (*CANInitialize*)

- Set the CAN module to CONFIG mode (*CANSetOperationMode*)

- Set the mask registers (*CANSetMask*)

- Set the filter registers (*CANSetFilter*)

- Set the CAN module to normal mode (*CANSetOperationMode*)

- Write/read data (*CANWrite/CANRead*)

PROJECT 9.1—Temperature Sensor CAN Bus Project

The following is a simple two-node CAN bus–based project. The block diagram of the project is shown in Figure 9.15. The system is made up of two CAN nodes. One node (called DISPLAY node) requests the temperature every second and displays it on an LCD. This process is repeated continuously. The other node (called COLLECTOR node) reads the temperature from an external semiconductor temperature sensor.

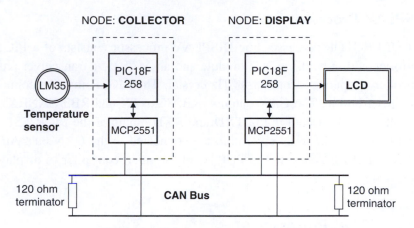

Figure 9.15: Block diagram of the project

The project's circuit diagram is given in Figure 9.16. Two CAN nodes are connected together using a two-meter twisted pair cable, terminated with a 120-ohm resistor at each end.

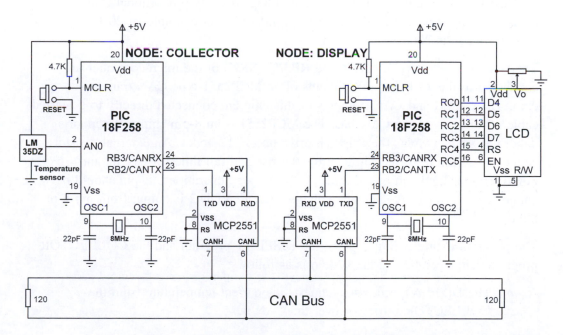

Figure 9.16: Circuit diagram of the project

The DISPLAY Processor

Like the COLLECTOR processor, the DISPLAY processor consists of a PIC18F258 microcontroller with a built-in CAN module and an MCP2551 transceiver chip. The microcontroller is operated from an 8MHz crystal. The MCLR input is connected to an external reset button. The CAN outputs (RB2/CANTX and RB3/CANRX) of the microcontroller are connected to the TXD and RXD inputs of the MCP2551. Pins CANH and CANL of the transceiver chip are connected to the CAN bus. An HD44780-type LCD is connected to PORTC of the microcontroller to display the temperature values.

The COLLECTOR Processor

The COLLECTOR processor consists of a PIC18F258 microcontroller with a built-in CAN module and an MCP2551 transceiver chip. The microcontroller is operated from an 8MHz crystal. The MCLR input is connected to an external reset button. Analog input AN0 of the microcontroller is connected to a LM35DZ-type semiconductor temperature sensor. The sensor can measure temperature in the range of 0°C to 100°C and generates an analog voltage directly proportional to the measured temperature (i.e., the output is 10mV/°C). For example, at 20°C the output voltage is 200mV.

The CAN outputs (RB2/CANTX and RB3/CANRX) of the microcontroller are connected to the TXD and RXD inputs of an MCP2551-type CAN transceiver chip. The CANH and CANL outputs of this chip are connected directly to a twisted cable terminating at the CAN bus. The MCP2551 is an 8-pin chip that supports data rates up to 1Mb/s. The chip can drive up to 112 nodes. An external resistor connected to pin 8 of the chip controls the rise and fall times of CANH and CANL so that EMI can be reduced. For high-speed operation this pin should be connected to ground. A reference voltage equal to VDD/2 is output from pin 5 of the chip.

The program listing is in two parts: the DISPLAY program and the COLLECTOR program. The operation of the system is as follows:

- The DISPLAY processor requests the current temperature from the COLLECTOR processor over the CAN bus

- The COLLECTOR processor reads the temperature, formats it, and sends to the DISPLAY processor over the CAN bus

- The DISPLAY processor reads the temperature from the CAN bus and then displays it on the LCD

- This process is repeated every second

DISPLAY Program

Figure 9.17 shows the program listing of the DISPLAY program, called DISPLAY.C. At the beginning of the program PORTC pins are configured as outputs, RB3 is configured as input (CANRX), and RB2 is configured as output (CANTX). In this project the CAN bus bit rate is selected as 100Kb/s. With a microcontroller clock frequency of 8MHz, the Baud Rate Calculator program (see Figure 9.14) is used to calculate the timing parameters as:

```
SJW = 1
BRP = 1
Phase_Seg1 = 6
Phase_Seg2 = 7
Prop_Seg = 6
```

The mikroC CAN bus function *CANInitialize* is used to initialize the CAN module. The timing parameters and the initialization flag are specified as arguments in this function. The initialization flag is made up from the bitwise AND of:

```
init_flag = CAN_CONFIG_SAMPLE_THRICE &
            CAN_CONFIG_PHSEG2_PRG_ON &
            CAN_CONFIG_STD_MSG &
            CAN_CONFIG_DBL_BUFFER_ON &
            CAN_CONFIG_VALID_XTD_MSG &
            CAN_CONFIG_LINE_FILTER_OFF;
```

Where sampling the bus three times is specified, the standard identifier is specified, double buffering is turned on, and the line filter is turned off.

Then the operation mode is set to CONFIG and the filter masks and filter values are specified. Both mask 1 and mask 2 are set to all 1's (-1 is a shorthand way of writing hexadecimal FFFFFFFF, i.e., setting all mask bits to 1's) so that all filter bits match up with incoming data.

```
/************************************************************************
                    CAN BUS EXAMPLE - NODE: DISPLAY
                    ===============================

This is the DISPLAY node of the CAN bus example. In this project a PIC18F258
type microcontroller is used. An MCP2551 type CAN bus transceiver is used to
connect the microcontroller to the CAN bus. The microcontroller is operated from
an 8MHz crystal with an external reset button.

Pin CANRX and CANTX of the microcontroller are connected to pins RXD
and TXD of the transceiver chip respectively. Pins CANH and CANL of
the transceiver chip are connected to the CAN bus.

An LCD is connected to PORTC of the microcontroller. The ambient
temperature is read from another CAN node and is displayed on the LCD.

The LCD is connected to the microcontroller as follows:

Microcontroller    LCD

     RC0           D4
     RC1           D5
     RC2           D6
     RC3           D7
     RC4           RS
     RC5           EN

CAN speed parameters are:

     Microcontroller clock:      8MHz
     CAN Bus bit rate:           100Kb/s
     Sync_Seg:                   1
     Prop_Seg:                   6
     Phase_Seg1:                 6
     Phase_Seg2:                 7
     SJW:                        1
     BRP:                        1
     Sample point:               65%

Author:      Dogan Ibrahim
Date:        October 2007
File:        DISPLAY.C
************************************************************************/

void main()
{
    unsigned char temperature, data[8];
    unsigned short init_flag, send_flag, dt, len, read_flag;
    char SJW, BRP, Phase_Seg1, Phase_Seg2, Prop_Seg, txt[4];
    long id, mask;
```

Figure 9.17: DISPLAY program listing

```
    TRISC = 0;                          // PORTC are outputs (LCD)
    TRISB = 0x08;                       // RB2 is output, RB3 is input
//
// CAN BUS Parameters
//
    SJW = 1;
    BRP = 1;
    Phase_Seg1 = 6;
    Phase_Seg2 = 7;
    Prop_Seg = 6;

    init_flag = CAN_CONFIG_SAMPLE_THRICE  &
            CAN_CONFIG_PHSEG2_PRG_ON  &
            CAN_CONFIG_STD_MSG        &
            CAN_CONFIG_DBL_BUFFER_ON  &
            CAN_CONFIG_VALID_XTD_MSG  &
            CAN_CONFIG_LINE_FILTER_OFF;

    send_flag = CAN_TX_PRIORITY_0        &
            CAN_TX_XTD_FRAME          &
            CAN_TX_NO_RTR_FRAME;

    read_flag = 0;
//
// Initialize CAN module
//
    CANInitialize(SJW, BRP, Phase_Seg1, Phase_Seg2, Prop_Seg, init_flag);
//
// Set CAN CONFIG mode
//
    CANSetOperationMode(CAN_MODE_CONFIG, 0xFF);

    mask = -1;
//
// Set all MASK1 bits to 1's
//
    CANSetMask(CAN_MASK_B1, mask, CAN_CONFIG_XTD_MSG);
//
// Set all MASK2 bits to 1's
//
    CANSetMask(CAN_MASK_B2, mask, CAN_CONFIG_XTD_MSG);
//
// Set id of filter B2_F3 to 3
//
    CANSetFilter(CAN_FILTER_B2_F3,3,CAN_CONFIG_XTD_MSG);
//
// Set CAN module to NORMAL mode
//
    CANSetOperationMode(CAN_MODE_NORMAL, 0xFF);
```

Figure 9.17: (Cont'd)

```
//
// Configure LCD
//
    Lcd_Config(&PORTC,4,5,0,3,2,1,0);          // LCD is connected to PORTC
    Lcd_Cmd(LCD_CLEAR);                          // Clear LCD
    Lcd_Out(1,1,"CAN BUS");                      // Display heading on LCD
    Delay_ms(1000);                              // Wait for 2 seconds

//
// Program loop. Read the temperature from Node:COLLECTOR and display
// on the LCD continuously
//
    for(;;)                                      // Endless loop
    {
        Lcd_Cmd(LCD_CLEAR);                      // Clear LCD
        Lcd_Out(1,1,"Temp = ");                  // Display "Temp = "
        //
        // Send a message to Node:COLLECTOR and ask for data
        //
        data[0] = 'T';                           // Data to be sent
        id = 500;                                // Identifier
        CANWrite(id, data, 1, send_flag);        // send 'T'
        //
        // Get temperature from node:COLLECT
        //
        dt = 0;
        while(!dt)dt = CANRead(&id, data, &len, &read_flag);
        if(id == 3)
        {
            temperature = data[0];
            ByteToStr(temperature,txt);          // Convert to string
            Lcd_Out(1,8,txt);                    // Output to LCD
            Delay_ms(1000);                      // Wait 1 second
        }
    }

}
```

Figure 9.17: (Cont'd)

Filter 3 for buffer 2 is set to value 3 so that identifiers having values 3 are accepted by the receive buffer.

The operation mode is then set to NORMAL. The program then configures the LCD and displays the message "CAN BUS" for one second on the LCD.

The main program loop executes continuously and starts with a *for* statement. Inside this loop the LCD is cleared and text "TEMP =" is displayed on the LCD. Then character "T" is sent over the bus with the identifier equal to 500 (the COLLECTOR

```
/************************************************************************
                  CAN BUS EXAMPLE - NODE: COLLECTOR
                  ===================================
```

This is the COLLECTOR node of the CAN bus example. In this project a
PIC18F258 type microcontroller is used. An MCP2551 type CAN bus transceiver
is used to connect the microcontroller to the CAN bus. The microcontroller is
operated from an 8MHz crystal with an external reset button.

Pin CANRX and CANTX of the microcontroller are connected to pins RXD
and TXD of the transceiver chip respectively. Pins CANH and CANL of the
transceiver chip are connected to the CAN bus.

An LM35DZ type analog temperature sensor is connected to port AN0 of the
microcontroller. The microcontroller reads the temperature when a request is
received and then sends the temperature value as a byte to Node:DISPLAY on
the CAN bus.

CAN speed parameters are:

```
    Microcontroller clock:      8MHz
    CAN Bus bit rate:           100Kb/s
    Sync_Seg:                   1
    Prop_Seg:                   6
    Phase_Seg1:                 6
    Phase_Seg2:                 7
    SJW:                        1
    BRP:                        1
    Sample point:               65%
```

```
Author:      Dogan Ibrahim
Date:        October 2007
File:        COLLECTOR.C
************************************************************************/
```

```c
void main()
{
    unsigned char temperature, data[8];
    unsigned short init_flag, send_flag, dt, len, read_flag;
    char SJW, BRP, Phase_Seg1, Phase_Seg2, Prop_Seg, txt[4];
    unsigned int temp;
    unsigned long mV;
    long id, mask;

    TRISA = 0xFF;                        // PORTA are inputs
    TRISB = 0x08;                        // RB2 is output, RB3 is input
//
// Configure A/D converter
//
    ADCON1 = 0x80;
```

Figure 9.18: COLLECTOR program listing

(Continued)

```
//
// CAN BUS Timing Parameters
//
    SJW = 1;
    BRP = 1;
    Phase_Seg1 = 6;
    Phase_Seg2 = 7;
    BRP = 1;
    Prop_Seg = 6;

    init_flag = CAN_CONFIG_SAMPLE_THRICE   &
            CAN_CONFIG_PHSEG2_PRG_ON  &
            CAN_CONFIG_STD_MSG        &
            CAN_CONFIG_DBL_BUFFER_ON  &
            CAN_CONFIG_VALID_XTD_MSG  &
            CAN_CONFIG_LINE_FILTER_OFF;

    send_flag = CAN_TX_PRIORITY_0      &
            CAN_TX_XTD_FRAME        &
            CAN_TX_NO_RTR_FRAME;

    read_flag = 0;
//
// Initialise CAN module
//
    CANInitialize(SJW, BRP, Phase_Seg1, Phase_Seg2, Prop_Seg, init_flag);
//
// Set CAN CONFIG mode
//
    CANSetOperationMode(CAN_MODE_CONFIG, 0xFF);

    mask = -1;
//
// Set all MASK1 bits to 1's
//
    CANSetMask(CAN_MASK_B1, mask, CAN_CONFIG_XTD_MSG);
//
// Set all MASK2 bits to 1's
//
    CANSetMask(CAN_MASK_B2, mask, CAN_CONFIG_XTD_MSG);
//
// Set id of filter B1_F1 to 3
//
    CANSetFilter(CAN_FILTER_B2_F3,500,CAN_CONFIG_XTD_MSG);
//
// Set CAN module to NORMAL mode
//
    CANSetOperationMode(CAN_MODE_NORMAL, 0xFF);

//
```

Figure 9.18: (Cont'd)

```
// Program loop. Read the temperature from analog temperature
// sensor
//
    for(;;)                                              // Endless loop
    {
        //
        // Wait until a request is received
        //
        dt = 0;
        while(!dt) dt = CANRead(&id, data, &len, &read_flag);
        if(id == 500 && data[0] == 'T')
        {
            //
            // Now read the temperature
            //
            temp = Adc_Read(0);                          // Read temp
            mV = (unsigned long)temp * 5000 / 1024;      // in mV
            temperature = mV/10;                         // in degrees C
            //
            // send the temperature to Node:Display
            //
            data[0] = temperature;
            id = 3;                                      // Identifier
            CANWrite(id, data, 1, send_flag);            // send temperature
        }
    }
}
```

Figure 9.18: (Cont'd)

Node: DISPLAY

Initialize CAN module
Set mode to CONFIG
Set Mask bits to 1's
Set Filter value to 3
Set mode to NORMAL

DO FOREVER
    Send character "T" with *identifier* 500
    Read temperature with *identifier* 3
    Convert temperature to string
    Display temperature on LCD
    Wait 1 second
ENDDO

Node: COLLECTOR

Initialize CAN module
Set mode to CONFIG
Set Mask bits to 1's
Set Filter value to 500
Set mode to NORMAL

DO FOREVER
    Read a character
    IF character is "T"
        Read temperature
        Convert to digital
        Convert to °C
        Send with *identifier* 3
    ENDIF
ENDDO

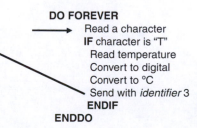

Figure 9.19: Operation of both nodes

node filter is set to accept identifier 500). This is a request to the COLLECTOR node to send the temperature reading. The program then reads the temperature from the CAN bus, converts it to a string in array *txt*, and displays it on the LCD. This process repeats after a one-second delay.

COLLECTOR Program

Figure 9.18 shows the program listing of the COLLECTOR program, called COLLECTOR.C. The initial part of this program is the same as the DISPLAY program. The receive filter is set to 500 so that messages with identifier 500 are accepted by the program.

Inside the program loop, the program waits until it receives a request to send the temperature. Here the request is identified by the reception of character "T". Once a valid request is received, the temperature is read and converted into °C (stored in variable *temperature*) and then sent to the CAN bus as a byte with an identifier value equal to 3. This process repeats forever.

Figure 9.19 summarizes the operation of both nodes.

Multi-Tasking and Real-Time Operating Systems

Nearly all microcontroller-based systems perform more than one activity. For example, a temperature monitoring system is made up of three tasks that normally repeat after a short delay, namely:

- Task 1 Reads the temperature
- Task 2 Formats the temperature
- Task 3 Displays the temperature

More complex systems may have many complex tasks. In a multi-tasking system, numerous tasks require CPU time, and since there is only one CPU, some form of organization and coordination is needed so each task has the CPU time it needs. In practice, each task takes a very brief amount of time, so it seems as if all the tasks are executing in parallel and simultaneously.

Almost all microcontroller-based systems work in real time. A real-time system is a time responsive system that can respond to its environment in the shortest possible time. Real time does not necessarily mean the microcontroller should operate at high speed. What is important in a real-time system is a fast response time, although high speed can help. For example, a real-time microcontroller-based system with various external switches is expected to respond immediately when a switch is activated or some other event occurs.

A real-time operating system (RTOS) is a piece of code (usually called the kernel) that controls task allocation when the microcontroller is operating in a multi-tasking environment. RTOS decides, for instance, which task to run next, how to coordinate the task priorities, and how to pass data and messages among tasks.

This chapter explores the basic principles of multi-tasking embedded systems and gives examples of an RTOS used in simple projects. Multi-tasking code and RTOS are complex and wide topics, and this chapter describes the concepts pertaining to these tools only briefly. Interested readers should refer to the many books and papers available on operating systems, multi-tasking systems, and RTOS.

There are several commercially available RTOS systems for PIC microcontrollers. At the time of writing, mikroC language did not provide a built-in RTOS. Two popular high-level RTOS systems for PIC microcontrollers are Salvo (www.pumpkin.com), which can be used from a Hi-Tech PIC C compiler, and the CCS (Customer Computer Services) built-in RTOS system. In this chapter, the example RTOS projects are based on the CCS (www.ccsinfo.com) compiler, one of the popular PIC C compilers developed for the PIC16 and PIC18 series of microcontrollers.

10.1   State Machines

State machines are simple constructs used to perform several activities, usually in a sequence. Many real-life systems fall into this category. For example, the operation of a washing machine or a dishwasher is easily described with a state machine construct.

Perhaps the simplest method of implementing a state machine construct in C is to use a *switch-case* statement. For example, our temperature monitoring system has three tasks, named Task 1, Task 2, and Task 3 as shown in Figure 10.1. The state machine implementation of the three tasks using *switch-case* statements is shown in Figure 10.2. The starting state is 1, and each task increments the state number by one to select the next state to be executed. The last state selects state 1, and there is a delay at the end of the *switch-case* statement. The state machine construct is executed continuously inside an endless *for* loop.

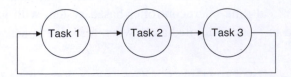

Figure 10.1: State machine implementation

```
for(;;)
{
        state = 1;
        switch (state)
        {
                CASE 1:
                                implement TASK 1
                                state++;
                                break;
                CASE 2:
                                implement TASK 2
                                state++;
                                break;
                CASE 3:
                                implement TASK 3
                                state = 1;
                                break;
        }
        Delay_ms(n);
}
```

Figure 10.2: State machine implementation in C

In many applications, the states need not be executed in sequence. Rather, the next state is selected by the present state either directly or based on some condition. This is shown in Figure 10.3.

```
for(;;)
{
        state = 1;
        switch (state)
        {
                CASE 1:
                                implement TASK 1
                                state = 2;
                                break;
                CASE 2:
                                implement TASK 2
                                state = 3;
                                break;
                CASE 3:
                                implement TASK 3
                                state = 1;
                                break;
        }
        Delay_ms(n);
}
```

Figure 10.3: Selecting the next state from the current state

State machines, although easy to implement, are primitive and have limited application. They can only be used in systems which are not truly responsive, where the task activities are well-defined and the tasks are not prioritized.

Moreover, some tasks may be more important than others. We may want some tasks to run whenever they become eligible. For example, in a manufacturing plant, a task that sets off an alarm when the temperature is too hot must be run. This kind of implementation of tasks requires a sophisticated system like RTOS.

10.2   The Real-Time Operating System (RTOS)

Real-time operating systems are built around a multi-tasking kernel which controls the allocation of time slices to tasks. A time slice is the period of time a given task has for execution before it is stopped and replaced by another task. This process, also known as context switching, repeats continuously. When context switching occurs, the executing task is stopped, the processor registers are saved in memory, the processor registers of the next available task are loaded into the CPU, and the new task begins execution. An RTOS also provides task-to-task message passing, synchronization of tasks, and allocation of shared resources to tasks.

The basic parts of an RTOS are:

- Scheduler

- RTOS services

- Synchronization and messaging tools

10.2.1   The Scheduler

A scheduler is at the heart of every RTOS, as it provides the algorithms to select the tasks for execution. Three of the more common scheduling algorithms are:

- Cooperative scheduling

- Round-robin scheduling

- Preemptive scheduling

Cooperative scheduling is perhaps the simplest scheduling algorithm available. Each task runs until it is complete and gives up the CPU voluntarily. Cooperative scheduling cannot satisfy real-time system needs, since it cannot support the prioritization of tasks according to importance. Also, a single task may use the CPU too long, leaving too little time for other tasks. And the scheduler has no control of the various tasks' execution time. A state machine construct is a simple form of a cooperative scheduling technique.

In *round-robin scheduling*, each task is assigned an equal share of CPU time (see Figure 10.4). A counter tracks the time slice for each task. When one task's time slice completes, the counter is cleared and the task is placed at the end of the cycle. Newly added tasks are placed at the end of the cycle with their counters cleared to 0. This, like cooperative scheduling, is not very useful in a real-time system, since very often some tasks take only a few milliseconds while others require hundreds of milliseconds or more.

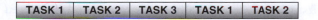

Figure 10.4: Round-robin scheduling

Preemptive scheduling is considered a real-time scheduling algorithm. It is priority-based, and each task is given a priority (see Figure 10.5). The task with the highest priority gets the CPU time. Real-time systems generally support priority levels ranging from 0 to 255, where 0 is the highest priority and 255 is the lowest.

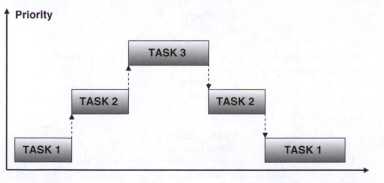

Figure 10.5: Preemptive scheduling

In some real-time systems where more than one task can be at the same priority level, preemptive scheduling is mixed with round-robin scheduling. In such cases, tasks at higher priority levels run before lower priority ones, and tasks at the same priority level run by round-robin scheduling. If a task is preempted by a higher priority task, its run time counter is saved and then restored when it regains control of the CPU.

In some systems a strict real-time priority class is defined where tasks above this class may run to completion (or run until a resource is not available) even if there are other tasks at the same priority level.

In a real-time system a task can be in any one of the following states (see Figure 10.6):

- Ready to run

- Running

- Blocked

When a task is first created, it is usually *ready to run* and is entered in the task list. From this state, subject to the scheduling algorithm, the task can become a *running* task. According to the conditions of preemptive scheduling, the task will run if it is the highest priority task in the system and is not waiting for a resource.

A running task becomes a *blocked* task if it needs a resource that is not available. For example, a task may need data from an A/D converter and is blocked until it is

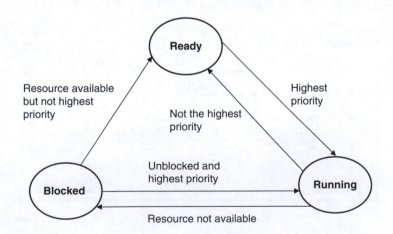

Figure 10.6: Task states

available. Once the resource can be accessed, the blocked task becomes a running task if it is the highest priority task in the system, otherwise it moves to the ready state. Only a running task can be blocked. A ready task cannot be blocked.

When a task moves from one state to another, the processor saves the running task's context in memory, loads the new task's context from memory, and then executes the new instructions as required.

The kernel usually provides an interface to manipulate task operations. Typical task operations are:

- Creating a task
- Deleting a task
- Changing the priority of a task
- Changing the state of a task

10.3   RTOS Services

RTOS services are utilities provided by the kernel that help developers create real-time tasks efficiently. For example, a task can use time services to obtain the current date and time. Some of these services are:

- Interrupt handling services
- Time services
- Device management services
- Memory management services
- Input-output services

10.4   Synchronization and Messaging Tools

Synchronization and messaging tools are kernel constructs that help developers create real-time applications. Some of these services are:

- Semaphores
- Event flags

- Mailboxes

- Pipes

- Message queues

Semaphores are used to synchronize access to shared resources, such as common data areas. Event flags are used to synchronize the intertask activities. Mailboxes, pipes, and message queues are used to send messages among tasks.

10.5   CCS PIC C Compiler RTOS

The CCS PIC C compiler is one of the popular C compilers for the PIC16 and PIC18 series of microcontrollers. In addition to their PIC compilers, Customer Computer Services offers PIC in-circuit emulators, simulators, microcontroller programmers, and various development kits. The syntax of the CCS C language is slightly different from that of the mikroC language, but readers who are familiar with mikroC should find CCS C easy to use.

CCS C supports a rudimentary multi-tasking cooperative RTOS for the PIC18 series of microcontrollers that uses their PCW and PCWH compilers. This RTOS allows a PIC microcontroller to run tasks without using interrupts. When a task is scheduled to run, control of the processor is given to that task. When the task is complete or does not need the processor any more, control returns to a dispatch function, which gives control of the processor to the next scheduled task. Because the RTOS does not use interrupts and is not preemptive, the user must make sure that a task does not run forever. Further details about the RTOS are available in the compiler's user manual.

The CCS language provides the following RTOS functions in addition to the normal C functions:

rtos_run() initiates the operation of RTOS. All task control operations are implemented after calling this function.

rtos_terminate() terminates the operation of RTOS. Control returns to the original program without RTOS. In fact, this function is like a return from *rtos_run*().

rtos_enable() receives the name of a task as an argument. The function enables the task so function *rtos_run*() can call the task when its time is due.

rtos_disable() receives the name of a task as an argument. The function disables the task so it can no longer be called by *rtos_run*() unless it is re-enabled by calling *rtos_enable*().

rtos_yield(), when called from within a task, returns control to the dispatcher. All tasks should call this function to release the processor so other tasks can utilize the processor time.

rtos_msg_send() receives a task name and a byte as arguments. The function sends the byte to the specified task, where it is placed in the task's message queue.

rtos_msg_read() reads the byte located in the task's message queue.

rtos_msg_poll() returns true if there is data in the task's message queue. This function should be called before reading a byte from the task's message queue.

rtos_signal() receives a semaphore name and increments that semaphore.

rtos_wait() receives a semaphore name and waits for the resource associated with the semaphore to become available. The semaphore count is then decremented so the task can claim the resource.

rtos_await() receives an expression as an argument, and the task waits until the expression evaluates to true.

rtos_overrun() receives a task name as an argument, and the function returns true if that task has overrun its allocated time.

rtos_stats() returns the specified statistics about a specified task. The statistics can be the minimum and maximum task run times and the total task run time. The task name and the type of statistics are specified as arguments to the function.

10.5.1 Preparing for RTOS

In addition to the preceding functions, the *#use rtos*() preprocessor command must be specified at the beginning of the program before calling any of the RTOS functions. The format of this preprocessor command is:

```
#use rtos(timer=n, minor_cycle=m)
```

where *timer* is between 0 and 4 and specifies the processor timer that will be used by the RTOS, and *minor_cycle* is the longest time any task will run. The number entered here must be followed by s, ms, us, or ns.

In addition, a *statistics* option can be specified after the *minor_cycle* option, in which case the compiler will keep track of the minimum and maximum processor times the task uses at each call and the task's total time used.

10.5.2   Declaring a Task

A task is declared just like any other C function, but tasks in a multi-tasking application do not have any arguments and do not return any values. Before a task is declared, a *#task* preprocessor command is needed to specify the task options. The format of this preprocessor command is:

#task(rate=n, max=m, queue=p)

where *rate* specifies how often the task should be called. The number specified must be followed by s, ms, us, or ns. *max* specifies how much processor time a task will use in one execution of the task. The time specifed here must be equal to or less than the time specified by *minor_cycle*. *queue* is optional and if present specifies the number of bytes to be reserved for the task to receive messages from other tasks. The default value is 0.

In the following example, a task called *my_ticks* is every 20ms and is expected to use no more than 100ms of processor time. This task is specified with no *queue* option:

```
#task(rate=20ms, max=100ms)
void my_ticks()
{
        . . . . . . . . . . .
        . . . . . . . . . . .
}
```

PROJECT 10.1—LEDs

In the following simple RTOS-based project, four LEDs are connected to the lower half of PORTB of a PIC18F452-type microcontroller. The software consists of four tasks, where each task flashes an LED at a different rate:

- Task 1, called *task_B0*, flashes the LED connected to port RB0 at a rate of 250ms.

- Task 2, called *task_B1*, flashes the LED connected to port RB1 at a rate of 500ms.

- Task 3, called *task_B2*, flashes the LED connected to port RB2 once a second.

- Task 4, called *task_B3*, flashes the LED connected to port RB3 once every two seconds.

Figure 10.7 shows the circuit diagram of the project. A 4MHz crystal is used as the clock. PORTB pins RB0–RB3 are connected to the LEDs through current limiting resistors.

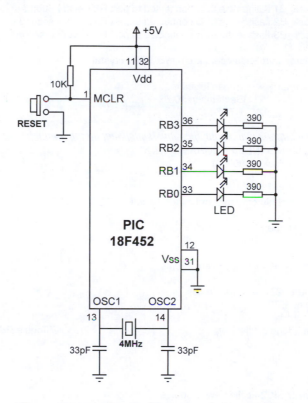

Figure 10.7: Circuit diagram of the project

The software is based on the CCS C compiler, and the program listing (RTOS1.C) is given in Figure 10.8. The main program is at the end of the program, and inside the main program PORTB pins are declared as outputs and RTOS is started by calling function *rtos_run()*.

The file that contains CCS RTOS declarations should be included at the beginning of the program. The preprocessor command *#use delay* tells the compiler that we are using

```
//////////////////////////////////////////////////////////////////////////////////////////////////////////////////////////
//
//                              SIMPLE RTOS EXAMPLE
//                          ---------------------------------
//
// This is a simple RTOS example. 4 LEDs are connected to lower half of
// PORTB of a PIC18F452 microcontroller. The program consists of 4
// tasks:
//
// Task task_B0 flashes the LED connected to port RB0 every 250ms.
// Task task_B1 flashes the LED connected to port RB1 every 500ms.
// Task task_B2 flashes the LED connected to port RB2 every second
// Task task_B3 flashes the LED connected to port RB3 every 2 seconds.
//
// The microcontroller is operated from a 4MHz crystal
//
// Programmer:        Dogan Ibrahim
// Date:              September, 2007
// File:              RTOS1.C
//
//////////////////////////////////////////////////////////////////////////////////////////////////////////////////////////
#include "C:\NEWNES\PROGRAMS\rtos.h"
#use delay (clock=4000000)


//
// Define which timer to use and minor_cycle for RTOS
//
  #use rtos(timer=0, minor_cycle=10ms)

//
// Declare TASK 1 - called every 250ms
//
  #task(rate=250ms, max=10ms)
  void task_B0()
  {
    output_toggle(PIN_B0);                              // Toggle RB0
  }


//
// Declare TASK 2 - called every 500ms
//
  #task(rate=500ms, max=10ms)
  void task_B1()
  {
    output_toggle(PIN_B1);                              // Toggle RB1
  }

//
// Declare TASK 3 - called every second
```

Figure 10.8: Program listing of the project

```
//
   #task(rate=1s, max=10ms)
   void task_B2()
   {
      output_toggle(PIN_B2);                        // Toggle RB2
   }

//
// Declare TASK 4 - called every 2 seconds
//
   #task(rate=2s, max=10ms)
   void task_B3()
   {
      output_toggle(PIN_B3);                        // Toggle RB3
   }



//
// Start of MAIN program
//
void main()
{
   set_tris_b(0);                                   // Configure PORTB as outputs
   rtos_run();                                      // Start RTOS
}
```

Figure 10.8: (Cont'd)

a 4MHz clock. Then the RTOS timer is declared as Timer 0, and *minor_cycle* time is declared as 10ms using the preprocessor command *#use rtos*.

The program consists of four similar tasks:

- *task_B0* flashes the LED connected to RB0 at a rate of 250ms. Thus, the LED is ON for 250ms, then OFF for 250ms, and so on. CCS statement *output_toggle* is used to change the state of the LED every time the task is called. In the CCS compiler *PIN_B0* refers to port pin RB0 of the microcontroller.

- *task_B1* flashes the LED connected to RB1 at a rate of 500ms as described.

- *task_B2* flashes the LED connected to RB2 every second as described.

- Finally, *task_B3* flashes the LED connected to RB3 every two seconds as described.

The program given in Figure 10.8 is a multi-tasking program where the LEDs flash independently of each other and concurrently.

PROJECT 10.2—Random Number Generator

In this slightly more complex RTOS project, a random number between 0 and 255 is generated. Eight LEDs are connected to PORTB of a PIC18F452 microcontroller. In addition, a push-button switch is connected to bit 0 of PORTD (RD0), and an LED is connected to bit 7 of PORTD (RD7).

Three tasks are used in this project: *Live*, *Generator*, and *Display*.

- Task *Live* runs every 200ms and flashes the LED on port pin RD7 to indicate that the system is working.

- Task *Generator* increments a variable from 0 to 255 continuously and checks the status of the push-button switch. When the push-button switch is pressed, the value of the current count is sent to task *Display* using a messaging queue.

- Task *Display* reads the number from the message queue and sends the received byte to the LEDs connected to PORTB. Thus, the LEDs display a random pattern every time the push button is pressed.

Figure 10.9 shows the project's block diagram. The circuit diagram is given in Figure 10.10. The microcontroller is operated from a 4MHz crystal.

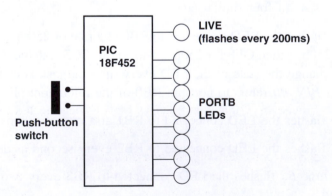

Figure 10.9: Block diagram of the project

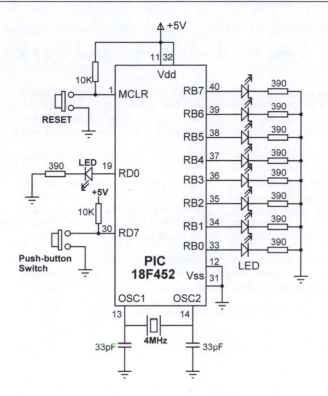

Figure 10.10: Circuit diagram of the project

The program listing of the project (RTOS2.C) is given in Figure 10.11. The main part of the program is in the later portion, and it configures PORTB pins as outputs. Also, bit 0 of PORTD is configured as input and other pins of PORTD are configured as outputs.

Timer 0 is used as the RTOS timer, and the *minor_cycle* is set to 1s. The program consists of three tasks:

- Task *Live* runs every 200ms and flashes the LED connected to port pin RD7. This LED indicates that the system is working.

- Task *Generator* runs every millisecond and increments a byte variable called *count* continuously. When the push-button switch is pressed, pin 0 of PORTD (RD0) goes to logic 0. When this happens, the current value of *count* is sent to task *Display* using RTOS function call *rtos_msg_send(display, count)*, where

```
///////////////////////////////////////////////////////////////////////////////////////////////////////////
//
//          SIMPLE RTOS EXAMPLE - RANDOM NUMBER GENERATOR
//          -------------------------------------------------------------------------------
//
// This is a simple RTOS example. 8 LEDs are connected to PORTB
// of a PIC18F452 microcontroller. Also, a push-button switch is
// connected to port RC0 of PORTC, and an LED is connected to port
// RC7 of the microcontroller. The push-button switch is normally at logic 1.
//
// The program consists of 3 tasks called "Generator", "Display", and "Live".
//
// Task "Generator" runs in a loop and increments a counter from 0 to 255.
// This task also checks the state of the push-button switch. When the
// push-button switch is pressed, the task sends the value of the count to the
// "Display" task using messaging passing mechanism. The "Display" task
// receives the value of count and displays it on the PORTB LEDs.
//
// Task "Live" flashes the LED connected to port RC7 at a rate of 250ms.
// This task is used to indicate that the system is working and is ready for
// the user to press the push-button.
//
// The microcontroller is operated from a 4MHz crystal
//
// Programmer:    Dogan Ibrahim
// Date:          September, 2007
// File:          RTOS2.C
//
///////////////////////////////////////////////////////////////////////////////////////
#include "C:\NEWNES\PROGRAMS\rtos.h"
#use delay (clock=4000000)
int count;


//
// Define which timer to use and minor_cycle for RTOS
//
   #use rtos(timer=0, minor_cycle=1ms)


//
// Declare TASK "Live" - called every 200ms
// This task flashes the LED on port RC7
//
   #task(rate=200ms, max=1ms)
   void Live()
   {
     output_toggle(PIN_D7);
   }


//
```

Figure 10.11: Program listing of the project

```
// Declare TASK "Display" - called every 10ms
//
   #task(rate=10ms, max=1ms, queue=1)
   void Display()
   {
     if(rtos_msg_poll() > 0)                          // Is there a message ?
     {
       output_b(rtos_msg_read());                     // Send to PORTB
     }
   }


//
// Declare TASK "Generator" - called every millisecond
//
   #task(rate=1ms, max=1ms)
   void Generator()
   {
     count++;                                         // Increment count
     if(input(PIN_D0) == 0)                           // Switch pressed ?
     {
       rtos_msg_send(Display,count);                  // send a message
     }
   }



//
// Start of MAIN program
//
void main()
{
   set_tris_b(0);                                     // Configure PORTB as outputs
   set_tris_d(1);                                     // RD0=input, RD7=output
   rtos_run();                                        // Start RTOS
}
```

Figure 10.11: (Cont'd)

Display is the name of the task where the message is sent and *count* is the byte
sent.

- Task *Display* runs every 10ms. This task checks whether there is a message
  in the queue. If so, the message is extracted using RTOS function call
  rtos_msg_read(), and the read byte is sent to the LEDs connected to PORTB.
  Thus, the LEDs display the binary value of *count* as the switch is pressed. The
  message queue should be checked by using function *rtos_msg_poll*(), as trying
  to read the queue without any bytes in the queue may freeze the program.

PROJECT 10.3—Voltmeter with RS232 Serial Output

In this RTOS project, which is more complex than the preceding ones, the voltage is read using an A/D converter and then sent over the serial port to a PC. The project consists of three tasks: *Live*, *Get_voltage*, and *To_RS232*.

- Task *Live* runs every 200ms and flashes an LED connected to port RD7 of the microcontroller to indicate that the system is working.

- Task *Get_voltage* reads channel 0 of the A/D converter where the voltage to be measured is connected. The read value is formatted and then stored in a variable. This task runs every two seconds.

- Task *To_RS232* reads the formatted voltage and sends it over the RS232 line to a PC every second.

Figure 10.12 shows the block diagram of the project. The circuit diagram is given in Figure 10.13. A PIC18F8520-type microcontroller with a 10MHz crystal is used in this project (though any PIC18F-series microcontroller can be used). The voltage to be measured is connected to analog port AN0 of the microcontroller. The RS232 TX output of the microcontroller (RC6) is connected to a MAX232-type RS232-level converter chip and then to the serial input of a PC (e.g., COM1) using a 9-pin D-type connector. Port pin RD7 is connected to an LED to indicate whether the project is working.

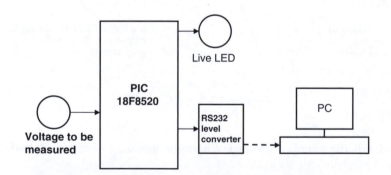

Figure 10.12: Block diagram of the project

The program listing (RTOS3.C) of the project is given in Figure 10.14. At the beginning of the program, the A/D is defined as 10 bits, the clock is defined as 10MHz, and the RS232 speed is defined as 2400 baud. The RTOS timer and the *minor_cycle* are then defined using the *#use rtos* preprocessor command.

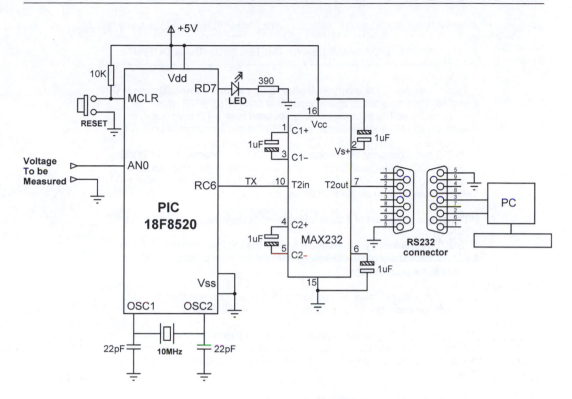

Figure 10.13: Circuit diagram of the project

In the main part of the program PORTD is configured as output and all PORTD pins are cleared. Then PORTA is configured as input (RA0 is the analog input), the microcontroller's analog inputs are configured, the A/D clock is set, and the A/D channel 0 is selected (AN0). The RTOS is then started by calling function *rtos_run()*.

The program consists of three tasks:

- Task *Live* runs every 200ms and flashes an LED connected to port pin RD7 of the microcontroller to indicate that the project is working.

- Task *Get_voltage* reads the analog voltage from channel 0 (pin RA0 or AN0) of the microcontroller. The value is then converted into millivolts by multiplying by 5000 and dividing by 1024 (in a 10-bit A/D there are 1024 quantization levels, and when working with a reference voltage of +5V, each quantization level corresponds to 5000/1024mV). The voltage is stored in a global variable called *Volts*.

```
//////////////////////////////////////////////////////////////////////////////////////////////////////////
//
//          SIMPLE RTOS EXAMPLE - VOLTMETER WITH RS232 OUTPUT
//          ------------------------------------------------------------------------------------
//
// This is a simple RTOS example. Analog voltage to be measured (between 0V
// and +5V) is connected to analog input AN0 of a PIC18F8520 type
// microcontroller. The microcontroller is operated from a 10MHz crystal. In
// addition, an LED is connected to port in RD7 of the microcontroller.
//
// RS232 serial output of the mirocontroller (RC6) is connected to a MAX232
// type RS232 voltage level converter chip. The output of this chip can be
// connected to the serial input of a PC (e.g., COM1) so that the measured
// voltage can be seen on the PC screen.
//
// The program consists of 3 tasks called "live", "Get_voltage", and "To_RS232".
//
// Task "Live" runs every 200ms and it flashes the LED conencted to port pin
// RD7 of the microcontroller to indicate that the program is running and is
// ready to measure voltages.
//
// task "Get_voltage" reads analog voltage from port AN0 and then converts
// the voltage into millivolts and stores in a variable called Volts.
//
// Task "To_RS232" gets the measured voltage, converts it into a character
// array and then sends to the PC over the RS232 serial line. The serial line
// is configured to operate at 2400 Baud (higher Baud rates can also be used if
// desired).
//
// Programmer:      Dogan Ibrahim
// Date:            September, 2007
// File:            RTOS3.C
//
//////////////////////////////////////////////////////////////////////////////////////////////////////////

#include <18F8520.h>
#device adc=10
#use delay (clock=10000000)
#use rs232(baud=2400,xmit=PIN_C6,rcv=PIN_C7)

unsigned int16 adc_value;
unsigned int32 Volts;

//
// Define which timer to use and minor_cycle for RTOS
//
   #use rtos(timer=0, minor_cycle=100ms)

//
// Declare TASK "Live" - called every 200ms
```

Figure 10.14: Program listing of the project

```
// This task flashes the LED on port RD7
//
  #task(rate=200ms, max=1ms)
  void Live()
  {
    output_toggle(PIN_D7);                          // Toggle RD7 LED
  }


//
// Declare TASK "Get_voltage" - called every 10ms
//
  #task(rate=2s, max=100ms)
  void Get_voltage()
  {
    adc_value = read_adc();                         // Read A/D value
    Volts = (unsigned int32)adc_value*5000;
    Volts = Volts / 1024;                           // Voltage in mV
  }


//
// Declare TASK "To_RS232" - called every millisecond
//
  #task(rate=2s, max=100ms)
  void To_RS232()
  {
    printf("Measured Voltage = %LumV\n\r",Volts);   // send to RS232
  }



//
// Start of MAIN program
//
void main()
{
  set_tris_d(0);                                    // PORTD all outputs
  output_d(0);                                      // Clear PORTD
  set_tris_a(0xFF);                                 // PORTA all inputs
  setup_adc_ports(ALL_ANALOG);                      // A/D ports
  setup_adc(ADC_CLOCK_DIV_32);                      // A/D clock
  set_adc_channel(0);                               // Select channel 0 (AN0)
  delay_us(10);
  rtos_run();                                       // Start RTOS
}
```

Figure 10.14: (Cont'd)

- Task *To_RS232* reads the measured voltage from common variable *Volts* and sends it to the RS232 port using the C *printf* statement. The result is sent in the following format:

$$\text{Measured voltage} = \text{nnnn mV}$$

The HyperTerminal program is run on the PC to get an output from the program. A typical screen output is shown in Figure 10.15.

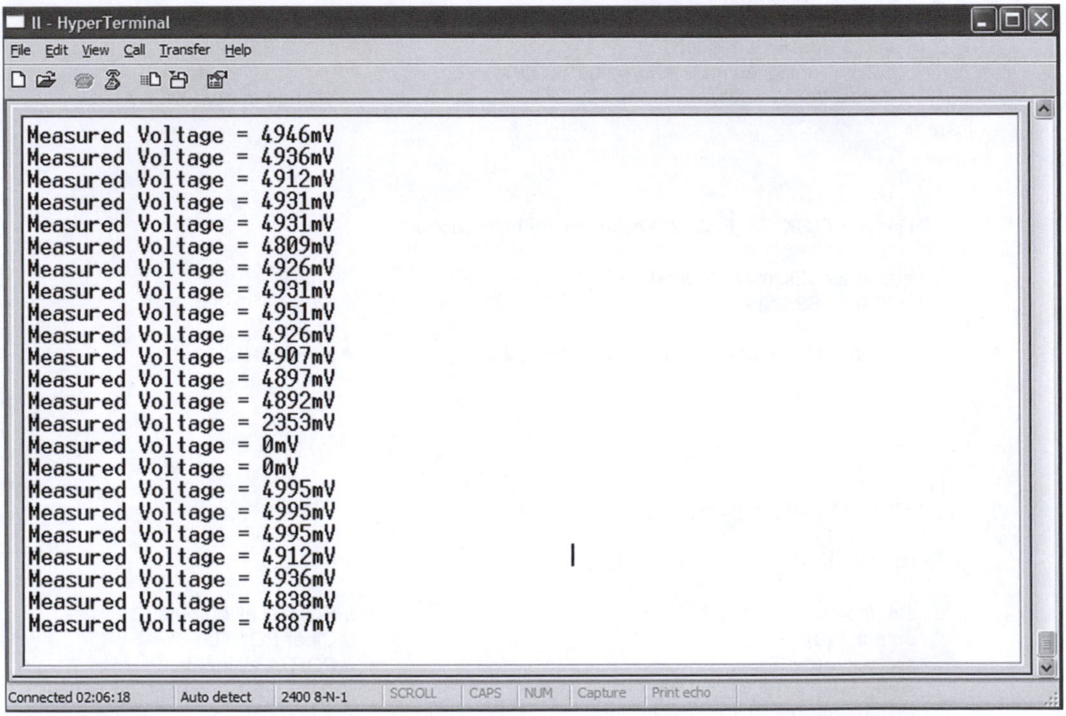

Figure 10.15: Typical output from the program

Using a Semaphore

The program given in Figure 10.14 is working and displays the measured voltage on the PC screen. This program can be improved slightly by using a semaphore to synchronize the display of the measured voltage with the A/D samples. The modified

```
/////////////////////////////////////////////////////////////////////////////////////////////////////
//
//          SIMPLE RTOS EXAMPLE - VOLTMETER WITH RS232 OUTPUT
//          ------------------------------------------------------------------------------------------
//
// This is a simple RTOS example. Analog voltage to be measured (between 0V
// and +5V) is connected to analog input AN0 of a PIC18F8520 type
// microcontroller. The microcontroller is operated from a 10MHz crystal. In
// addition, an LED is connected to port in RD7 of the microcontroller.
//
// RS232 serial output of the mirocontroller (RC6) is connected to a MAX232
// type RS232 voltage level converter chip. The output of this chip can be
// connected to the serial input of a PC (e.g., COM1) so that the measured
// voltage can be seen on the PC screen.
//
// The program consists of 3 tasks called "live", "Get_voltage", and "To_RS232".
//
// Task "Live" runs every 200ms and it flashes the LED connected to port RD7
// of the microcontroller to indicate that the program is running and is ready to
// measure voltages.
//
// task "Get_voltage" reads analog voltage from port AN0 and then converts the
// voltage into millivolts and stores in a variable called Volts.
//
// Task "To_RS232" gets the measured voltage  and then sends to the PC over
// the RS232 serial line. The serial line is configured to operate at 2400 Baud
// (higher Baud rates can also be used if desired).
//
// In this modified program, a semaphore is used to synchronize
// the display of the measured value with the A/D samples.
//
// Programmer:      Dogan Ibrahim
// Date:            September, 2007
// File:            RTOS4.C
//
/////////////////////////////////////////////////////////////////////////////////////////////////////

#include <18F8520.h>
#device adc=10
#use delay (clock=10000000)
#use rs232(baud=2400,xmit=PIN_C6,rcv=PIN_C7)

unsigned int16 adc_value;
unsigned int32 Volts;
int8 sem;


//
// Define which timer to use and minor_cycle for RTOS
//
    #use rtos(timer=0, minor_cycle=100ms)
```

Figure 10.16: Modified program listing

(Continued)

```
//
// Declare TASK "Live" - called every 200ms
// This task flashes the LED on port RD7
//
  #task(rate=200ms, max=1ms)
  void Live()
  {
    output_toggle(PIN_D7);                          // Toggle RD7 LED
  }


//
// Declare TASK "Get_voltage" - called every 10ms
//
  #task(rate=2s, max=100ms)
  void Get_voltage()
  {
    rtos_wait(sem);                                 // decrement semaphore
    adc_value = read_adc();                         // Read A/D value
    Volts = (unsigned int32)adc_value*5000;
    Volts = Volts / 1024;                           // Voltage in mV
    rtos_signal(sem);                               // increment semaphore
  }


//
// Declare TASK "To_RS232" - called every millisecond
//
  #task(rate=2s, max=100ms)
  void To_RS232()
  {
    rtos_wait(sem);                                 // Decrement semaphore
    printf("Measured Voltage = %LumV\n\r",Volts);   // Send to RS232
    rtos_signal(sem);                               // Increment semaphore
  }



//
// Start of MAIN program
//
void main()
{
  set_tris_d(0);                                    // PORTD all outputs
  output_d(0);                                      // Clear PORTD
  set_tris_a(0xFF);                                 // PORTA all inputs
  setup_adc_ports(ALL_ANALOG);                      // A/D ports
  setup_adc(ADC_CLOCK_DIV_32);                      // A/D clock
  set_adc_channel(0);                               // Select channel 0 (AN0)

  delay_us(10);
  sem = 1;                                          // Semaphore is 1
  rtos_run();                                       // Start RTOS
}
```

Figure 10.16: (Cont'd)

program (RTOS4.C) is given in Figure 10.16. The operation of the new program is as follows:

- The semaphore variable (*sem*) is set to 1 at the beginning of the program.

- Task *Get_voltage* decrements the semaphore (calls *rtos_wait*) variable so that task *To_RS232* is blocked (semaphore variable *sem* = 0) and cannot send data to the PC. When a new A/D sample is ready, the semaphore variable is incremented (calls *rtos_signal*) and task *To_RS232* can continue. Task *To_RS232* then sends the measured voltage to the PC and increments the semaphore variable to indicate that it had access to the data. Task *Get_voltage* can then get a new sample. This process is repeated forever.

Index